WITHDRAWN

169

Strength of materials for engineering technology

Strength of materials for engineering technology

Second edition

Irving Granet, P.E.
Queensborough Community College
New York Institute of Technology

Reston Publishing Company, Inc.
A Prentice-Hall Company
Reston, Virginia

This book is dedicated to
my wife
Arlene,
my children
Ellen, Kenny, and **David**
and to the beloved memory
of my father-in-law
Samuel A. Wertheim

Library of Congress Cataloging in Publication Data

Granet, Irving
 Strength of materials for engineering technology.

 Includes bibliographies and index.
 1. Strength of materials. I. Title.
TA405.G72   1980   620.1'12   79-15761
ISBN: 0-8359-7074-4

©1980 by
Reston Publishing Company, Inc.
A Prentice-Hall Company
Reston, Virginia

All rights reserved. No part of this book may be reproduced in any way, or by any means, without permission in writing from the publisher.

10 9 8 7 6 5 4 3 2 1

Printed in the United States of America

Preface

Wherever possible, the features and arrangement that made the original edition of this book successful have been retained in this revision. The most notable change has been the adoption of the SI (metric) system of units. The adoption of this system throughout the world has made its use in the United States inevitable. Many industries in this country have already converted to it, and it is necessary for technically trained personnel to become thoroughly versed in the use of this system. There are two chapters in the book where the use of SI units proved to be very difficult. The first of these was Chapter 2, Mechanical Properties of Materials. The present ANSI/ASTM Standard E8-77a notes, "Dimensions of specimens are given in U.S. customary units and metric units. The metric units are not the exact arithmetic equivalents but have been adjusted to provide practical equivalents for critical dimensions while retaining geometric proportionality." Since almost all existing data as well are found in U.S. customary units, most of Chapter 2 has been retained in these units.

The second chapter where difficulty was encountered was Chapter 9, Columns. Communications with both the American Iron and Steel Institute and the American Institute of Steel Construction indicate that the steel industry in this country will not convert to any major extent to the SI system in the near future. It therefore becomes necessary to design columns using the standard sizes and the design standards that exist in U.S. customary units. Care must be exercised in this matter since one is dealing with more than just a conversion of units.

Approximately 150 new problems have been included in this edition, making the total number of problems nearly 700. These problems are graded in order of difficulty and by topic, and their large number should enable the user to have a much wider latitude in the choice of assignments.

I am indebted to Prof. D. C. Klingensmith of Bluefield State College for his in-depth review of the book and his suggestions for its improvement. Also, Prof. G. Thomas Burch, Jr., of Olympic College, Prof. O. Paradzick of Springfield Technical Community College, and Prof. R. E. White of Nashville State Technical Institute were most helpful and cooperative during the preparation of this edition. Mr. William C. Benzer of the American Iron and Steel Institute and Mr. Robert D. Bauer of the ASTM graciously gave their time to help clarify several matters concerning standards and testing. My colleagues at Queensborough Community College and New York Institute of Technology were most supportive of me during this revision of the book.

To my Wife, Arlene, and our children, Ellen, Kenny, and David I owe a debt beyond words. The few words that dedicate this book to them cannot express the magnitude of the love and patience they have so willingly and freely given me so that this book could be completed.

Irving Granet

Contents

Preface, v

Introduction to the SI system of units, 1

1 Concepts of stress and strain, 10

    1.1   Introduction, 10
    1.2   Structural Loads, 10
    1.3   Stress, 11
    1.4   Strain, 17
           1.4a Shearing strain, 18
    1.5   Hooke's Law, 19
    1.6   Poisson's Ratio, 23
    1.7   Statically Indeterminate Stresses, 29
    1.8   Temperature Stresses, 32
    1.9   Stresses in Cylinders and Spheres, 40
    1.10 Closure, 43
Problems, 43
References, 53

2 Mechanical properties of materials, 54

    2.1   Introduction, 54
    2.2   The Tensile Test, 54
    2.3   The Stress-Strain Curve in Tension, 56
    2.4   Other Tests, 68
    2.5   Closure, 72
Problems, 72
References, 76

3 Riveted and welded structures, 77

    3.1   Introduction, 77

- 3.2 Riveted Connections, 77
- 3.3 Design Stresses for Riveted Connections, 80
- 3.4 Failure in Riveted Joints, 83
- 3.5 Design of Riveted Joints, 87
- 3.6 Eccentrically Loaded Riveted Joints, 96
- 3.7 Welded Structures—General, 99
- 3.8 Weld Design, 104
- 3.9 Closure, 110

Problems, 111
References, 120

4 Center of gravity and moment of inertia—the first and second moment of area, 121

- 4.1 Introduction, 121
- 4.2 Center of Gravity, 121
- 4.3 Center of Gravity of Composite Areas, 136
- 4.4 The Second Moment of Area—Moment of Inertia, 140
- 4.5 The Parallel Axis Transfer Theorem, 142
- 4.6 The Moment of Inertia of Composite Areas, 144
- 4.7 The Radius of Gyration, 149
- 4.8 Closure, 151

Problems, 151
References, 163

5 Torsion, 165

- 5.1 Introduction, 165
- 5.2 The Horsepower-Torque Equation, 165
- 5.3 Torsion of Circular Shafts, 169
- 5.4 Shaft Couplings, 177
- 5.5 Springs, 180
- 5.6 Torsion of Noncircular Sections, 189
- 5.7 Closure, 196

Problems, 197
References, 204

6 Shear and bending moments in beams, 206

- 6.1 Introduction, 206
- 6.2 Beam Supports and Beam Loadings, 206
  - 6.2a Supports, 207
  - 6.2b Loads, 209
- 6.3 Shear and Bending Moment, 211

6.4 Shear and Bending Moment in Cantilever Beams, 215
6.5 General Relation Between Shear and Bending Moment: Applications, 224
6.6 Moving Loads, 235
6.7 Closure, 239
Problems, 239
References, 254

7 Stresses in beams, 255

7.1 Introduction, 255
7.2 Pure Bending—The Flexure Formula, 255
    7.2a Flooring, 264
7.3 Beams of Several Materials, 266
7.4 Reinforced Concrete Beams, 273
7.5 Shear Stresses in Beams, 278
7.6 Closure, 288
Problems, 289
References, 302

8 The deflection of beams, 304

8.1 Introduction, 304
8.2 The Elastic Curve, 305
8.3 The Moment Area Propositions, 309
    8.3a Cantilever beams, 313
    8.3b Simply supported beams, 320
8.4 The Superposition Method and Its Application, 325
8.5 Statically Indeterminate Beams, 334
8.6 Continuous Beams—The Three Moment Equation, 341
8.7 Closure, 345
Problems, 346
References, 359

9 Columns, 360

9.1 Introduction, 360
9.2 Columns—General, 361
9.3 Long Columns, 363
9.4 Intermediate Steel Columns—Historical, 368
9.5 Intermediate Steel Columns—AISC Design Procedure, 376
9.6 Eccentrically Loaded Columns, 381
9.7 Closure, 384
Problems, 385
References, 389

10 Combined stress, 391

    10.1   Introduction, 391
    10.2   Bending and Axial Loads Combined, 392
    10.3   Pure Shear, 394
    10.4   Combined Normal Stresses, 399
    10.5   General Case of Plane Stress, 408
    10.6   Combined Stresses in Circular Shafts, 413
    10.7   Closure, 419
    Problems, 420
    References, 427

Answers to even numbered problems, 428

Appendix A, Average mechanical properties of selected materials, 434

Appendix B, Properties of structural shapes, 436

Appendix C, Column tables, 456

Appendix D, Properties of pipe, 467

Appendix E, Miscellaneous tables, 474

Introduction to the SI system of units

At the time of the French Revolution, the systems of weights and measures used throughout the world were an incoherent and almost hopeless jumble. International trade as well as the interchange of scientific information suffered greatly due to this condition. French scientists and scholars of this era developed a rational system of weights and measures, which was called the metric system and which was adopted by most countries of the world. In 1960, the General Conference of Weights and Measures extensively revised and simplified the older metric system and gave it the French title, System International d' Unites (International System of Units), commonly abbreviated SI. The latest revisions and additions were made in an international conference in 1971, and work still continues on these standards.

The SI system consists of three classes of units, namely:

1. Base Units
2. Supplementary Units
3. Derived Units
   (a) With Special Names
   (b) Without Special Names

Table I-1 gives the seven base units of the SI system. Several observations concerning this table should be noted. The unit of length is the metre (not meter), and the kilogram is a unit of mass, not weight. Also, symbols are never pluralized, never written with a period; and the use of upper- and lowercase symbols *must* be used as shown *without exception.*

Table I-2 gives the supplementary units of the SI system. These units can be regarded as either base units or derived units.

Table I-3 gives the derived units (with and without symbols) often used in engineering mechanics. These derived units are formed by the algebraic combination of base and supplementary units. It is noted that where the name is named for a person the first letter of the symbol appears as a

Table I-1 Base SI Units

Quantity	Name of Base SI Unit	Symbol
length	metre	m
mass	kilogram	kg
time	second	s
electric current	ampere	A
thermodynamic temperature	kelvin	K
amount of substance	mole	mol
luminous intensity	candela	cd

Table I-2 Supplementary SI Units

Quantity	Supplementary SI Unit	Symbol
plane angle	radian	rad
solid angle	steradian	sr

Table I-3 Derived SI Units

Quantity	Name	Symbol	Formula	Expressed in Terms of Base Units
acceleration	acceleration	m/s^2	m/s^2	m/s^2
area	square metre	m^2	m^2	m^2
density	kilogram per cubic metre	—	kg/m^3	$kg \cdot m^{-3}$
energy or work	joule	J	$N \cdot m$	$m^2 \cdot kg \cdot s^{-2}$
force	newton	N	$m \cdot kg \cdot s^{-2}$	$m \cdot kg \cdot s^{-2}$
length	metre	m	m	m
mass	kilogram	kg	kg	kg
moment	newton-metre	$N \cdot m$	$N \cdot m$	$m^2 \cdot kg \cdot s^{-2}$
moment of inertia of area	—	m^4	m^4	m^4
plane angle	radian	rad	rad	rad
power	watt	W	J/s	$m^2 \cdot kg \cdot s^{-3}$
pressure or stress	pascal	Pa	N/m^2	$N \cdot m^{-2}$
rotational frequency	revolutions per second	rev. per sec.	s^{-1}	s^{-1}
temperature	degree celsius	°C	°C	1 °C = 1 K
time	second	s	s	s
torque (see moment)	newton-metre	$N \cdot m$	$N \cdot m$	$m^2 \cdot kg \cdot s^{-2}$
velocity (speed)	metre per second	metre per sec.	m/s	$m \cdot s^{-1}$
volume	cubic metre	—	m^3	m^3

capital; e.g., newton is N, etc. Otherwise the convention is to make the symbol lowercase.

For the engineer, the greatest confusion has been the units for mass and weight. The literature abounds with units such as slugs, pounds mass, pound force, poundal, kilogram force, kilogram mass, dyne, etc. In the SI system, the base unit for *mass* (not weight or force) is the kilogram, which is equal to the mass of the international standard kilogram located at the International Bureau of Weights and Measures. It is used to specify the quantity of matter in a body. The mass of a body never varies, and it is independent of gravitational force.

The SI *derived* unit for force is the newton (N). The unit of force is defined from Newton's Law of Motion, namely: force is equal to mass times acceleration (F = ma). Thus, by this definition, one newton applied to a mass of one kilogram gives the mass an acceleration of one metre per second squared (N = kg·m/s^2). The newton is used in all combination of units that include force, i.e., pressure or stress (N/m^2), energy (N·m), power (N·m/s = W), etc. By this procedure, the unit of force is not related to gravity as was the older kilogram-force.

Weight is defined as a measure of gravitational force acting on a material object at a specified location. Thus weight is a force that has both a mass component and an acceleration component (gravity). Gravitational forces vary by about 0.5 percent over the earth's surface. For non-precision measurements, these variations can normally be ignored. Thus, a constant mass has an approximate constant weight on the surface of the earth. The agreed standard value (standard acceleration) of gravity is 9.806 650 m/s^2. Figure I-1 illustrates the difference between mass (kilogram) and force (newton).

The term, "mass" or "unit mass," should only be used to indicate the quantity of matter in an object, and the old practice of using weight in such cases should be avoided in engineering and scientific practice. However, since the determination of an object's mass will be accomplished by the use of a weighing process, the common usage of the term weight instead of mass is expected to continue, but should be avoided.

In order for the SI system to be universally understood, it is most important that the symbols for the SI units and the conventions governing their use be strictly adhered to. Care should be taken to use the correct case for symbols, units, and their multiples (for example, K for kelvin, k for kilo, m for milli, M for mega). As noted earlier, unit *names* are never capitalized except at the beginning of a sentence. SI unit *symbols* derived from proper names are written with the first letter in uppercase; all other symbols are written in lowercase. For example, m (metre), s (second), K (kelvin), Wb (weber). Also, unit names form their plurals in the usual manner. Unit symbols are always written in singular form, for example: 350 megapascals, or 350 MPa; 50 milligrams, or 50 mg. Since the unit symbols

4 Strength of Materials for Engineering Technology

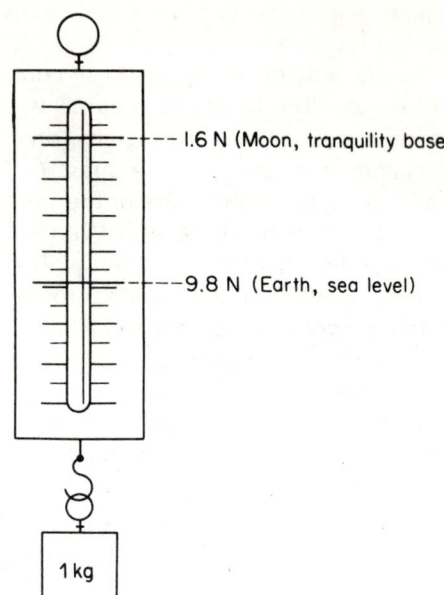

Fig I-1   Mass and force

are standardized, the symbols should always be used in preference to the unit names. An exception is made when a number written out in words precedes the unit, e.g., seven metres, not seven m. Unit symbols are not followed by a period unless they occur at the end of a sentence, and the numerical value associated with a symbol should be separated from that symbol by a space, e.g., 1.81 mm, *not* 1.81mm. The period is only to be used as a decimal marker. Since the comma is used by some countries as a decimal marker, the SI system does not use the comma. A space is used to separate large numbers in groups of threes starting from the decimal in either direction. Thus, 3 807 747.0 and 0.030 704 254 indicate this type of grouping. Notice that for numerical values less than one, the decimal point is preceded by a zero. For a number of four digits, the space can be omitted.

In addition, certain style rules should also be adhered to:

1. When a product is to be indicated, use a space between unit names, e.g., newton metre.
2. When a quotient is indicated, use the word "per," e.g., metre per second.
3. When a product is indicated, use the word "square," "cubic," etc., e.g., square metre.
4. In designating the product of units, use a centered dot, e.g., N·s, kg·m.

5. For quotients, use a solidus (/) or a negative exponent, e.g., m/s or m·s^{-1}. The solidus (/) should not be repeated in the same expression unless ambiguity is avoided by using parentheses. Thus, one should use m/s^2 or m·s^{-2}, but *not* m/s/s; also, use m·kg/(s^3·A) or m·kg·s^{-3}·A^{-1}, but *not* m·kg/s^3/A.

One of the features of the older metric system and the current SI system that is most useful is the fact that multiples and submultiples of the units are in terms of factors of 10. Thus the prefixes given in Table I-4 are used in conjunction with SI units to form names and symbols of multiples of SI units. Certain general rules apply to the use of these prefixes:

1. The prefix becomes part of the name or symbol with no separation, e.g., kilometre, megagram, etc.
2. Compound prefixes should not be used: use GPa, not kMPa.
3. In calculations, use powers of ten in place of prefixes.
4. Try to select a prefix where the numerical value will fall between 0.1 and 1000. This rule may be disregarded when it is better to use the same multiple for all items. It is also recommended that prefixes representing 10 raised to a power that is a multiple of 3 be used; e.g., 100 mg not 10 cg.
5. The prefix is combined with the unit to form a new unit, which can be provided with a positive or negative exponent. Therefore, mm^3 is (10^{-3}m)^3 or 10^{-9}m^3.
6. Where possible, avoid the use of prefixes in the denominator of compound units. The exception to this rule is the prefix k in the base unit kg (kilogram).

Table I-4 Factors of Ten for SI Units

Prefix	Symbol	Factor	
tera	T	10^{12}	1 000 000 000 000
giga	G	10^9	1 000 000 000
mega	M	10^6	1 000 000
kilo	k	10^3	1 000
hecto	h	10^2	100
deka	da	10^1	10
deci	d	10^{-1}	0.1
centi	c	10^{-2}	0.01
milli	m	10^{-3}	0.001
micro	μ	10^{-6}	0.000 001
nano	n	10^{-9}	0.000 000 001
pico	p	10^{-12}	0.000 000 000 001
femto	f	10^{-15}	0.000 000 000 000 001
atto	a	10^{-18}	0.000 000 000 000 000 001

There are certain units outside the SI that may be used together with the SI units and their multiples. These are recognized by the International Committee for Weights and Measures as having to be retained because of their practical importance. These are listed in Table I-5.

Table I-5 Retained Common Units

Quantity	Name of Unit	Unit Symbol	Definition
time	minute	min	1 min = 60 s
	hour	h	1 h = 60 min = 3600 s
	day	d	1 d = 24 h = 86 400 s
plane angle	degree	°	$1° = 1/(\pi/180)$ rad
	minute	'	$1' = (1/60)° = 2.909 \times 10^{-4}$ rad
	second	"	$1" = (1/60)' = 4.848 \times 10^{-6}$ rad
volume	litre	l	$1 = 1$ dm^3 $= 10^{-3}$ m^3
mass	tonne	t	1 t = 1 Mg = 10^3 kg

It is almost universally agreed that when a new language is to be learned the student should be completely immersed and made to "think" in the new language. This technique has been proven most effective by the Berlitz language schools and the Ulpan method of language teaching. A classical joke about this is of the American traveling in Europe who was amazed that two-year-old children were able to speak "foreign" languages. In dealing with the SI system, the student should not "think" in terms of customary units and then perform a mental conversion. It is better to learn to "think" in terms of the SI system, which will then become a second language. However, there will be times when it may be necessary to convert from customary U.S. units to SI units. In order to facilitate such conversions, Table I-6 gives some commonly used conversion factors.

Table I-6 Conversion Factors

Multiply	By	To Obtain
atmospheres	2.992×10^1	inches mercury (32 deg. F)
atmospheres	1.033×10^4	kilogram/sq metre
atmospheres (760 torr)	1.013×10^2	kilopascals
bars	9.869×10^{-1}	atmospheres
bars	1.000×10^2	kilopascals
British thermal units (Btu)	3.927×10^{-4}	horsepower-hours
British thermal units (Btu)	1.056	kilojoules
British thermal units (Btu)	2.928×10^{-4}	kilowatt-hours
British thermal units (Btu)	1.221×10^{-8}	megawatt-days

Introduction 7

Table I-6 (*continued*)

Multiply	By	To Obtain
Btu/hr-square foot	3.153×10^{-4}	Watts/sq centimeter
Btu/hr-sq ft-deg. F	5.676×10^{-4}	W/sq cm-degree Celsius
Btu/minute	2.356×10^{-2}	horsepower
Btu/minute	1.757×10^{1}	Watts
calories	4.190	Joules
cubic feet	2.832×10^{-2}	cubic metres
cubic feet	2.832×10^{1}	litres
cubic feet/min	4.720×10^{-4}	cubic metres/sec
cubic metres	8.107×10^{-4}	acre-feet
cubic metres	3.531×10^{1}	cubic feet
cubic metres	2.642×10^{2}	gallons (US)
cubic metres/sec	2.119×10^{3}	cubic feet/min
cubic metres/sec	1.585×10^{4}	gallons/min
degrees Celsius	$(9/5)\,C + 32$	degrees Fahrenheit
degrees Fahrenheit	$5/9\,(F - 32)$	degrees Celsius
feet	3.048×10^{-1}	metres
feet of H_2O (39.2 deg. F)	3.048×10^{2}	kilogram/sq metre
feet of H_2O (39.2 deg. F)	4.335×10^{-1}	pounds/sq inch
feet/sec	3.048×10^{-1}	metres/sec
foot-pound (force)	1.356	Joules
foot-pounds (force)/min	2.260×10^{-2}	Watts
gallons	3.785×10^{-3}	cubic metres
gallons/min	6.309×10^{-5}	cubic metres/sec
horsepower	4.244×10^{1}	British thermal units/min
horsepower	7.457×10^{-1}	kilowatts
horsepower-hours	2.547×10^{3}	British thermal units
horsepower-hours	7.457×10^{-1}	kilowatt-hrs
inches of H_2O (39.2 deg. F)	2.491×10^{-1}	kilopascals
inches mercury (32 deg. F)	3.342×10^{-2}	atmospheres
inches mercury (32 deg. F)	3.453×10^{2}	kilograms/sq metre
inches mercury (32 deg. F)	3.386	kilopascals
inches mercury (32 deg. F)	4.912×10^{-1}	pounds/sq inch
Joules	7.376×10^{-1}	foot-pounds (force)
Joules	1.000	Watt-seconds
Joules	2.387×10^{-1}	calories
kilograms	2.205	pounds
kilograms	1.102×10^{-3}	tons (short)
kilograms/cubic metre	6.243×10^{-2}	pounds/cubic foot

Table I-6 (continued)

Multiply	By	To Obtain
kilograms/square metre	9.678×10^{-5}	atmospheres
kilograms/square metre	3.281×10^{-3}	ft H_2O (at 39.2 deg. F)
kilograms/square metre	2.896×10^{-3}	inches mercury (32 deg. F)
kilograms/square metre	1.422×10^{-3}	pounds/sq inch
kilojoules	9.471×10^{-1}	British thermal units
kilopascals	4.015	inches H_2O (at 39.2 deg. F)
kilopascals	1.450×10^{-1}	pounds (force)/sq inch
kilopascals	2.953×10^{-1}	inches mercury (32 deg. F)
kilopascals	1.000×10^{-2}	bars
kilopascals	9.869×10^{-3}	atmospheres (760 torr)
kilowatts	1.341	horsepower
kilowatt-hours	3.413×10^{3}	British thermal units
kilowatt-hours	1.341	horsepower-hours
kilowatt-hours	4.167×10^{-5}	megawatt-days
litre	3.531×10^{-2}	cubic feet
megawatt-days	8.189×10^{7}	British thermal units
megawatt-days	2.400×10^{4}	kilowatt-hours
metres	3.281	feet
Newtons	2.248×10^{-1}	pounds (force)
pounds	4.536×10^{-1}	kilograms
pounds (force)	4.448	Newtons
pounds/cubic feet	1.602×10^{1}	kilograms/cu metre
pounds/square inch	2.307	ft H_2O (at 39.2 deg. F)
pounds/square inch	2.036	inches mercury (32 deg. F)
pounds/square inch	7.031×10^{2}	kilograms/sq metre
pounds/square inch	6.895	kilopascals
square feet	9.290×10^{-2}	square metres
square metres	2.471×10^{-4}	acres
square metres	1.076×10^{1}	square feet
tonnes	2.205×10^{3}	pounds
tons (short)	9.072×10^{2}	kilograms
Watts	5.688×10^{-2}	Btu/minute
Watts	4.427×10^{1}	foot-pounds (force)/minute
Watt-seconds	1.000	Joules
Watts/sq centimeter	3.171×10^{3}	Btu/hr-sq ft
Watts/sq cm-deg. C	1.762×10^{3}	Btu/hr-sq ft-deg. F

REFERENCES

American Iron and Steel Institute. *AISI Metric Practice Guide—SI Units and Conversion Factors for the Steel Industry.* Washington, DC: American Iron and Steel Institute, 1975.

American Society of Mechanical Engineers. *ASME Orientation and Guide for Use of SI (Metric) Units.* 5th ed., 1974.

American Society of Mechanical Engineers. *ASME Text Booklet—SI Units in Strength of Materials,* 1975.

Meriam, J. L. *Statics.* 2nd ed., SI version. New York: John Wiley & Sons, 1975.

Walker, K. M. *Applied Mechanics for Engineering Technology.* Reston, VA: Reston Publishing Co., 1974.

1

Concepts of stress and strain

1.1 INTRODUCTION

The objectives of any course in "Strength of Materials" are to study how materials behave when subjected to differing loadings and restraints and to be able to predict this behavior in a given situation. The ultimate goal of this study is to relate the external forces on a body to the internal forces and deformations of that body. In most instances we will not be able to calculate the exact distribution of the internal resisting forces in the body, but by the use of the equations of equilibrium, we can usually obtain the resultant of these forces. A review of statics, especially that portion concerning the equilibrium of bodies subjected to various external forces, is strongly suggested at this time.

The purposes of this chapter are to define certain terms such as stress and strain, to illustrate these definitions, and to apply the free body concept and these definitions to some common simple situations.

1.2 STRUCTURAL LOADS

Since all structures, structural members, machine elements, etc. can be subjected to various types of loading such as static loads, dynamic loads, and repeated loads, all of which may act over wide temperature ranges, let us first discuss load categories. For our present purposes we will separate loadings into two broad classifications: *static loading* and *dynamic loading*.

The category of static loading can be further broken down into three subdivisions, namely: *continuous loading, gradually* or *slowly applied loading,* and *repeated gradually applied loading.* The term *continuous load* is used to characterize a load that remains on a member for a long time period. The weight of a structure or a tank subjected to internal pressure for a long period of time is a common example of this type of loading. A *gradually* or *slowly applied load* is one that slowly builds up to its maximum value and does not cause shock or vibration when it is applied. The last type of static load is the *repeated gradually applied load.* In this load classification are included those loads that are gradually applied but repeated a large number of times. The repeated gradually applied load is important because it can cause failure under a load that would be safe if the load were applied once or only a few times. Failure of structures due to repeated loads usually occurs catastrophically.

When the application of a load is much more rapid than it is for those loads that are classified as static loads, appreciable shock or vibration can occur. This type of load is known as a *dynamic* or *impact load,* and its application can cause internal forces in the structure to momentarily exceed those that occur when the same load is applied gradually. In addition, the deflection of the structure will also momentarily exceed the deflection caused by an equal gradually applied load. As an example, the gradual placing of a weight on the end of a cantilever beam will cause the beam to deflect and gradually come to maximum value. However, if a weight of the same magnitude is dropped on the end of the beam, the maximum deflection will be found to be several times greater than for the same gradually applied load even if the height of drop is small.

In the following chapters of this book, we shall be concerned with static loads and, in particular, continuous or slowly applied loads. Structures or machine members that are subjected to either repeated or dynamic loads will not be part of our study. The reactions of structures to dynamic loads is a study area involving considerations beyond the scope of this text.

1.3 STRESS

The application of an external force to a solid causes internal resisting forces to exist within the body whose resultant will be equal in magnitude but opposite in direction to the applied force. A bar subjected to a longitudinal axial force that tends to elongate the bar is said to be *in tension,* while a bar subject to a longitudinal axial force that tends to decrease the axial dimension of the bar is said to be *in compression.* Figure 1.1 shows short bars in tension and compression. Due to the applied axial forces, internal resisting forces that are continuously distributed over the cross section of the bar are shown on the *free-body diagrams.* If the applied force passes

Strength of Materials for Engineering Technology

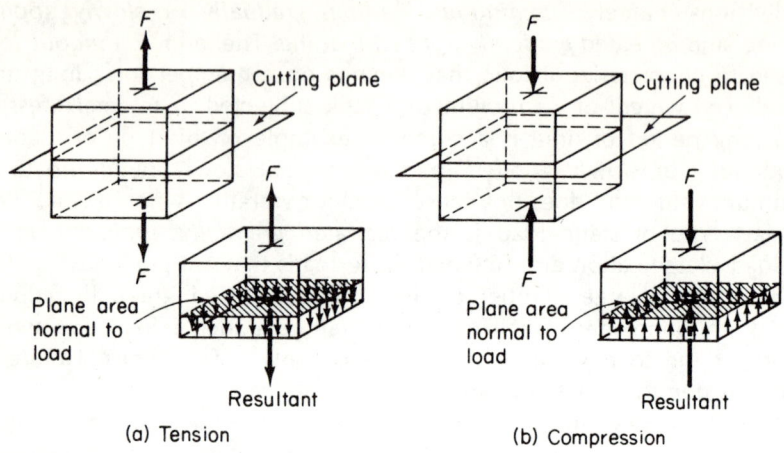

Fig 1.1 Tension and compression loads

through the *centroid* (center of gravity) of the cross section of the bar, the resisting forces will be uniformly distributed over the cross section of the bar. This rule is quite general and holds with only minor exceptions.

When comparing the resisting loads of geometrically similar members, we will find it more convenient to use as a basis of comparison the resistive load per unit area. For a uniform distribution of internal resistive forces, we simply express the resistive force per unit area in terms of the applied loading as

$$S = \frac{F}{A} \tag{1.1}$$

where S is the stress, A is the cross-sectional area of the bar, and F is the applied force. With force expressed in pounds, and area expressed in square inches, the stress is expressed in pounds per square inch (psi). For the loading conditions shown in Fig. 1.1, we note that the area is perpendicular (*normal*) to the direction of the applied force, and the stress defined by equation (1.1) is called a *normal stress*. If the distribution of the internal resistive force in a body is not uniform, it is necessary to consider a small element of area ΔA having a resultant resistive force ΔF acting on it. As ΔA is made very small, we approach the conditions that exist at a point in the body. For this situation the stress at a point can be expressed as

Chapter 1, Concepts of Stress and Strain   13

$$S = \left(\frac{\Delta F}{\Delta A}\right)_{(\Delta A \to 0)} \tag{1.2}$$

where the symbol $\Delta A \to 0$ reads "as ΔA approaches zero."

Illustrative problem 1.1

The body shown in Fig. 1.2 is subjected to an axial tensile load of 40 kN. Determine the stress at sections A-A, B-B, and C-C.

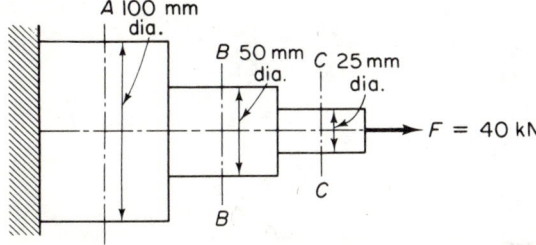

Fig 1.2   Illustrative problem 1.1

Solution:

Each of the sections must be capable of carrying the full load of 40 kN. The area at each of the sections is tabulated, and the stress at each section is F/A or 40 kN/A or 40 000/A Pa.

Section	Diameter (m)	Area (m²)	Stress = $\frac{40\,000}{\text{Area}}$
A-A	100×10^{-3}	7.85×10^{-3}	5.1 MPa
B-B	50×10^{-3}	1.96×10^{-3}	20.4 MPa
C-C	25×10^{-3}	4.91×10^{-4}	81.5 MPa

Illustrative problem 1.2

A uniform circular rod is hung vertically with one end free and the other end fixed. If the rod weighs w N/m, determine the stress at sections A-A, B-B, C-C, D-D, and E-E if the cross-sectional area of the rod is A.

Solution:

The stress in the bar arises from the weight of the portion of the bar hanging below the section in question, shown in Fig. 1.3(b). Thus we can tabulate the loads at each section as follows to obtain the stress on the section. As will be seen, the stress varies from zero at the free end to a maximum value at the fixed end. This type of stress variation in a member occurs frequently.

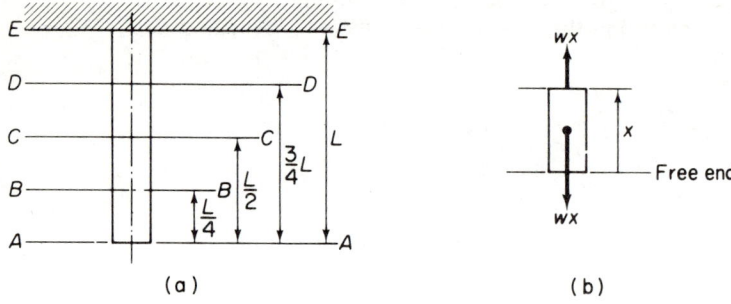

Fig 1.3   Illustrative problem 1.2

Section	Weight hanging on Section	Stress $= \dfrac{F}{A}$
A-A	0	0
B-B	$\dfrac{wL}{4}$	$\dfrac{wL}{4A}$
C-C	$\dfrac{wL}{2}$	$\dfrac{wL}{2A}$
D-D	$\dfrac{3wL}{4}$	$\dfrac{3wL}{4A}$
E-E	wL	$\dfrac{wL}{A}$

The tensile and compressive stresses discussed in the previous paragraphs arise from internal resistive forces within the body when the body is subjected to external forces. Another direct stress occurs when there is contact between two bodies. This external force is known as *bearing* and the contact pressure between the two bodies is known as a *bearing stress*. Some examples of bearing stresses are shown in the situation illustrated in Fig. 1.4. Bearing occurs between the post and plate, the plate and footing, and between the footing and the soil. The bearing stress in each case is the load F divided by the normal area carrying the load.

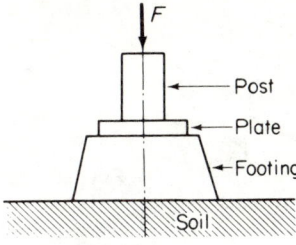

Fig 1.4 Bearing stress

Illustrative problem 1.3

If the post in Fig. 1.4 is a 150-mm × 150-mm member, the plate is 250 mm × 250 mm, and the base of the concrete footing is 750 mm × 750 mm, determine the bearing stresses when the structure must carry 450 kN.

Solution:

Bearing occurs at the three sections noted in the preceding paragraph. The bearing areas are noted and the bearing stress is 450 kN/A.

Location	Dimensions (m)	Bearing Area m²	Bearing Stress F/A
Post-Plate	$150(10)^{-3} \times 150(10)^{-3}$	$2.25(10)^{-2}$	20.0 MPa
Plate-Footing	$250(10)^{-3} \times 250(10)^{-3}$	$6.25(10)^{-2}$	7.2 MPa
Footing-Soil	$750(10)^{-3} \times 750(10)^{-3}$	$5.625(10)^{-1}$	800 kPa

Tensile and compression loads are resisted by areas that are perpendicular to the direction of the applied load. Shearing stresses occur when the force being resisted acts in the plane of the reacting area. Several representative situations that give rise to shear stresses are shown in Fig. 1.5. In each case, the shearing stress is the shear force divided by the area being sheared. Mathematically

$$S_s = \frac{F}{A} \qquad (1.1a)$$

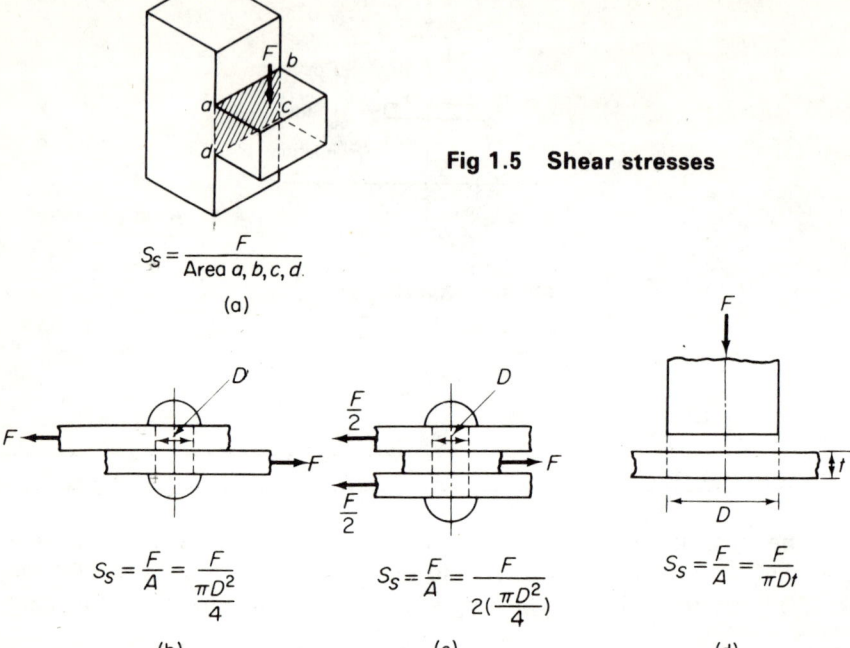

Fig 1.5 Shear stresses

Figures 1.5(a) and (b) show the shearing force being resisted by a single shear area. The situation shown in Fig. 1.5(c) illustrates a case of double shear, i.e., two areas resist the shear in the bolt being sheared. Figure 1.5(d) shows a punch shearing out a cylindrical disk from a plate with the shear area being the outside surface of the disk.

Illustrative problem 1.4

The eyebar shown in Fig. 1.6 must carry a load of 45 kN. Calculate the shear stress in the bolt.

Solution:

The bolt is in double shear and the shear area $= 2(\pi D^2/4) = 2(\pi/4)(10 \times 10^{-3})^2 = 1.57 (10)^{-4}$ m^2. The shear stress is given by

$$S_s = \frac{F}{A} = \frac{45\,000}{1.57\,(10)^{-4}} = 286.6 \text{ MPa}$$

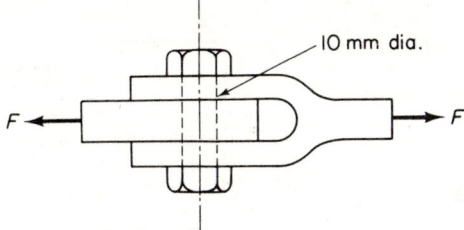

Fig 1.6   Illustrative problem 1.4

Illustrative problem 1.5

A punch is used to punch a 25-mm-diameter hole in a plate 6 mm thick. If the ultimate shearing stress in the material is 350 MPa, determine the necessary shearing force.

Solution:

Referring to Fig. 1.5(d), the shearing area is $\pi Dt = \pi(25)(10)^{-3}(6)(10)^{-3}$ = 471.2 $(10)^{-6}$ sq metres. Since $S_s = F/A$: $F = 471.2\,(10)^{-6} \times 350\,(10)^6 = 164.92$ kN.

1.4   STRAIN

The application of an external force gives rise not only to internal resistive forces in a body, but also to internal displacements in the body, which in the aggregate give rise to an elongation of the body if the external force is a tensile force. Similarly, a compressive force causes a shortening of the dimension of the body in the direction of the force. When discussing the elongation (or shortening) of a body, we will use several terms whose meaning must be clearly understood. The *deformation* will be designated by the symbol δ, and it will be taken to be the total change in a dimension due to an applied force. The *elongation* of the bar per unit of original length is designated by the symbol ε and is expressed as

$$\varepsilon = \frac{\delta}{L} \qquad (1.3)$$

where L is the original length. The elongation per unit length is called the *strain*. Strain is a dimensionless quantity usually expressed as inches per inch or metres per metre. These definitions are shown in Fig. 1.7

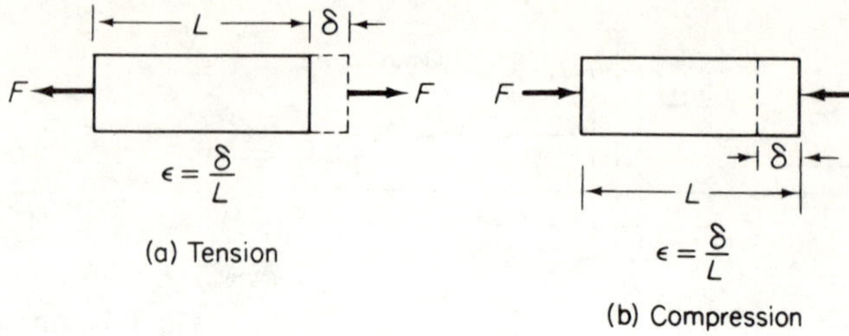

Fig 1.7 Strain

(a) Tension $\epsilon = \dfrac{\delta}{L}$

(b) Compression $\epsilon = \dfrac{\delta}{L}$

for tension and compression. It should be noted that the strain defined by equation (1.3) is an average value. The strain at any position is more correctly written as

$$\varepsilon = \left(\frac{\Delta \delta}{\Delta L}\right)_{(\Delta L \to 0)} \tag{1.4}$$

Illustrative problem 1.6

During the test of a specimen in a tensile testing machine, it is found that the specimen elongates 0.0024 inch between two punch marks that are initially 2.0 inches apart. Evaluate the strain.

Solution:

By definition the strain is

$$\varepsilon = \frac{\delta}{L} = \frac{0.0024}{2.0} = 0.0012 \text{ inches/inch},$$

which is the same as metres/metre

1.4a Shearing strain

When a member is subjected to a shear stress, it undergoes a deformation that is somewhat different from that described for the cases of tension and compression. Rather than an elongation or shortening, shearing stress causes an angular deformation of the body. Thus, a rectangular section becomes a parallelogram. The *shearing strain,* as shown in Fig. 1.8, can be defined as the total deformation divided by the height of the element L when the angle φ is small. Thus

Chapter 1, Concepts of Stress and Strain   19

$$\frac{\delta_s}{L} = \tan \varphi \cong \varphi \qquad (1.5)$$

(Since the angle φ is small, tan $\varphi \cong \varphi$, where φ is the angle expressed in radians.)
Therefore

$$\varphi = \frac{\delta_s}{L} \qquad (1.6)$$

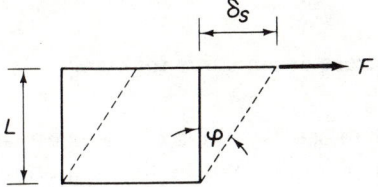

Fig 1.8   Shear strain

Illustrative problem 1.7

A short bar is subjected to a shearing force whose magnitude causes a total deformation of the top of the block with respect to its undeformed position of 0.005 metres. If the block is 2.5 metres high, what is the shearing strain?

Solution:

Using equation (1.6)

$$\varphi = \frac{\delta_s}{L} = \frac{0.005}{2.5} = 0.002 \text{ radians} \left(\frac{\text{metres}}{\text{metre}}\right)$$

1.5   HOOKE'S LAW

If a member is subjected to a specific stress, it will undergo a specific strain. If this strain vanishes, i.e., the body returns to its original dimensions upon the removal of the stress, the action is said to be *elastic*. However, if upon the removal of the stress the body does not return to its original dimensions and there is a residual strain, the action is said to be *inelastic*.

Almost three hundred years ago Robert Hooke observed that stress is directly proportional to strain. Figure 1.9 shows a typical stress-strain curve for

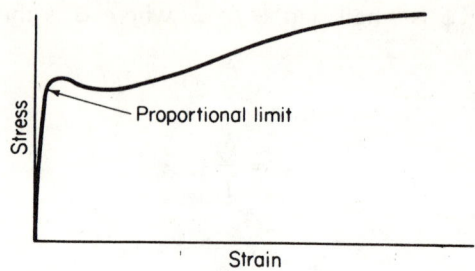

Fig 1.9 Typical stress-strain curve for steel

a material such as steel. As we can see from this curve, stress and strain are linearly proportional up to a point, known as the *proportional limit,* and past that point additional stress causes inelastic action to occur. All of the material in the subsequent chapters of this book will utilize the proportionality of stress and strain, which is known as *Hooke's Law*. We can mathematically express the proportionality of stress and strain as

$$S = E\varepsilon \tag{1.7}$$

where E is the constant of proportionality relating stress and strain. The proportionality constant E is sometimes known as *Young's modulus,* or more commonly the *modulus of elasticity*. Since stress is expressed in units of force per unit area and ε is non-dimensional (inches/inch or metres/metre), E has the same units as S, namely, psi or Pa. It is sometimes more convenient to express equation (1.7) in terms of the total deformation (δ), the length of the member (L), the cross-sectional area (A), and the applied load (F). Noting that $S = F/A$ and $\varepsilon = \delta/L$

$$\frac{F}{A} = E\frac{\delta}{L} \tag{1.8}$$

or

$$\delta = \frac{FL}{EA} \tag{1.9}$$

A similar proportionality exists between shear stress and shear strain. It can be written as

$$S_s = G\varphi \tag{1.10}$$

where G is the shear modulus of elasticity (more commonly known as the *modulus of rigidity*). Table 1.1 gives some typical values of E and G for some common materials.

Table 1.1 Modulus of Elasticity and Rigidity for Some Common Materials

Material	E (GPa)	E (psi)	G (GPa)	G (psi)
Carbon Steel	210	30×10^6	85	12×10^6
Alloy Steel	210	30×10^6	85	12×10^6
Cast Iron (grey)	105	15×10^6	40	6×10^6
Aluminum	70	10×10^6	28	4×10^6
Brass	105	14×10^6	40	6×10^6
Copper	115	17×10^6	40	6×10^6

Illustrative problem 1.8

A steel rod is used to support a weight of 2 tons. If the allowable stress is 20,000 psi, determine the diameter of the rod and its elongation if it is 2 ft long. The modulus of elasticity is 30×10^6 psi for steel.

Solution:

The tensile stress is given by $S = F/A$. Therefore, $A = F/S$. Using the numerical data of the problem in consistent units

$$A = \frac{2000 \times 2}{20,000} = 0.2 \text{ sq. in. } (1.29 \times 10^{-4} \text{ m}^2)$$

$$\frac{\pi D^2}{4} = 0.2 \text{ and } D = 0.505 \text{ in. } (1.283 \times 10^{-2} \text{ m})$$

The elongation may be found using either equation (1.7) or (1.9)

$$\varepsilon = \frac{S}{E} = \frac{20,000}{30 \times 10^6} = 0.67 \times 10^{-3} \text{ in./in. (m/m)}$$

$$\delta = \varepsilon L = 0.67 \times 10^{-3} \times 2 \times 12 = 16 \times 10^{-3}$$
$$= 0.016 \text{ in. } (4.06 \times 10^{-4} \text{ m})$$

Alternately, $\quad \delta = \dfrac{FL}{EA} = \dfrac{2000 \times 2 \times 24}{30 \times 10^6 \times 0.2} = 0.016 \text{ in. } (4.06 \times 10^{-4} \text{ m})$

Illustrative problem 1.9

A long circular uniform rod is hung vertically with one end free and the other end fixed. Determine the elongation of the rod if it weighs w lb/ft, and its total weight is W lb.

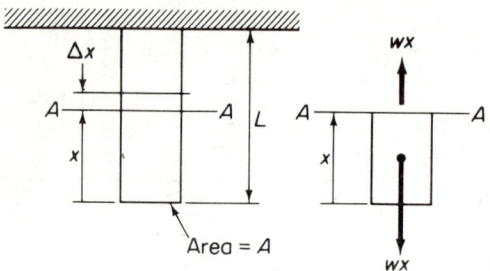

Fig 1.10 Illustrative problem 1.9

Solution:

Let us consider the load on section A-A located x feet from the free end. The tensile force on this section is equal to the weight of the bar below it, i.e., wx and the stress is wx/A. Notice that at the free end of the bar the stress is zero, and at the fixed end it is wL/A. The strain at section A-A will be the stress divided by the modulus of elasticity. Thus $\varepsilon_x = wx/EA$, and the strain will vary from zero at the free end of the bar to a maximum value of wL/EA at the built-in end of the bar. The total elongation of an element of height Δx located x feet from the free end will be $\varepsilon_x(\Delta x)$ or $(wx/EA)(\Delta x)$. In order to determine the total elongation of the bar, it is necessary to sum all such terms from the free end to the fixed end of the bar. This summation can best be performed by plotting the curve shown in Fig. 1.11. The small cross-hatched area is $x\Delta x$. Thus, the total area under this curve is the summation of all

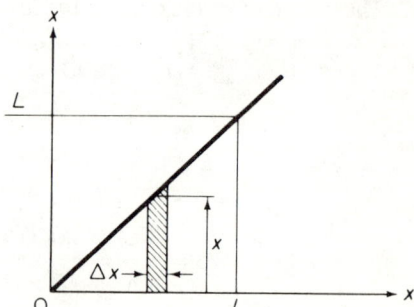

Fig 1.11 Evaluation of $x(\Delta x)$

of the $x \Delta x$ terms. The area under the curve for the triangle is $\frac{1}{2}(L)(L) = L^2/2$, which yields the total elongation of the bar as $(w/EA)(L^2/2)$. Since W, the total weight, is equal to wL, the total elongation of the bar can be written as $\frac{1}{2}(WL/EA)$. It is interesting to note that the elongation of the bar due to its own weight is one-half the value that would be obtained due to a force numerically equal to W acting at the end of the bar.

1.6 POISSON'S RATIO

As was noted earlier, when a bar or member is subjected to a direct stress, that bar changes its axial dimension by an amount directly proportional to the magnitude of the applied stress—if the stress is tensile, then a bar elongates; if the stress is compressive, a bar contracts. In addition to this direct effect, it is found that the bar simultaneously undergoes a change in its lateral dimension. Thus for the bar shown in Fig. 1.12, the length will increase from L to $L + \varepsilon L$, while the sides contract from their original dimensions of W and h. If we now denote the ratio of the lateral strain to the longitudinal strain as μ we have

$$\mu = \frac{\varepsilon_w}{\varepsilon_y} \quad \text{and} \quad \mu = \frac{\varepsilon_h}{\varepsilon_y} \tag{1.11}$$

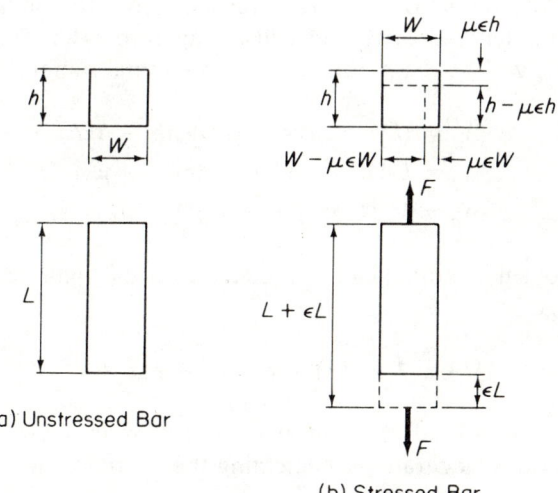

(a) Unstressed Bar

(b) Stressed Bar

Fig 1.12 Lateral strain due to longitudinal stress

This ratio is known as *Poisson's ratio*, and for steel it is usually taken to be approximately 0.3. Typical values of μ for some materials are given in Table 1.2.

Table 1.2 Value of Poisson's Ratio

Material	μ
Aluminum	0.34
Aluminum Alloys	0.32
Copper	0.35
Iron	0.28
Carbon Steel	0.28
Magnesium	0.33
Titanium	0.34

It is important to note that if the longitudinal strain causes extension, the lateral strain will be compressive; i.e., if the applied force causes the bar to elongate in the direction of the applied force, the bar will contract in the other two directions. Similarly, if a bar is compressed (shortened) in the longitudinal direction, it will undergo an increase in its lateral dimensions.

For the conditions shown in Fig. 1.12, the bar will increase its length from L to $L + \varepsilon L$, and the W and h dimensions will decrease from W to $W - \mu\varepsilon W$ and h to $h - \mu\varepsilon h$, respectively. The initial volume of the bar will be denoted as V_o and will be equal to LWh. The final volume of the bar, V_f, will be the product of the three final dimensions, namely,

$$V_f = (L + \varepsilon L)(W - \mu\varepsilon W)(h - \mu\varepsilon h) \tag{1.12}$$

Factoring,
$$V_f = LWh(1 + \varepsilon)(1 - \mu\varepsilon)(1 - \mu\varepsilon) \tag{1.12a}$$

or
$$V_f = V_o(1 + \varepsilon)(1 - \mu\varepsilon)(1 - \mu\varepsilon) \tag{1.12b}$$

If we carry out the multiplication indicated on the right side of equation (1.12b), we get

$$V_f = V_o[1 + \varepsilon(1 - 2\mu) + \varepsilon^2(\mu^2 - 2\mu) + \varepsilon^3 \mu^2] \tag{1.13}$$

Since ε is usually a small number, we can obtain an approximate relation for V_f that is quite accurate by neglecting the ε^2 and ε^3 terms in equation (1.13). Thus

$$V_f = V_o[1 + \varepsilon(1 - 2\mu)] \tag{1.14}$$

The *volumetric strain* is defined to be the ratio of the change in volume of the element divided by its initial volume. Using equation (1.14)

$$\frac{V_f - V_o}{V_o} = \frac{V_o[1 + \varepsilon(1 - 2\mu)] - V_o}{V_o} = \varepsilon(1 - 2\mu) \qquad (1.15)$$

Illustrative problem 1.10

A steel bar 25 mm × 25 mm × 250 mm is subjected to an axial tensile stress of 140 MPa (Fig. 1.13). Determine the axial elongation, the lateral contraction, the final volume, and the volumetric strain. Use $E = 210$ GPa and $\mu = 0.3$.

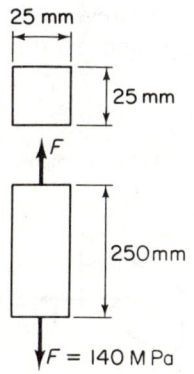

Fig 1.13   Illustrative problem 1.10

Solution:

The axial stress is 140 MPa. Therefore the axial strain, ε, is $S/E = 140 \times 10^6 / 210 \times 10^9 = 0.000\ 667$ metres/metre. The lateral strain is $\mu\varepsilon = 0.3 \times (0.000\ 667) = 0.000\ 20$ metres/metre. The elongation of the 250-mm length is $250 \times 0.000\ 667 = 0.166\ 75$ mm. The contraction of each side is $25 \times 0.000\ 20 = 0.005$ mm. The final dimensions of the bar are 250.166 75 mm, 24.995 mm and 24.995 mm. The final volume is $250.166\ 75 \times 24.995 \times 24.995 = 156\ 291.68$ mm³. If we had used the approximation given by equation (1.14), we would have $V_f = 250 \times 25 \times 25\ [1 + 0.000\ 667(0.4)] = 156\ 291.68$ mm³.

$$\frac{V_f - V_o}{V_o} = \frac{156\ 291.68 - 156\ 250}{156\ 250} = 0.000\ 266\ 8$$

Remember that even though the strains appear as small numbers, they are significant and are not to be ignored. Also, the foregoing discussion presumes that the material being deformed is not stressed beyond its proportional limit.

When a body is subjected to stresses in more than one direction, the strains may be found by considering each stress separately and determining the effect of each stress separately. The total effect is then the algebraic sum of the separate effects with attention to the fact that, by convention, tension causes a positive deformation, and compression causes a negative deformation.

Illustrative problem 1.11

A homogeneous, rectangular parallelepiped is subjected to a tensile stress of 35 MPa in the x direction and a compressive stress of 28 MPa in the y direction. Determine the strain in the x, y, and z directions if $E = 210$ GPa and $\mu = 0.3$.

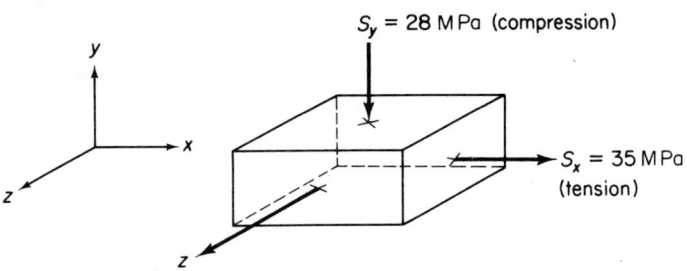

Fig 1.14 Illustrative problem 1.11

Solution:

Let us first consider the effect of the x-directed stress. The strain in the x direction is S_x/E. In the y and z directions, the x-directed stress causes a compression of $-\mu\varepsilon_x$ or $-\mu S_x/E$. Due to the y-directed compressive stress, there is a strain in the y direction equal to $-S_y/E$ (negative because stress is compression), and this stress causes a strain of $\mu S_y/E$ in both the x and z directions. We may tabulate these effects and use the data of the problem to numerically evaluate the strains.

Direction	Direct Stress	Direct Strain	Induced Strain due to x Stress	Induced Strain due to y Stress	Induced Strain due to z Stress
x	$S_x = 35$ MPa	$\dfrac{S_x}{E}$ $= \dfrac{35(10)^6}{210(10)^9}$	0	$\dfrac{\mu S_y}{E}$ $= \dfrac{0.3(28)10^6}{210(10)^9}$	0
y	$S_y = -28$ MPa	$\dfrac{S_y}{E}$ $= \dfrac{-28(10)^6}{210(10)^9}$	$-\dfrac{\mu S_x}{E}$ $= \dfrac{-0.3(35)10^6}{210(10)^9}$	0	0
z	$S_z = 0$	$\dfrac{S_z}{E} = 0$	$-\dfrac{\mu S_x}{E}$ $= \dfrac{-0.3 \times 35(10)^6}{210(10)^9}$	$\dfrac{\mu S_y}{E}$ $= \dfrac{0.3(28)10^6}{210(10)^9}$	0

The total strain in each direction is,

$$\varepsilon_x = \dfrac{35(10)^6}{210(10)^9} + \dfrac{0.3(28)10^6}{210(10)^9} = 0.000\ 167 + 0.000\ 04$$
$$= 0.000\ 21\ \text{m/m}$$
$$\varepsilon_y = \dfrac{-28(10)^6}{210(10)^9} - \dfrac{0.3(35)10^6}{210(10)^9} = -0.000\ 133 - 0.000\ 05$$
$$= -0.000\ 183\ \text{m/m}$$
$$\varepsilon_z = \dfrac{-0.3(35)10^9}{210(10)^9} + \dfrac{0.3(28)10^6}{210(10)^9} = -000\ 05 + 0.000\ 04$$
$$= -0.000\ 01\ \text{m/m}$$

We can generalize the results of Illustrative Problem 1.11 by considering the case of a parallelepiped subjected to three mutually perpendicular tensile stresses, as shown in Fig. 1.15. Using the same reasoning as for Illustrative Problem 1.11, we can construct the Table 1.3.

28   Strength of Materials for Engineering Technology

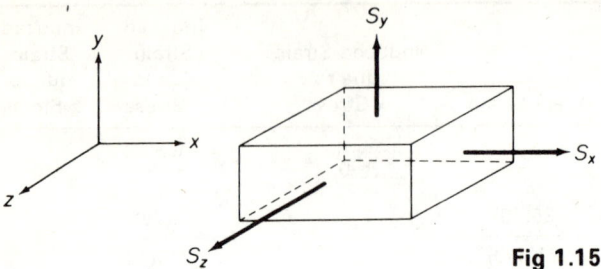

Fig 1.15   Generalized stresses

Table 1.3   Generalized Stresses

Direction	Direct Stress	Direct Strain	Induced Strain due to x Stress	Induced Strain due to y Stress	Induced Strain due to z Stress
x	S_x	$\dfrac{S_x}{E}$	0	$-\dfrac{\mu S_y}{E}$	$-\dfrac{\mu S_z}{E}$
y	S_y	$\dfrac{S_y}{E}$	$-\dfrac{\mu S_x}{E}$	0	$-\dfrac{\mu S_z}{E}$
z	S_z	$\dfrac{S_z}{E}$	$-\dfrac{\mu S_x}{E}$	$-\dfrac{\mu S_y}{E}$	0

Therefore,

$$\varepsilon_x = \frac{S_x}{E} - \frac{\mu S_y}{E} - \frac{\mu S_z}{E} \tag{1.16}$$

$$\varepsilon_y = \frac{S_y}{E} - \frac{\mu S_x}{E} - \frac{\mu S_z}{E} \tag{1.17}$$

$$\varepsilon_z = \frac{S_z}{E} - \frac{\mu S_x}{\mu} - \frac{\mu S_y}{E} \tag{1.18}$$

Equations (1.16), (1.17), and (1.18) are known as the *generalized Hooke's Law equations* and are the basic equations used in advanced studies of the theory of elasticity.

An important relation that connects the three constants, E, G, and μ can be derived from theoretical considerations. Such a relation is important since the direct determination of μ is quite difficult due to the necessity

Chapter 1, Concepts of Stress and Strain   29

of measuring small lateral contractions accurately. This connecting equation is

$$G = \frac{E}{2(1 + \mu)} \tag{1.19}$$

Illustrative problem 1.12

Using the values of E and G from Table 1.1, calculate μ for carbon steel.

Solution:

From equation (1.19), $\mu = (E/2G) - 1$. Using the values of $E = 210$ GPa and $G = 85$ GPa

$$\mu = \frac{210(10)^9}{2(85)(10)^9} - 1$$

$$\mu = 0.24$$

The value for carbon steel from Table 1.2 is 0.28.

1.7 STATICALLY INDETERMINATE STRESSES

Structures or members whose loadings and supporting conditions give rise to more conditions than can be solved using the three equations of statics for equilibrium are classified as being *statically indeterminate.* In order to solve this type of situation, it is necessary to have additional information, such as the conditions at a support, etc., to fully solve the problem. The method of approach to the solution of problems of this type is best illustrated by examining a specific problem.

Illustrative problem 1.13

A steel rod 25 mm in diameter is placed inside of a brass cylinder as shown in Fig. 1.16. Both are initially 75 mm long. A 22-kN weight is placed on the assembly. Determine the force, stress, and contraction of each member.

Strength of Materials for Engineering Technology

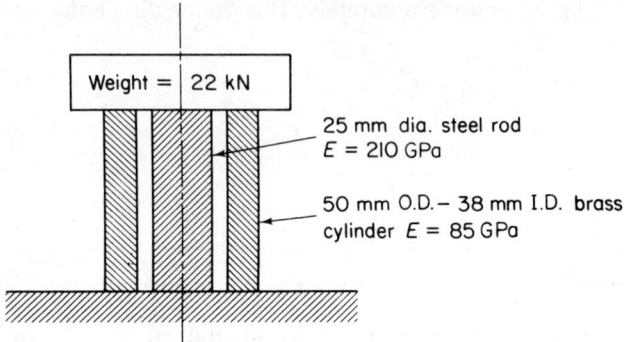

Fig 1.16 Illustrative problem 1.13

Solution:

Let us designate the force in the steel as F_S and the force in the brass as F_B. The necessary equation from statics for equilibrium is

$$F_S + F_B = 22\,000$$

The second condition of the problem is that the deflection of each member is the same. Therefore using δ_S as the contraction of the steel and δ_B as the contraction of the brass

$$\delta_S = \delta_B$$

In terms of Hooke's Law

$$\frac{F_S L_o}{A_S E_S} = \frac{F_B L_o}{A_B E_B}$$

or

$$F_S = F_B \left(\frac{A_S}{A_B}\right)\left(\frac{E_S}{E_B}\right) = F_B \left[\frac{\frac{\pi}{4}(25)^2 \times 10^{-6}}{\frac{\pi}{4}(50^2 - 38^2) \times 10^{-6}}\right] \cdot \left[\frac{210 \times 10^9}{85 \times 10^9}\right] = 1.462\, F_B$$

Using this information in the static equilibrium condition

$$1.462 F_B + F_B = 22\ 000$$
$$F_B = 8936 \text{ N}$$
$$F_S = 13\ 064 \text{ N}$$

The corresponding stresses are

$$S_B = 8936 \bigg/ \frac{\pi}{4}(0.050^2 - 0.038^2) = 10.77 \text{ MPa}$$

and

$$S_S = 13\ 064 \bigg/ \frac{\pi}{4}(0.025)^2 = 26.61 \text{ MPa}.$$

Since both members are originally and finally the same length, $\varepsilon_S = \varepsilon_B$. As a check

$$\varepsilon_S = \frac{S_S}{E_S} = -\frac{26.61 \times 10^6}{210 \times 10^9} = 0.1267 \times 10^{-3}$$

and

$$\varepsilon_B = \frac{10.77 \times 10^6}{85 \times 10^9} = 0.1267 \times 10^{-3}$$

which is satisfactory. The total deflection of each member $\delta = \varepsilon L_o = 0.1267 \times 10^{-3} \times 75 = 0.0095$ mm (contraction).

The procedure outlined above is for a specific type of statically indeterminate structure. The next section further illustrates this type of situation, and a portion of Chapter 8 is also devoted to statically indeterminate beams. Note that the equations of equilibrium are necessary in order to obtain a solution of statically indeterminate problems; however, these equations are not sufficient in themselves and must be augmented with further information, such as the stress-strain relation, boundary conditions, etc.

1.8 TEMPERATURE STRESSES

If a bar is not restrained in any way, it will be found that an increase in temperature will cause an increase in its dimensions, and conversely, a decrease in temperature will cause a decrease in its dimensions. It is usual to describe the dimensional change due to temperature changes in terms of the change in a linear dimension. Thus, the change in length of a bar, ΔL, is directly proportional to both the temperature change of the bar (Δt) and the original length of the bar, L_o. The constant of proportionality is called the *linear coefficient of expansion* and is usually denoted by the symbol α. Referring to Fig. 1.17, we can write

$$\Delta L = \alpha L_o (\Delta t) \tag{1.20}$$

or
$$L_f = L_o + \alpha L_o (\Delta t) = L_o(1 + \alpha \Delta t) \tag{1.21}$$

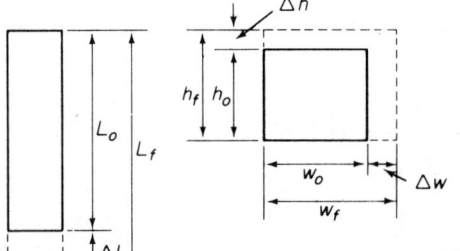

Fig 1.17 Thermal expansion of a bar

The coefficient of linear expansion (α) is therefore the change in length, per unit length for a one-degree change in temperature. Units for α are in./in. °F or m/m °C or mm/mm °C. Some typical values of the linear coefficient of expansion for several materials are given in Table 1.4, and a more complete table will be found in the Appendix.

Table 1.4 Linear Coefficients of Expansion

Material	Coefficient (α) in./in. °F	mm/mm °C
Aluminum	12.8×10^{-6}	23.0×10^{-6}
Brass	10.4×10^{-6}	18.7×10^{-6}
Copper	9.3×10^{-6}	16.7×10^{-6}
Iron	6.7×10^{-6}	12.1×10^{-6}
Steel (Mild)	6.5×10^{-6}	11.7×10^{-6}
Steel (Stainless)	9.9×10^{-6}	17.8×10^{-6}

Chapter 1, Concepts of Stress and Strain   33

It should be borne in mind that the values in Table 1.4 are representative values of α. Since α varies with temperature, the correct value of α to use in equations (1.20) and (1.21) is the average value for the temperature interval being considered.

Illustrative problem 1.14

Determine the changes in cross-sectional area and volume for the bar shown in Fig. 1.17 due to a temperature increase of Δt °F.

Solution:

The original cross-sectional area of the bar shown is $w_o h_o$. The final area is $w_f h_f$, where $w_f = w_o [1 + \alpha (\Delta t)]$ and $h_f = h_o [1 + \alpha (\Delta t)]$. The change in area is

$$\Delta \text{ area} = w_f h_f - w_o h_o = w_o [1 + \alpha (\Delta t)] [h_o (1 + \alpha (\Delta t))] - w_o h_o$$

Performing the algebraic operation indicated yields

$$\Delta \text{ area} = w_o h_o [1 + 2\alpha (\Delta t) + \alpha^2 (\Delta t)^2] - w_o h_o$$
$$= w_o h_o [2\alpha (\Delta t) + \alpha^2 (\Delta t)^2]$$

Since α is a small numerical quantity, α^2 is indeed very small when compared to α, and we can neglect the α^2 term. Therefore

$$\Delta \text{ area} = w_o h_o [2\alpha (\Delta t)]$$

Thus the coefficient of area expansion can be taken as twice the coefficient of linear expansion. In order to evaluate the change in volume of the bar, we proceed in a similar manner. The original volume is $L_o h_o w_o$ and the final volume is $L_f w_f h_f$. The change in volume, Δ volume, is

$$\Delta \text{ volume} = V_f - V_o = L_f w_f h_f - L_o w_o h_o$$
$$= L_o [1 + \alpha (\Delta t)] w_o [1 + \alpha (\Delta t)] h_o [1 + \alpha (\Delta t)] - L_o w_o h_o$$
$$= L_o w_o h_o [1 + \alpha (\Delta t)]^3 - L_o w_o h_o$$

Expanding the exponential term and neglecting α^2 and α^3 terms yields

$$\Delta \text{ volume} = L_o w_o h_o [3\alpha (\Delta t)]$$

The coefficient of volume expansion can therefore be closely approximated as being three times the coefficient of linear expansion.

If a bar is restrained and cannot move freely, stresses will be set up in the bar when it undergoes a change in temperature. Consider the bar in Fig. 1.18. If it were not restrained, it would expand a distance $\alpha L_o(\Delta t)$

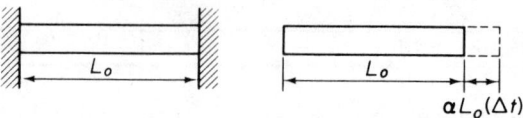

Fig 1.18 Restrained bar subject to a temperature increase

due to a temperature increase of Δt. However, the rigid walls prevent this from occuring and the final length of the bar must equal the initial length of the bar. We can consider the return of the bar from its expanded position to its initial position to have been accomplished by an axial compressive force on the bar, which caused a stress in the bar of magnitude S. From Hooke's Law

$$S = E\varepsilon = E \frac{\text{total deflection}}{\text{initial length}} \qquad (1.22)$$

and using $\alpha L_o(\Delta t)$ as the total deflection

$$S = E \frac{\alpha L_o(\Delta t)}{L_o} = E\alpha(\Delta t) \qquad (1.23)$$

From equation (1.23) we note that the stress in the bar, when the bar is restrained from freely expanding, is independent of the dimensions of the bar and is only a function of the temperature change, the modulus of elasticity, and the coefficient of linear expansion. Thus for a restrained, mild steel bar the stress for a temperature rise of 1 °C is $11.7 \times 10^{-6} \times 210 \times 10^9 = 2.457$ MPa; for a 100 °C rise, it is 245.7 MPa.

Illustrative problem 1.15

A 25-mm-diameter steel rod is used to brace two walls as shown. If the walls move 0.25 mm when the rod is subjected to a 50 °C temperature change, determine the stress in the rod. Assume $E = 210$ GPa and $\alpha = 11.7 \times 10^{-6}$.

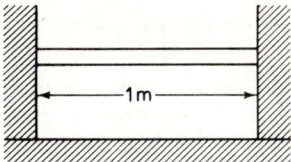

Fig 1.19   Illustrative problem 1.15

Solution:

The total elongation of the rod, if it is unrestrained, is

$$\Delta L = L_o \alpha (\Delta t) = 1 \times 11.7 \times 10^{-6} \times 50 = 585 \times 10^{-6} \text{ m} = 0.585 \text{ mm}$$

Since the walls move 0.25 mm, the rod must be compressed by a force causing a deflection of 0.585 − 0.25 = 0.335 mm. Therefore

$$S = E\varepsilon = 210 \times 10^9 \times \frac{0.335}{1000} = 70.35 \text{ MPa}$$

Thus far we have considered the case of temperature stresses that arise when a member is prevented from expanding freely due to imposed external conditions. When a structure is composed of more than one material, temperature stresses can occur due to the differences in the expansion between the materials. This condition is shown by the situation illustrated in Fig. 1.20. In this instance, we have bars of different materials whose upper ends are free to move. However, the very large rigid member, whose weight is W, joining the upper end of the bars, can only move parallel to itself. Due to this action, one bar will elongate due to a temperature rise less than the amount of expansion that would occur if it were unrestrained, while the other end will elongate more than the amount it would expand for the same temperature rise if it too were unrestrained. The final lengths of both bars will be the same due to the constraint imposed by the upper rigid body. The elongation of rod A, if it is free, is $\alpha_A L_o (\Delta t)$ and that of rod B is $\alpha_B L_o (\Delta t)$. The difference in these lengths is made zero by compressing bar A an amount that we shall denote as δ_1 and elongating bar B an amount δ_2. The force in rod A plus the force in rod B must support the weight W. Thus

$$\text{initially } F_A + F_B = W \quad (1.24)$$
$$\text{finally } F'_A + F'_B = W \quad (1.25)$$

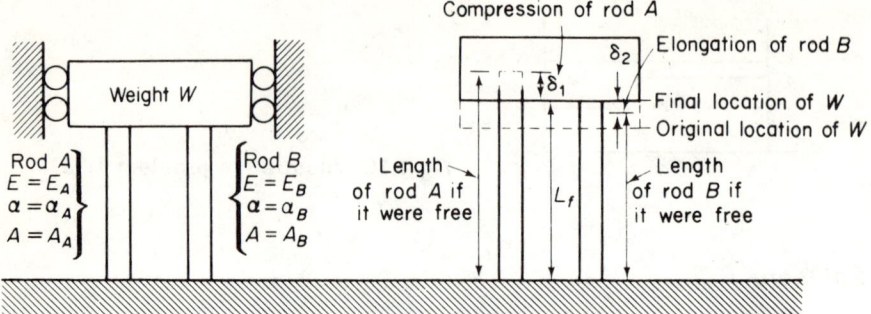

Fig 1.20 System with two different materials

Also, the extension of rod A due to the temperature increase minus the decrease in its length due to the increased compressive load in rod A must equal the extension of rod B plus the increase in its length due to the decrease in the compressive load in rod B. Therefore,

$$\alpha_A L_o (\Delta t) - \frac{(F'_A - F_A) L_o}{E_A A_A} = \alpha_B L_o (\Delta t) - \frac{(F'_B - F_B) L_o}{E_B A_B} \qquad (1.26)$$

Equations (1.25) and (1.26) yield the conditions necessary to solve the problem.

Illustrative problem 1.16

If rod A is a 25-mm-diameter stainless steel rod and rod B is a 50-mm-diameter mild steel rod, determine the stresses in each due to a 25 °C temperature rise if they support a weight W of 45 000 N. Assume both rods to have been the same length before W was placed on them, and that W was constrained to stay horizontal. Also take $E = 210$ GPa for both rods. Use α from Table 1.4.

Solution:

When the weight is initially placed on the rods, both rods will deflect by an equal amount. Thus $\delta_1 = \delta_2$ or

$$\frac{F_A L_o}{E_A A_A} = \frac{F_B L_o}{E_B A_B}$$

Since E_A equals E_B

$$F_A = \frac{(A_A)}{(A_B)} F_B$$

$$F_A = \left[\frac{\frac{\pi}{4}(25)^2}{\frac{\pi}{4}(50)^2}\right] F_B = \tfrac{1}{4} F_B$$

and both rods must support the total weight of 45 kN.

$$F_A + F_B = 45\,000$$

Therefore $\quad\tfrac{1}{4} F_B + F_B = 45\,000$

Thus, $\quad F_B = 36\,000$ N

$\quad F_A = 9000$ N

When the temperature increases by 25 °C the expansion of rod A is equal to $\alpha_A L_o (\Delta t)$ or

$$\alpha_A L_o (\Delta t) = 17.8 \times 10^{-6} \times L_o \times 25 = 445 \times 10^{-6} L_o$$

The mild steel rod, rod B, expands

$$\alpha_B (L_o)(\Delta t) = 11.7 \times 10^{-6} \times L_o \times 25 = 292.5 \times 10^{-6} L_o$$

Denoting the new forces in the rods as F'_A and F'_B

$$F'_A + F'_B = 45\,000 \text{ or } F'_A = 45\,000 - F'_B$$

Using the condition that the final length of both bars are equal [equation (1.26)],

$$\alpha_A L_o (\Delta t) - \frac{(F'_A - F_A) L_o}{E_A A_A} = \alpha_B L_o (\Delta t) - \frac{(F'_B - F_B) L_o}{E_B A_B}$$

Substituting numerical data and canceling out the common L_o term yields

$$445 \times 10^{-6} - \frac{(F'_A - 9000)}{E_A A_A} = 292.5 \times 10^{-6} + \frac{(36\,000 - F'_B)}{E_B A_B}$$

Substituting for F'_A and letting $E = E_A = E_B$

38 Strength of Materials for Engineering Technology

$$445 \times 10^{-6} - 292.5 \times 10^{-6} = \frac{36\,000 - F'_B}{E_A A_A} + \frac{36\,000 - F'_B}{E_B A_B}$$

$$= \frac{(36\,000 - F'_B)}{E}\left[\frac{1}{A_A} + \frac{1}{A_B}\right]$$

$$= \frac{4}{\pi E}(2000)(36\,000 - F'_B)$$

Simplifying $F'_B = 36\,000 - 152.5 \times 10^{-6} \times 210 \times 10^9$

$$\times \frac{\pi}{4(2000)} = 23\,424\text{ N}$$

and $F'_A = 21\,576\text{ N}$

Notice that the load in rod B has decreased since that rod has expanded less than rod A, while the load in rod A increased. The final stresses are F/A for each rod, which yield

$$S_A = \frac{F'_A}{A_A} = \frac{21\,576}{\frac{\pi}{4}(0.025)^2} = 43.95\text{ MPa}$$

$$S_B = \frac{F'_B}{A_B} = \frac{23\,424}{\frac{\pi}{4}(0.050)^2} = 11.93\text{ MPa}$$

Illustrative problem 1.17

A mild steel rod is brazed to a brass tube as shown in Fig. 1.21. Assuming E of brass = 70 GPa, E of steel = 210 GPa, $\alpha_{brass} = 18 \times 10^{-6}$, and $\alpha_{steel} = 11.7 \times 10^{-6}$, determine the stresses in each member due to α 50 °C temperature rise.

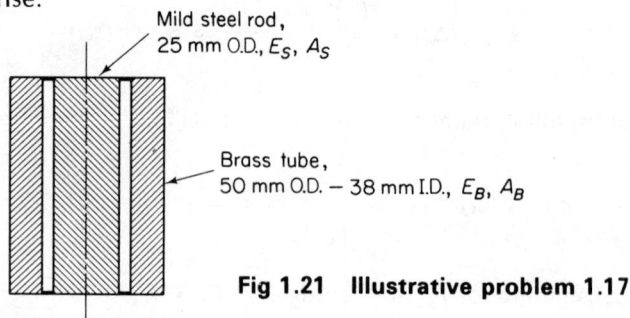

Fig 1.21 Illustrative problem 1.17

Chapter 1, Concepts of Stress and Strain

Solution:

In this problem, we note that the final lengths of both members must be the same and also that the forces within each member must be the same. Denoting the force in the steel as F_S and the force in the brass as F_B

$$F_S = F_B$$

If the brass were permitted to expand freely, it would have been extended by the amount $\alpha_B L_o (\Delta t)$. Due to the restraint of the steel tube, the compressive force F_B causes a contraction of the brass tube equal to $F_B L_o / E_B A_B$. The net change in length of the brass is therefore $\alpha_B L_o (\Delta t) - F_B L_o / E_B A_B$. The steel tube has a temperature expansion of $\alpha_S L_o (\Delta t)$ and an additional extension due the tensile force of $F_S L_o / E_S A_S$, for a net change of $\alpha_S L_o (\Delta t) + F_S L_o / E_S A_S$. Using the conditions that the net length change of both members must be equal and $F_S = F_B$, we obtain the following relation for F_S

$$F_S = F_B = \frac{(\alpha_B - \alpha_S)(\Delta t)}{1/E_B A_B + 1/E_S A_S}$$

Using the numerical data of the problem

$$F_B = F_S = \frac{(18 - 11.7) \times 10^{-6} \times 50}{\left[1/70 \times 10^9 \; \frac{\pi}{4} (0.050^2 - .038^2)\right] + \left[1/210 \times 10^9 \times \frac{\pi}{4} (0.025)^2\right]}$$

$$= 11.7 \text{ kN}$$

The stress in each member is the force of 11.7 kN divided by the respective cross-sectional areas. Thus

$$S_S = \frac{11.7 \times 10^3}{\frac{\pi}{4}(0.025)^2} = 23.83 \text{ MPa (tensile)}$$

$$S_B = \frac{11.7 \times 10^3}{\frac{\pi}{4}(0.050^2 - .038^2)} = 14.11 \text{ MPa (compressive)}$$

1.9 STRESSES IN CYLINDERS AND SPHERES

When a fluid is contained, it exerts forces on the walls of the container that stress the material of the container. As an extension of the principles discussed in this chapter, let us consider the problem of evaluating the stresses in thin cylinders and spheres subjected to an internal pressure. To evaluate these stresses, we shall assume that the ratio of the thickness to the diameter of the cylinder or sphere is small and that the stresses developed in these pressure vessels are uniform over the cross section of the vessel. By *thin* we mean that the ratio of thickness to diameter is 0.1 or less.

With these assumptions in mind, let us consider the cylinder shown in Fig. 1.22. Due to the internal pressure, there is a stress in the circumferential direction S_2 and a stress in the longitudinal direction S_1 [as shown in Fig. 1.22(a)]. If the pipe is cut perpendicular to the longitudinal axis (and far from the ends), the resultant free-body diagram will be the one shown in Fig. 1.22(b). The total resisting force in the cylinder will be the stress S_1 multiplied by the area over which the stress acts. Thus, the resisting force is $S_1(2\pi Rt)$ since the material's area is $2\pi Rt$. The applied load is due to the pressure in the tube over the tube area. The applied load is therefore $p\pi R^2$. For equilibrium, these forces must be equal. Therefore

$$S_1 2\pi Rt = p\pi R^2$$

or
$$S_1 = \frac{pR}{2t} \qquad (1.27)$$

By considering a diametrical cut through the cylinder shown in Fig. 1.22(c), S_2 can be evaluated in an analogous manner. The area is $2tL$, where L is the pipe length. Acting on the pipe is the internal force component

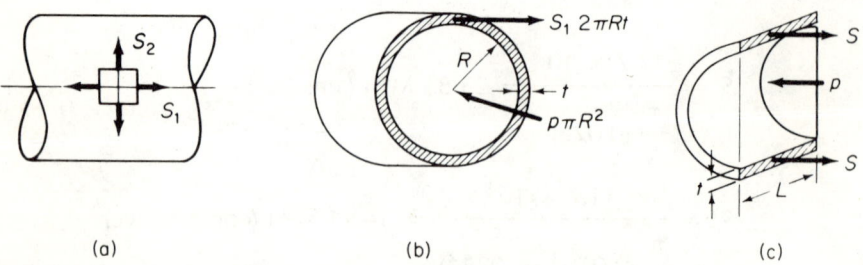

Fig 1.22 Thin cylinder under internal pressure

in the horizontal direction, which is p2RL, where 2RL is the projected area of the tube. Thus,

$$2tLS_2 = p2RL$$

and
$$S_2 = \frac{pR}{t} \tag{1.28}$$

From equations (1.27) and (1.28) we can see that S_2 is twice S_1. The student should also note that these stresses were arrived at by assuming the stress distribution in the material of the cylinder to be uniform.

Using considerations beyond the scope of this text, we will find it possible to calculate the stresses in thick cylinders. Figure 1.23 shows the ratio of the stresses calculated for thick cylinders to the results obtained from equation (1.28) as a function of the outer radius R_o and the inner radius R_i. Our earlier definition of thin was the ratio of thickness to diameter as 0.1 or less. This corresponds to $t/R_i = 0.2$, which from Fig. 1.23 indicates a 10% error in calculating the stress using the thin cylinder formula. As t/R_i increases, the error in using the thin cylinder formula increases, as shown in Fig. 1.23.

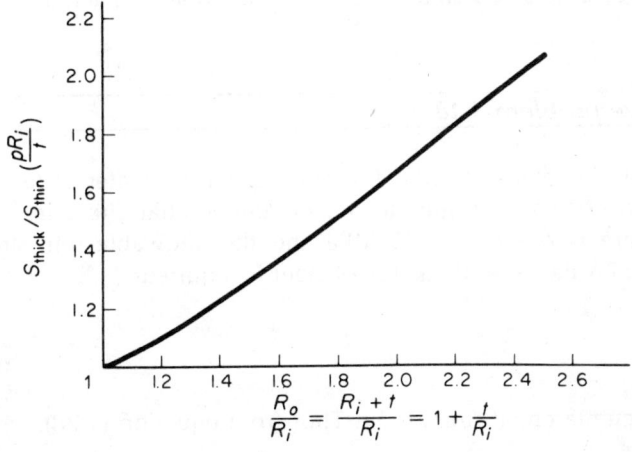

Fig 1.23 Ratio of stresses on a thick cylinder to stress calculated from thin cylinder formula

Let us now consider a thin sphere subjected to an internal pressure. In this sphere S_1 will be equal to S_2 by symmetry. As can be seen from Fig. 1.24, the stress S_2 acts over an area equal to $2\pi Rt$. The sum of the force components in the horizontal direction is once again just the product

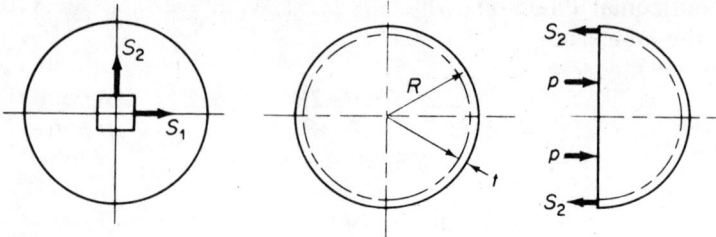

Fig 1.24 Thin sphere under internal pressure

of the pressure multiplied by the projected area, that is, $p\pi R^2$. Therefore

$$S_2 = 2\pi R t = p\pi R^2$$

or
$$S_2 = S_1 = \frac{pR}{2t} \tag{1.29}$$

Equation (1.29) yields the result that S_2 in a thin sphere is half of S_2 in a thin cylinder when both are subjected to the same internal pressure. The student is cautioned to consult applicable codes before attempting to design cylinders and spheres for either internal or external pressure.

Illustrative problem 1.18

A 1.5-m-diameter steel pipe, 6 mm thick, carries water under an internal pressure of 700 kPa. Compute the circumferential stress in the steel. If the pressure is raised to 1.75 MPa and the allowable unit stress in the steel is 105 MPa, what thickness of steel is required?

Solution:

The circumferential stress is S_2. Thus from equation (1.28),

$$S_2 = \frac{pR}{t} = \frac{700(10)^3 \times 1.5/2}{0.006} = 87.5 \text{ MPa}$$

$$t = \frac{pR}{S_2} = \frac{1.75(10)^6 \times 1.5/2}{105(10)^6} = 0.0125 \text{ m} = 12.5 \text{ mm}$$

1.10 CLOSURE

The concepts of stress and strain defined and illustrated in this chapter, Hooke's law, and the free-body diagram provide the foundations for the study of "Strength of Materials." It is therefore most important for the student to master this material fully before proceeding further into the subject. Similar precautionary statements will be found at the closure of other chapters, but no other chapter will contain more essential material than this chapter. We may use the old saying that "a child must crawl before it walks and it must walk before it runs" to place this chapter in its proper perspective—having mastered it, the student is now prepared to walk.

One further suggestion is in order. When doing problems, draw a sketch and put all the given pertinent data on that sketch. Then draw the necessary free-body diagram (or diagrams). Apply the conditions of equilibrium to the free-body diagrams and use the known boundary conditions (such as equal deflection) and the stress strain relations, to solve the problem. Such an organized approach to problem solving in this course will prove to be most helpful in both understanding and solving problems.

PROBLEMS

1.1 A steel bar having a 25-mm diameter is to support a load of 220 kN. Determine the stress in the bar.

1.2 A short solid cylinder is to support a compressive load of 105 kN. If the allowable stress of the material is 35 MPa, determine the required diameter of the cylinder.

1.3 A 19-mm-diameter bolt is used to connect two plates. If the load F is 22 kN, determine the shear stress in the bolt.

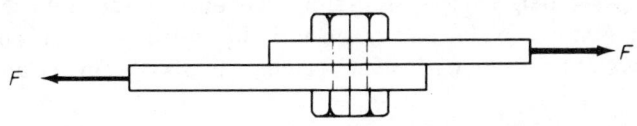

Fig P1.3

1.4 A 19-mm bolt has a root diameter of 15.2 mm. If the allowable tensile stress in the unthreaded material is 105 MPa, determine the maximum tensile force that the bolt can support.

1.5 How long should the threaded portion of the bolt in Problem 1.4 be made if the allowable shearing stress is 42 MPa?

1.6 Determine the average stress at sections A-A, B-B, and C-C.

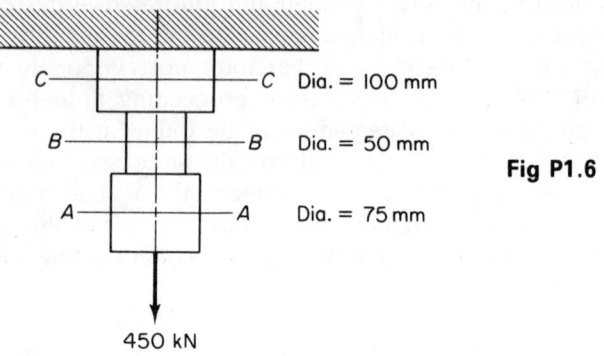

Fig P1.6

1.7 Two bars, one aluminum and the other steel, are arranged as shown. Determine the total change in length when these bars are subjected to an axial force of 45 kN. E_{al} = 70 GPa and E_{steel} = 210 GPa.

Fig P1.7

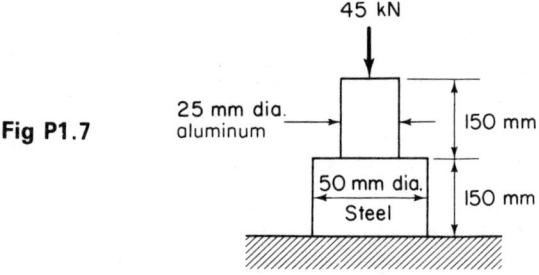

1.8 A steel bar, 3 m long, is to support a tensile force of 45 kN. Determine the stress in the bar at the built-in end of the bar—including the weight of the bar. Steel weighs 78 400 N/m³. Can the weight

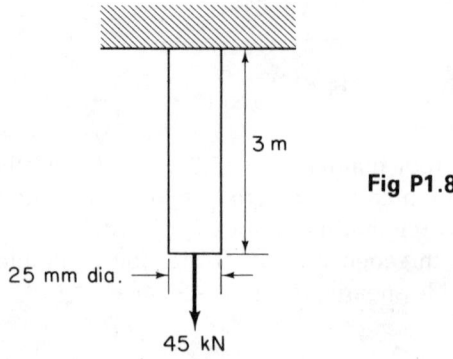

Fig P1.8

of the bar be neglected? What percentage of error would have resulted if the weight were neglected?

1.9 If the allowable shear stress in steel is taken to be 140 MPa, what load can the bolt carry in the configuration shown?

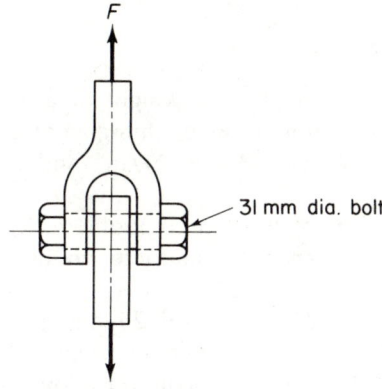

Fig P1.9

1.10 Determine the maximum tensile stress in the configuration shown in Fig. P1.10.

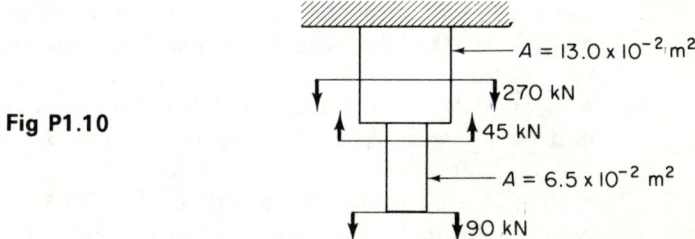

Fig P1.10

1.11 A steel bolt carries a tensile load of 22 kN. Determine the tensile stress in the unthreaded portion of the bolt and in the bottom of the threaded portion of the bolt, and also determine the shear stress at the root of the thread if the nut is 12 mm long.

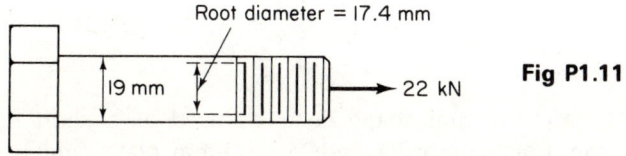

Fig P1.11

1.12 A glued joint is made between a 100-mm × 300-mm board and a large table as shown in Fig. P1.12. Determine the shear stress on the joint when it is subjected to the load shown.

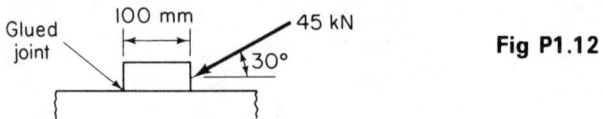

Fig P1.12

1.13 Determine the elongation due to its own weight of a 16-m-long steel bar that is 12 mm in diameter when hung vertically. Use the specific weight of steel to be 78 400 N/m^3 and $E = 210$ GPa.

1.14 Determine the total elongation of the bar in Problem 1.13 if the bar is stressed to 70 MPa by an external force on its free end as well as by its weight.

1.15 What is the maximum length of a steel rod 25 mm in diameter that can be hung from one end if the specific weight of steel is 78 400 N/m^3, $E = 210$ GPa and the maximum allowable stress is 70 MPa?

1.16 It is desired to punch a 25-mm-diameter hole in a steel plate 4.8 mm thick. Determine the force on the punch if the shear stress is 358 MPa.

1.17 It is desired to punch a plate that is 6 mm thick. If the punch can exert a force of 0.1 MN and the ultimate shear stress of the material is 200 MPa, what is the maximum diameter hole that can be sheared?

1.18 A steel plate 16 mm thick and 300 mm wide is to be sheared into two pieces. What shearing force is required if the shearing stress is 280 MPa?

1.19 A shaft is used to transmit a torque of 2100 N·m. A key 12 mm wide × 125 mm long is used to couple a wheel to the shaft. Determine the shear stress in the key.

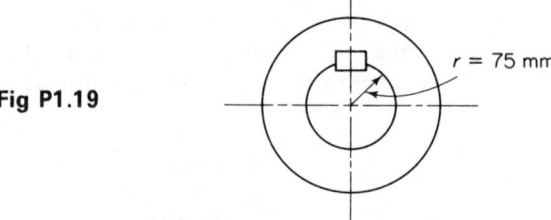

Fig P1.19

1.20 An S7 × 20 structural shape is to be used as a short column. If the column rests on a 300-mm × 300-mm plate, which in turn rests on a concrete footing, whose bearing surface is 0.6 m × 0.6

m, determine the stresses caused by an axial compressive load of 450 kN in each member. The area of a S7 × 20 member is 5.89 square inches.

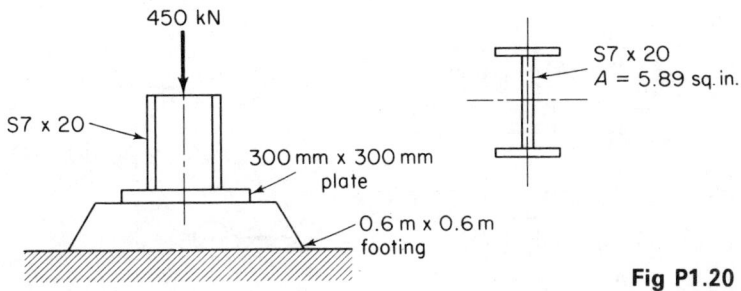

Fig P1.20

1.21 A 4 in. Schedule 40 pipe (standard weight pipe) having an O.D. of 4.500 in. and an I.D. of 4.026 in. is subjected to an axial compressive load of 300 kN. Using $E = 210$ GPa, calculate the shortening of the pipe if it is 3 m long.

1.22 A compression member consists of a short length of 4 in. Schedule 40 pipe having an O.D. of 4.5 in. and an I.D. of 4.026 in. If it is filled with concrete and supports an axial load of 45 kN, determine the load taken by the steel and concrete; also determine the stress in each material. $E_s = 210$ GPa and $E_c = 70$ GPa.

1.23 What is the change of length of the composite bar shown due to the axially applied load? All bars have the same diameter, 25 mm.

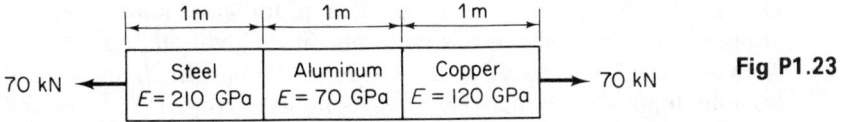

Fig P1.23

1.24 Determine the change in length of the steel shaft shown due to an axial tension load of 450 kN. Use $E = 210 \times 10^9$ Pa.

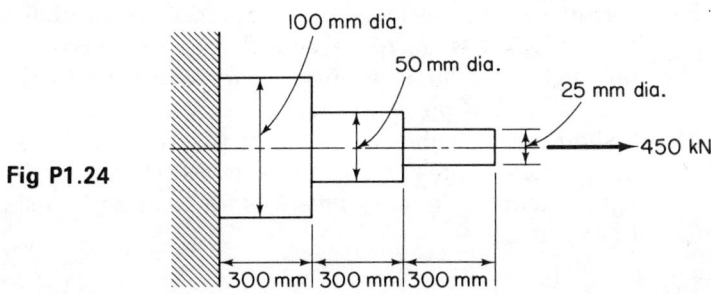

Fig P1.24

1.25 A flange coupling is designed to support a twist (torque) of 700 N·m. If the mean bolt radius is 125 mm and there are four, 12-mm-diameter bolts, determine the shearing stress in the bolts.

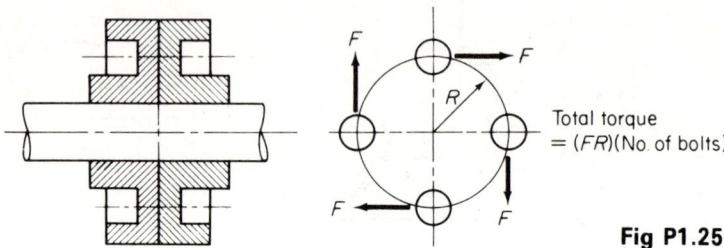

Fig P1.25

1.26 What load will cause a bar L m long and A m^2 in cross section to increase in length by 1%? Assume the bar to be elastic.

1.27 A bar 12 mm in diameter and 250 mm long elongates 0.2 mm when subjected to an axial load of 22 kN. Compute the stress, strain, and E for the bar.

1.28 If the total elongation of a bar is 0.5 mm when subjected to a load of 50 kN, determine the stress in the bar. $E = 210$ GPa and the bar had an initial length of 0.5 m.

1.29 The modulus of rigidity of a material is 80.0×10^9 Pa, and the modulus of elasticity is 202×10^9 Pa. Determine μ.

1.30 If the modulus of rigidity of a material is 11×10^6 psi and its modulus of elasticity is 30×10^6 psi, determine μ.

1.31 Using the data in Table 1.1 calculate μ for aluminum, iron, and copper. Compare your results with the values given in Table 1.2.

1.32 A steel bar ($E = 210$ GPa, $\mu = 0.3$) is 25 mm in diameter and 50 mm long. Determine the change in its diameter and volume when it is subjected to a 350 kN tensile load.

1.33 An aluminum bar, $E = 70$ GPa and $\mu = 0.34$, is 30 mm in diameter and 100 mm long. What is the change in its diameter and its volumetric strain when subjected to a 500 kN tensile load?

1.34 A short steel bar is subjected to a transverse shear load of 22 kN. If the block has a cross-sectional area of 0.0013 m^2 and is 25 mm high, determine its total shear deformation and its shearing strain. Use $G = 84 \times 10^9$ Pa.

1.35 A short bar is subjected to a transverse load of 30 kN. If the cross-sectional area of the block is 0.0018 m^2 and it is 40 mm high, determine its total shear deformation and its shearing strain if $G = 30$ GPa.

1.36 A steel rod having a diameter of 50 mm is subjected to an axial tensile load of 450 kN. If $\mu = 0.28$ and $E = 210$ GPa, determine the change in diameter of the rod.

1.37 Determine the change in cross-sectional area for the steel rod in Problem 1.36.

1.38 Determine the volumetric strain of the steel rod in Problem 1.36.

1.39 A bar 50 mm × 38 mm carries a compressive load of 400 kN. If its volumetric strain is found to be 0.0015, determine μ. Assume $E = 84$ GPa.

1.40 Determine the modulus of rigidity for the bar in Problem 1.39.

1.41 A bar 50 mm × 50 mm carries a compressive load of 500 kN. Its volumetric strain is found to be 0.0017. If $E = 84$ GPa, determine μ.

1.42 Determine the volumetric strain of a steel bar subjected to a stress of 105 MPa. Use $E = 210$ GPa and $G = 77$ GPa.

1.43 What is the volumetric strain in a bar of aluminum that is subjected to a stress of 40 MPa if $E = 70$ GPa and $G = 28$ GPa?

1.44 A steel rod, 25 mm in diameter, is placed between two rigid walls. If the rod is initially 1 m long, determine its compressive stress after a temperature increase of 45 °C. Use $E = 210$ GPa and $\alpha = 11.7 \times 10^{-6}$ m/m °C.

1.45 An aluminum rod 12.5 mm in diameter is placed between two rigid walls. If the rod is initially 500 mm long, determine its compressive stress after its temperature increases 20 °C. $E = 70$ GPa and $\alpha = 23.0 \times 10^{-6}$ m/m°C.

1.46 A steel wire having a diameter of 2.5 mm is found to be 8 m long when subjected to a tensile force of 350 N at 20 °C. How long will it be at 38 °C when subjected to a force of 140 N in tension? Use $E = 210$ GPa and $\alpha = 12.6 \times 10^{-6}$ m/m °C.

1.47 A 4 in.-Schedule 40 pipe is used as a short compression member. At 20 °C it is 250 mm long. It is loaded by a compressive load of a magnitude to cause it to be 250 mm. long at 38 °C. Determine the magnitude of the compressive load if $E = 210 \times 10^9$ Pa and $\alpha = 12.6 \times 10^{-6}$ m/m °C.

1.48 The composite bar shown in Fig. P1.23 just fits snuggly between two rigid walls. If the temperature is raised 50 °C, determine the stress in each of the bars. $\alpha_{steel} = 11.7 \times 10^{-6}$ m/m°C, $\alpha_{copper} = 16.7 \times 10^{-6}$ m/m°C and $\alpha_{al} = 23.0 \times 10^{-6}$ m/m°C.

1.49 Solve Problem 1.48 if the steel bar is made 2 m long.

1.50 What total elongation will occur in a steel rail initially 8 m long due to a 56 °C temperature change? Use $E = 210$ GPa and $\alpha = 11.7 \times 10^{-6}$ m/m°C.

1.51 If the axial stress in the rail described in Problem 1.50 is to be limited to 35 MPa, determine the required clearance between the rails.

1.52 Determine the stresses in each of the bars if the temperature increases 50 °C. The bars are each 25 mm in diameter.

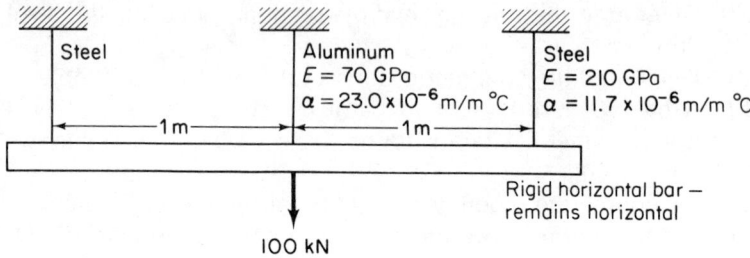

Fig P1.52

1.53 If the diameter of the bars in Problem 1.52 is made to be 50 mm, determine the stresses in each bar.
1.54 If the length of the bars in Problem 1.52 is doubled, determine the stresses in each bar.
1.55 Two rods, one of steel and the other of an unknown material, are initially at the same temperature. An increase of temperature of 42 °C causes the unknown material to expand so that a line drawn across the tops of the rods makes an angle of 0.000 188 radians with the horizontal. If the rods were initially 300 mm long and were separated by 300 mm determine α for the unknown material. α for steel can be taken as 11.7×10^{-6} m/m °C.

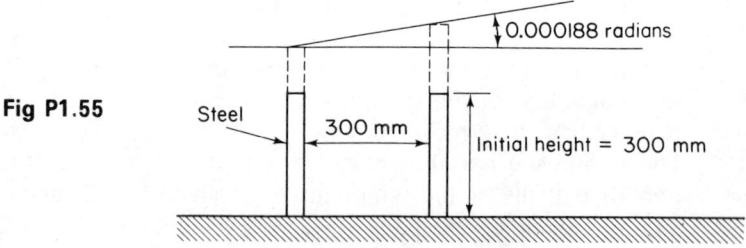

Fig P1.55

1.56 A weight of 450 kN is to be carried by a steel rod in an aluminum cylinder. Determine the forces and stresses in each member. $E_{steel} = 210$ GPa and $E_{al} = 70$ GPa.

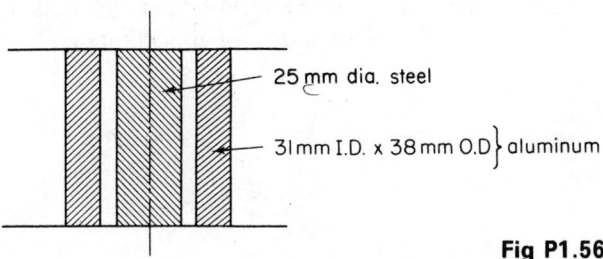

Fig P1.56

1.57 If the steel is replaced by copper in Problem 1.56, determine the forces and stresses in each member. $E_{copper} = 120$ GPa.

1.58 A 25-mm-diameter steel bolt has 10 threads per 25 mm and an aluminum sleeve around it as shown in Fig. P1.58. If the nut on the bolt is turned $\frac{1}{10}$ of a turn, determine the stress in each material. $E_{steel} = 210 \times 10^9$ Pa, $E_{al} = 70 \times 10^9$ Pa.
(HINT: The final force in each member is the same. Also, the elongation of the bolt must equal the shortening of the sleeve due to the turning of the nut minus the elongation of the sleeve due to the elongation of the bolt.)

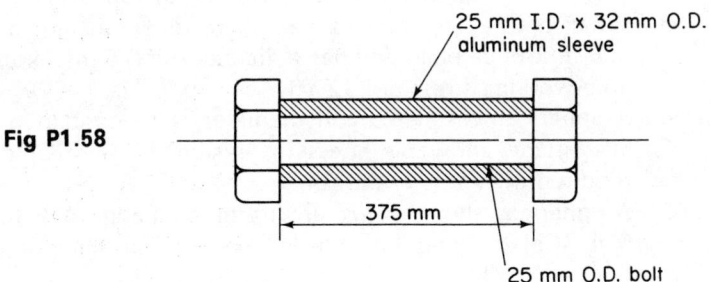

Fig P1.58

52 Strength of Materials for Engineering Technology

1.59 Two rods are mounted between rigid walls as shown. Determine the stress in each bar due to a temperature drop of 47 °C. α_{steel} = 11.7 × 10^{-6} m/m °C, E_{steel} = 210 × 10^9 Pa, α_{copper} = 16.7 × 10^{-6} m/m °C, E_{copper} = 112 GPa.
(HINT: The force in each member is the same. Also the sum of the final lengths of the members must equal 300 mm.)

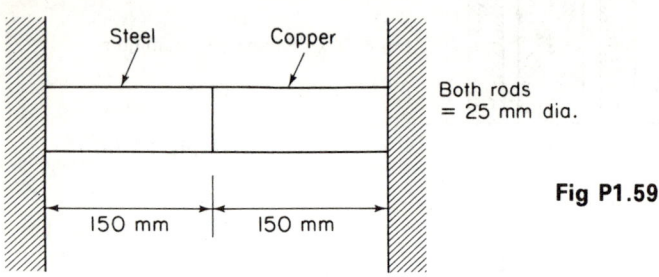

Fig P1.59

1.60 If the steel rod in Problem 1.59 is replaced with an aluminum rod, determine the stress in each rod. Use E_{al} = 70 GPa and α_{al} = 23 × 10^{-6} m/m°C.

1.61 If the stress in the steel bars in Problem 1.52 is not to exceed 70 MPa, what is the maximum temperature increase that the assembly can be subjected to?

1.62 A thin-walled cylinder is subjected to an internal pressure of 2.8 MPa. If the internal diameter is 500 mm and the allowable working stress is 126 MPa, calculate the required wall thickness.

1.63 What internal pressure can a sphere be subjected to if it is made of 6-mm-thick plate and has a diameter of 0.6 m? Assume an allowable working stress of 112 MPa.

1.64 A spherical vessel 6 m in diameter is to contain a fluid at 700 kPa. If the allowable stress in tension is 70 MPa, determine the required thickness of the sphere.

1.65 A spherical shell having an inside diameter of 6 m is made of 9 mm plate. Find the tensile stress in the shell if it contains a fluid at 525 kPa.

1.66 A cylindrical tank, 450 mm in diameter, has a flat cover plate held on by eight 12-mm-diameter bolts. If the allowable stress is 105 MPa in the bolts, what pressure can the plate be subjected to? Assume bolt tension governs.

1.67 Using the results of Problem 1.66, determine the thickness of the tank if the allowable stress in the tank is the same as the bolts.

1.68 A cylindrical tank is made with a 0.5 m I.D. and a 0.625 m O.D. Calculate the stress in this vessel using the mean diameter and a pressure of 7 MPa. Compare your results with the value obtained from Fig. 1.23.

REFERENCES

American Institute of Steel Construction. *Manual of Steel Construction.* 7th ed., 1970.

Bassin, M. E. and S. M. Brodsky, *Statics and Strength of Materials.* 2nd ed. New York: McGraw-Hill Book Co., 1969.

Bruhn, E. F. *Analysis and Design of Flight Vehicle Structures.* Cincinnati: Tri-State Offset Co.

Case, John and A. H. Chilver. *Strength of Materials.* London: Edward Arnold, Ltd.

Granet, Irving. *Fluid Mechanics for Engineering Technology.* Englewood-Cliffs, NJ: Prentice-Hall, Inc., 1971.

Jensen, A. and H. H. Chenoweth. *Applied Strength of Materials.* 3rd ed. New York: McGraw-Hill Book Co., 1971.

Levinson, Irving J. *Mechanics of Materials.* 2nd ed. Englewood Cliffs, NJ: Prentice-Hall, Inc., 1970.

Panlilio, F. *Elementary Theory of Structural Strength.* New York: John Wiley & Sons, 1963.

Seely, F. B. and James O. Smith. *Advanced Mechanics of Materials.* 2nd ed. New York: John Wiley & Sons, 1952.

Singer, F. L. *Strength of Materials.* 2nd ed. New York: Harper and Row, 1962.

Timoshenko, S. and Donovan H. Young. *Elements of Strength of Materials.* 5th ed. New York: D. Van Nostrand Co., 1968.

2

Mechanical properties of materials

2.1 INTRODUCTION

The process of selecting a material for a given application is complex and, therefore, it is quite difficult simply to write down a prescription for selection that would be generally applicable. The purpose of this chapter is to briefly discuss material properties that are generally needed, common test methods used to obtain data, and the results of such tests for certain selected materials. The study of the internal structure of a particular material, the effects of alloying, heat treating, or cold working, as well as other methods of tailoring the properties of materials, are most properly the concern and province of a course in physical metallurgy. However, the first consideration in selecting a material is usually that the strength of the material is adequate for the specified situation. Since more than one material can usually be found that would be suitable for a given application it becomes the designer's function to select the most suitable material, using considerations of cost, availability, ease of handling, life, corrosion resistance, machinability, etc.

The mechanical properties of materials and tests that will be discussed briefly in this chapter represent only a portion of the information that a designer needs to know. However, they are important in that they usually represent the starting point in the selection process.

2.2 THE TENSILE TEST

The *tensile test* is the most common test applied to materials. It is also the most important of the mechanical tests used to obtain data on the properties of materials. The test is usually performed by slowly and steadily applying a tensile load to a test specimen at room temperature. The test is usually performed on a standard test specimen in order to standardize the results. One common type of test specimen is shown in Fig. 2.1.

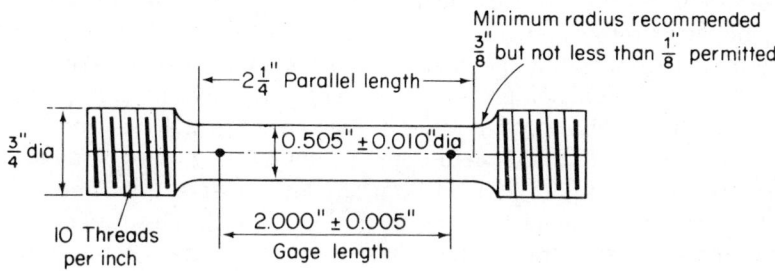

Fig 2.1 Standard tensile test specimens

Other types of test specimens are specified by the American Society for Testing and Materials (ASTM) for machines having different types of grips, for wire, for sheet metal, etc. The ASTM standards should be consulted for more detailed information on standardized test specimens. The test specimen must be accurately machined so that it is symmetrical, and extreme care must be exercised to mount it in the test machine so that only axial loads, as differentiated from bending loads, will be imposed on the specimen. For the test specimen shown in Fig. 2.1, it will be noted that two light prick point marks are shown located 2.000 ± 0.005 in. apart. This distance is known as the *gage length*. After the sample has been loaded in tension and is broken, the two halves of the test specimen are held firmly together and the distance between the marks is again measured. The *percent elongation* is defined as the change in length divided by the original length × 100. Thus

$$\text{percent elongation} = \left(\frac{L_f - L_o}{L_o}\right) \times 100 \qquad (2.1)$$

where L_f is the final gage length and L_o is the initial gage length. Percent elongation is also used as a measure of the *ductility* of a material, i.e., the ability of a material to be drawn into a wire or tube, or to be forged. In referring to ductility in terms of percent elongation, we must also necessarily state the gage length since percent elongation varies with gage length. This is due to the fact that a large part of the total strain occurs in the necked down portion of the gage length just before fracture (see Fig. 2.9) and within one inch of the fracture. For example, a specimen of material having a gage length of 6.000 inches might have a final length of 7.8 inches, which would give a percent elongation of $(7.8 - 6)/6 \times 100 = 30\%$. A 2.000-inch gage length of the same material might have as a final length 2.8 inches or a percent elongation of $(2.8 - 2)/2 \times 100 = 40\%$.

Figure 2.2 shows a modern universal testing machine with an automatic stress-strain recorder. The testing machine is a device in which the specimen can be accurately loaded at rates that are standardized. One standard requirement is that the speed of the testing machine cross-head should not exceed $\frac{1}{16}$ inch per inch of gage length per minute up to the yield point of the material, and it should not exceed $\frac{1}{2}$ inch per inch of gage length per minute from the yield point to the rupturing point of the material.

The reliability of data derived from any test depends upon the accuracy of the test instrumentation, the recorder, and the test machine. The measurement of the strain of the test specimen is made with an instrument known as an *extensometer*. Figure 2.3 shows six types of extensometers, all of which can be used to obtain the specimen strain. The extensometer is clamped to the test specimen and, as the specimen stretches or compresses, the core of a differential transformer moves proportionally to this displacement to produce a displacement voltage and a direction signal. This signal is subsequently amplified and is used to activate a servo motor, which rotates the recorder drum (2) and a cam (3), as shown in Fig. 2.4. As the cam turns, it moves the cam follower and the attached cores of the three differential transformers (one for each magnification range). This core motion produces an opposite and equal signal. The signal for the selected magnification range travels to the servo amplifier. Thus the pen line drawn by the rotation of the recorder drum indicates the exact specimen strain at any instant. The signal restores the null balance when the opposing signals are of equal intensity.

While the recorder drum is being rotated in direct proportion to specimen strain, the load-activated pen mechanism moves the pen point across the drum a distance that is in direct proportion to the increasing or decreasing stress (load) on the specimen. For all practical purposes, both of these indicating processes are instantaneous. As a result of the simultaneous rotation of the recorder drum and the movement of the recorder pen, a highly magnified and extremely accurate stress-strain diagram, which can be studied in detail without further transposition, is produced.

2.3 THE STRESS-STRAIN CURVE IN TENSION

Having tested a specimen, we find it necessary to reduce the observed data to meaningful engineering results. The most useful engineering presentation of the data obtained is the *engineering stress-strain diagram*. This diagram is a plot of strain as the abscissa, and stress as the ordinate. The *stress* is defined as the load in pounds divided by the cross-sectional

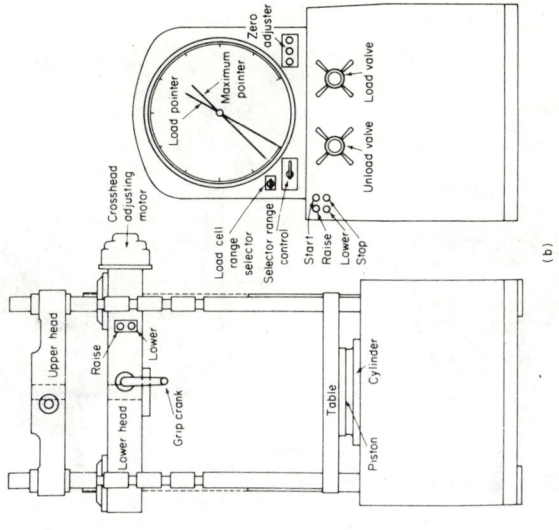

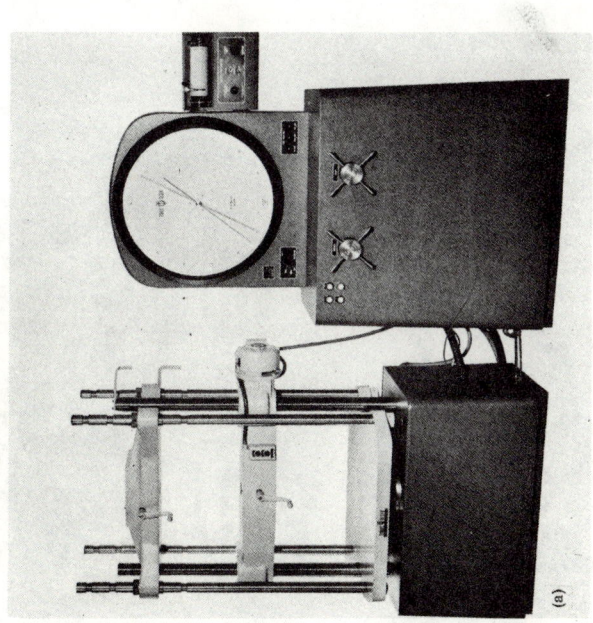

Fig 2.2 Universal testing machine (Courtesy of Tinius Olsen Testing Machine Company)

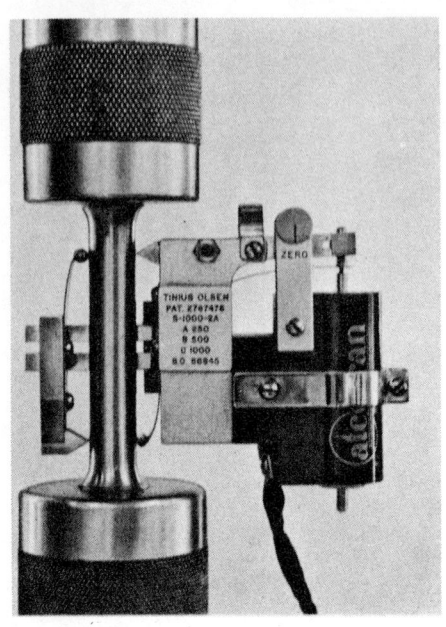

1

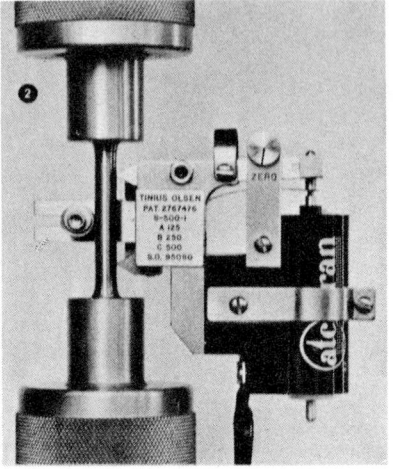

2

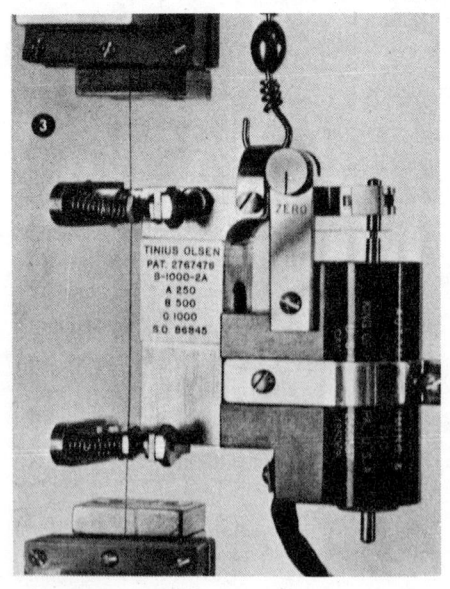

3

4

Chapter 2, Mechanical Properties of Materials 59

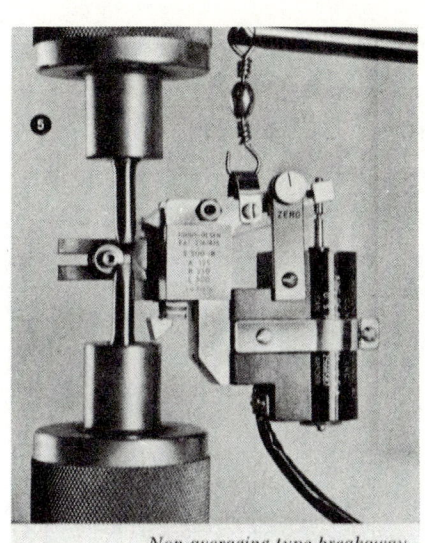
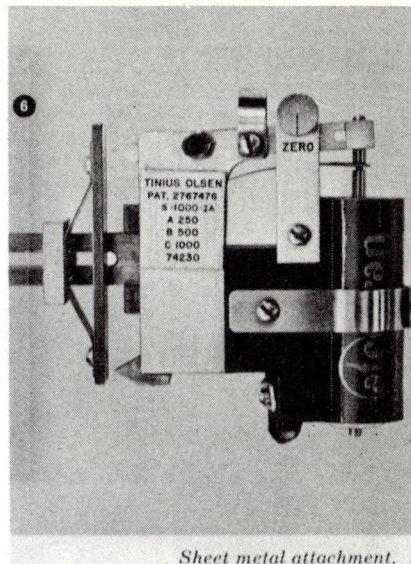

Non-averaging type breakaway.   *Sheet metal attachment.*

5   6

Fig 2.3 Extensometers: 1. Averaging type—Knife edges on opposite sides of specimen provide a measure of the average amount of strain between gage points; 2. Non-averaging type—Provides measurement of strain between knife edges on same side of specimen; 3. Film clamps (non-averaging only)—Spring activated film clamps permit strain measurements of thin materials; 4. Breakaway extensometer, averaging type—Lower knife edge pivots free of specimen when elongation exceeds measuring range or specimen breaks; 5. Breakaway extensometer, non-averaging type—Same as part 4 but measures elongation between knife edges on the same side of specimen; 6. Sheet metal attachment (non-averaging only)—Used for thin flat specimens. (All photos courtesy of Tinius Olsen Testing Machine Company)

area of the specimen at the start of the test. As the test proceeds, the actual cross-sectional area decreases, and at high stresses this reduction in area becomes appreciable. It should be noted that the stress based upon the initial area is not the true stress, but it is generally used and the calculated stress in load-carrying members is almost universally based on this original area. The strain used is the elongation of a unit length of the test specimen taken over the gage length. Figure 2.5 shows a typical engineering stress-strain diagram and the diagram obtained for the same material when based upon the actual area of the specimen. That the actual stress curve is one that continually rises indicates an ever increasing applied

60   Strength of Materials for Engineering Technology

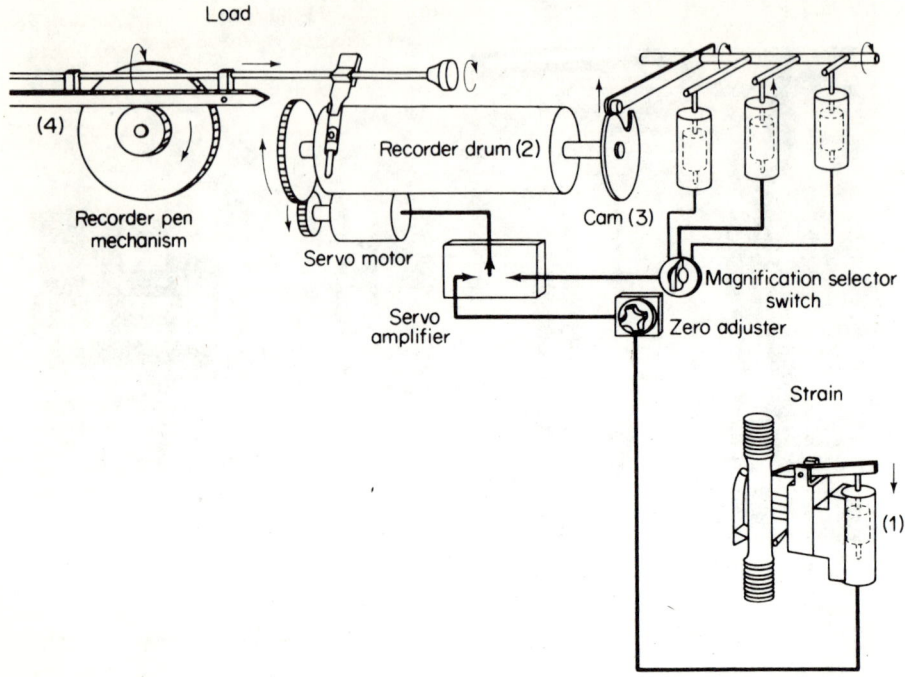

Fig 2.4 Automatic stress-strain recording machine (Courtesy of Tinius Olsen Testing Machine Company)

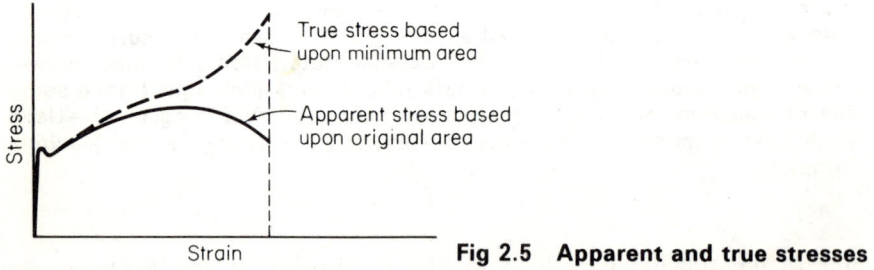

Fig 2.5 Apparent and true stresses

stress up to the point of rupture, while the curve based upon the original area of the specimen shows a maximum stress and then a *decrease* in stress up to the rupture point. We shall examine all of these occurrences in some detail later on in this section.

Table 2.1* is the data recorded during a tension test conducted on

---

*Table 2.1, and Figs. 2.6 and 2.7 are reproduced from *Engineering Materials and Processes* by D. S. Clark, International Textbook Co., with permission.

TABLE 2.1 Log of a Tension Test of Mild Steel
(Initial Diameter 0.502 in.; Initial Area of Section, A_o 0.198 sq. in.;
Initial Gage Length 8 in.)

Applied Load		Elongation				
Total (F) lb	Stress $\left(\dfrac{F}{A_o}\right)$ psi	In Gage Length in.	Strain $\left(\dfrac{\Delta L}{L}\right)$ in./in.	Least Diameter in.	A Area sq. in.	True Stress psi
1100	5500	0.001	0.000125	0.5020	0.1979	5550
1950	9850	0.002	0.000250	0.5015	0.1975	9870
2570	12,480	0.003	0.000375	0.5012	0.1973	13,020
3230	16,310	0.004	0.000500	0.5012	0.1973	16,370
3800	19,200	0.005	0.000625	0.5012	0.1973	19,240
4600	23,200	0.006	0.000750	0.5012	0.1973	23,300
5300	26,800	0.007	0.000875	0.5012	0.1973	26,820
6150	31,050	0.008	0.001000	0.5012	0.1973	31,170
6600	33,300	0.009	0.001125	0.5012	0.1973	33,420
7300	36,850	0.010	0.001250	0.5012	0.1973	37,000
7600	38,400	0.011	0.001375	0.5012	0.1973	38,500
7500	37,850	0.012	0.001500	0.5012	0.1973	38,000
7450	37,600	0.0135	0.001688	0.5012	0.1973	37,700
7560	38,200	0.014	0.001750	0.5012	0.1973	38,300
7600	38,400	0.015	0.001875	0.5011	0.1972	38,500
7780	39,300	0.016	0.002000	0.5010	0.1971	39,420
7700	38,900	0.0505	0.006310	0.4990	0.1956	39,350
8100	40,900	0.100	0.023500	0.4965	0.1936	41,800
8000	40,400	0.150	0.018750	0.4960	0.1932	41,400
8700	43,900	0.200	0.025000	0.4960	0.1932	45,000
9300	46,900	0.250	0.031250	0.4940	0.1917	48,500
9860	49,800	0.300	0.037500	0.4925	0.1909	51,700
10,330	52,200	0.360	0.045000	0.4900	0.1886	54,800
10,980	55,400	0.460	0.057500	0.4870	0.1863	58,800
11,850	59,800	0.600	0.082500	0.4810	0.1817	65,200
12,340	62,300	0.860	0.107500	0.4740	0.1765	69,800
12,450	62,900	1.060	0.132500	0.4660	0.1706	73,000
12,620	63,700	1.260	0.157500	0.4570	0.1640	76,900
12,760	63,900*†	1.460	0.182500	0.4390	0.1514	83,600
9980	50,400	1.660	0.207500			
9840	49,700‡	1.670	0.208750	0.3350	0.0881	111,500

*Specimen begins to neck. †Ultimate tensile strength. ‡Failure.

a mild steel specimen. The specimen had an initial diameter of 0.502 in., an initial area (A_o) of 0.196 sq. in. and a gage length of 8.00 in.

The total applied load and the elongation of the specimen in the gage length are tabulated in Table 2.1. From the test data and the known size of the test specimen the rest of the table is completed. The stress-strain

62   Strength of Materials for Engineering Technology

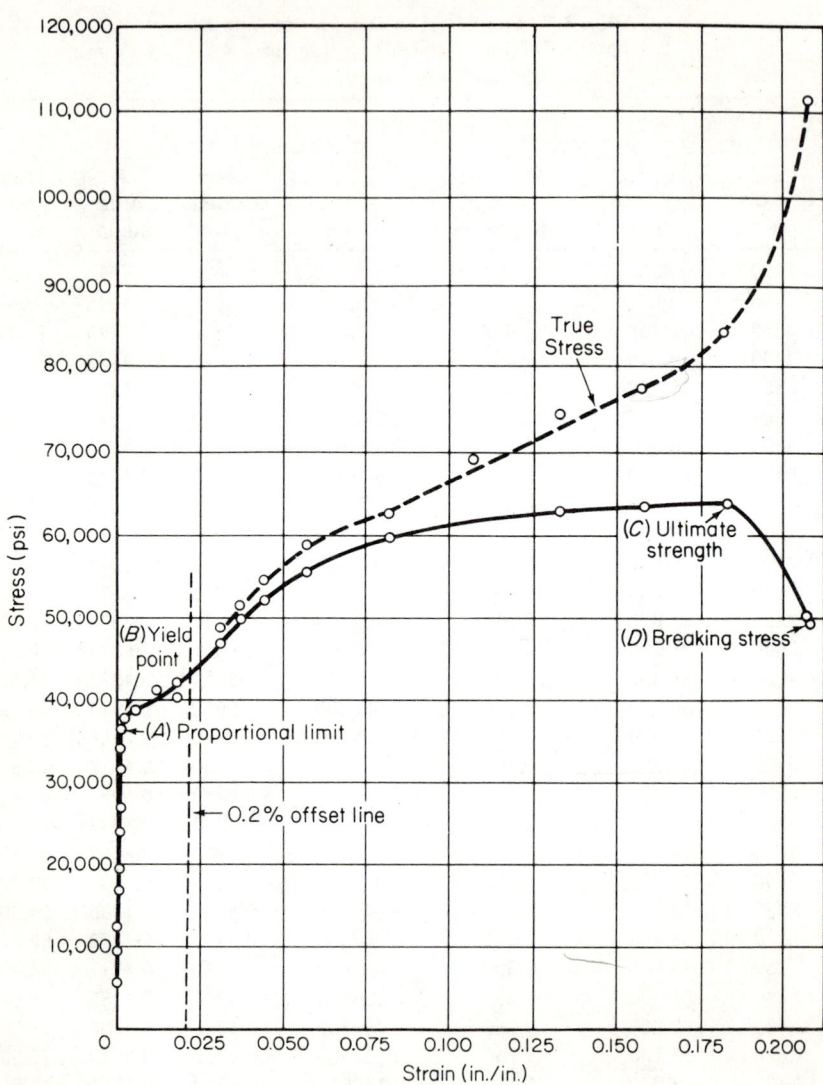

Fig 2.6   Stress-strain diagram for mild steel

diagram has been plotted in Fig. 2.6 with both the engineering stress and true stress curves shown. Note that the stress-strain curve begins with an almost vertical straight line when plotted on a scale to show the strain up to the breaking point. In order to show this initial section in detail, Fig. 2.7 has been plotted. This curve shows that the extensometer had a zero correction of 0.001 in./in.; that is, all strains should be corrected by adding 0.001 in./in. to the tabulated value. Basing our calculations

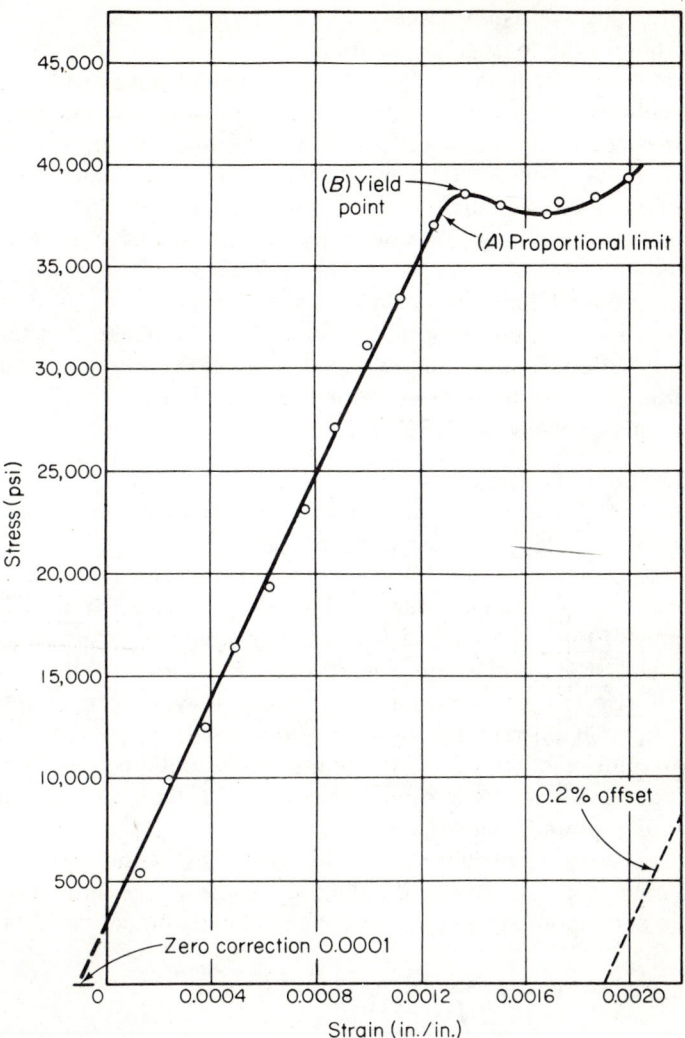

Fig 2.7 Portion of stress-strain diagram for mild steel

upon Table 2.1 and Figs. 2.6 and 2.7, we can obtain much significant information. These are defined and illustrated for the data of Table 2.1.

1. Proportional Limit

Hooke's Law is a statement of the proportionality of stress and strain. The proportional limit is defined as the greatest stress that a material is capable of developing without deviation from Hooke's Law of stress-strain proportionality. In order to determine the proportional limit, it is necessary to use very sensitive extensometers to detect the slightest deviation from a straight line in the tensile test diagram.

Other methods of defining the proportional limit have been proposed. These usually define the proportional limit as that stress at which a certain defined permanent set takes place. Thus a permanent strain of 0.001% and a permanent strain of 0.01% have been used to obtain the proportional limit. Certain materials, such as cast iron and copper, do not have the straight line portion of the stress-strain curve as does steel. For these materials a value known as *Johnson's apparent elastic limit* is sometimes used instead of the proportional limit. Johnson's apparent elastic limit is defined as that stress at which the rate of deformation is 50% greater than the initial rate of deformation. Figure 2.8 shows the construction for Johnson's apparent elastic limit. The proportional limit for the data shown in Fig. 2.7 occurs at point A and its value is 37,500 psi.

2. Modulus of Elasticity

The mechanical property that defines resistance of a material in the elastic range is called *stiffness;* and for ductile materials it is measured by the value called the *modulus of elasticity,* (or Young's modulus) which is designated by the capital letter *E*. Referring to Fig. 2.7, we note that the first part of the diagram is a straight line, which indicates a constant ratio between stress and strain over this range. The numerical value of this ratio is referred to as the modulus of elasticity, *E*. *E* is therefore the slope of the initial straight portion of the stress-strain diagram, and its numerical value is obtained by dividing stress in pounds per square inch by the strain, which in non-dimensional; thus *E* has the same units as stress, namely, pounds per square inch.

When the straight line portion of the curve passes through the origin, *E* can be determined simply by dividing any stress value along the straight line by the corresponding strain. As we have already noted, the data shown

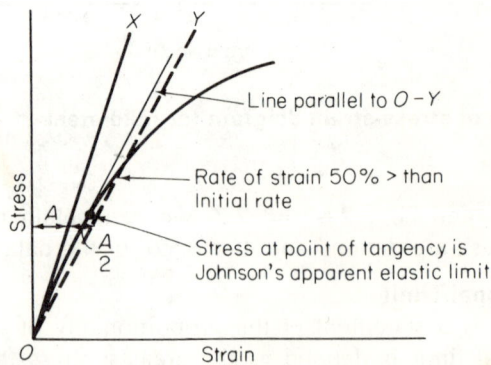

Fig 2.8 Construction for Johnson's apparent elastic limit

in Fig. 2.7 show that a zero correction is needed due to the offset of the intercept. For cases such as this, E can be found by taking two values of stress and two values of strain far enough from the origin to eliminate any errors. Thus

$$E = \frac{S_1 - S_2}{\varepsilon_1 - \varepsilon_2} \tag{2.2}$$

For Fig. 2.7 we can obtain E from two points C and D as

$$E = \frac{30,000 - 13,750}{0.0010 - 0.0004} = 27.1 \times 10^6 \text{ psi}$$

We can check this by using the data at a strain (apparent) of 0.0010. The stress corresponds to 30,000 psi, and the true strain is 0.0010 + 0.0001. Thus, E is

$$E = \frac{30,000}{0.0010 + 0.0001} = 27.3 \times 10^6 \text{ psi}$$

It should be noted that the value of 27.3×10^6 psi for E is from a single test of one specific steel specimen.

3. Yield Strength

When the load on the test specimen is increased beyond the proportional limit, a stress level is reached where the material continues to elongate without an increase of load. The *yield point* is defined as the stress at which a marked increase in strain occurs without a concurrent increase in applied stress. Point B on Fig. 2.7 is the yield point. More correctly, point B is known as the *upper yield point*, and, after it is reached, the force resisting deformation decreases due to the yielding of the material. After the initial yielding, the stress reaches a relatively constant lower value, while the deformation process continues. This latter value of stress is known as the *lower yield point*. The lower yield point is usually taken to be the true material characteristic to be used as the basis for the determination of working stresses.

Many materials do not exhibit well-defined yield points, and the *yield strength* is defined as the stress at which the material exhibits a specified limiting permanent set. The specified set (or offset) most commonly used is 0.2%, which corresponds to a strain of 0.002 inches/inch. The yield strength is therefore the stress corresponding to the intersection of a line parallel to the straight line portion of the stress-strain curve, and the stress-strain curve.

The defined yield strength does not represent a physical property of the material, and its value is a function of the defined offset. Actually, this yield strength of a material is of importance in determining when a material is stressed beyond its ability to act elastically. For most applications, the onset of inelastic action is considered to be the point at which a member can no longer perform its structural function.

4. Ultimate Strength (Tensile Strength)

The *ultimate strength* of a material is defined as the stress obtained by dividing the maximum load reached before the specimen breaks by the initial cross-sectional area of the specimen. For the test data shown in Fig. 2.6, this corresponds to a stress of approximately 64,000 psi at point C. It will be noted from Fig. 2.6 that beyond the yield point, the stress continues to increase until it reaches the ultimate strength. This is true when the initial area of the test specimen is used to define the stress. The true stress-strain curve based upon the actual area of the specimen shows a continual increase in stress until the breaking point of the specimen is reached. The ultimate strength (often called the *tensile strength*) of the material is commonly used as a basis for establishing working stresses for a material.

5. Elongation

We earlier noted that *elongation* is a measure of the ability of a material to undergo deformation without rupture. Percentage elongation, defined by equation (2.1) is a measure of the ductility of a material. This property, ductility, is a desirable and necessary property, and a member must possess it to prevent failure due to local overstressing. From the data of Table 2.1, percent elongation is

$$\left(\frac{9.67 - 8.00}{8.00}\right) \times 100 = 20.9\% \text{ in 8 inches}$$

6. Breaking Strength (Rupture Strength, Fracture Strength)

The *breaking strength* of a material is the load on the material at the time of failure divided by the original cross-sectional area of the specimen. As we see from Fig. 2.6, it is the final ordinate on this stress-strain curve. For the data of Fig. 2.6, the breaking strength is very close to 50,000 psi. It should be re-emphasized that this definition of the breaking strength is based upon the original area of the test specimen during the test. The breaking stress on this basis is less than the ultimate stress of the data of Fig. 2.6, while the true stress at failure is the maximum stress on the material. For the test interval between the ultimate stress and the breaking stress, the specimen continues to elongate even though the resisting stress based on the original area decreases.

7. Reduction of Area

As the load on the material undergoing testing is increased, the original cross-sectional area decreases until it is at a minimum at the instant of fracture. It is usual to express this reduction in area as the ratio of the change in area to the original specimen cross-sectional area, expressed as a percentage. As Fig. 2.9 shows, the failed specimen exhibits a local

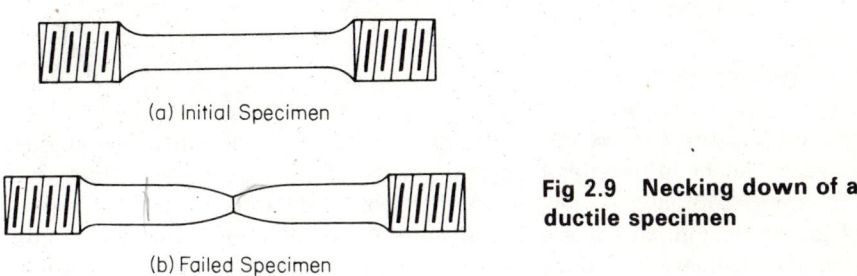

Fig 2.9 Necking down of a ductile specimen

decrease in diameter known as *necking down* in the region where failure occurs. It is very difficult to determine the onset of necking down and to differentiate it from the uniform decrease in diameter of the specimen. The percent reduction in area is

$$\% \text{ reduction in area} = \left(\frac{A_o - A_f}{A_o} \right) \times 100 \qquad (2.3)$$

where A_f is the final cross-sectional area at the point of fracture and A_o is the initial cross-sectional area of the specimen. The percent reduction in area is also a measure of the ductility of a material. Brittle materials exhibit almost no reduction in area, while ductile materials exhibit a high percent reduction in area.

The reduction in area is less dependent on the dimensions of the test specimen than is percent elongation, and it is also independent of the gage length of the test specimen. For the data of Table 2.1, the percent reduction in area is

$$\left(\frac{0.1979 - 0.0881}{0.1979} \right) \times 100 = 55.5\%$$

8. Toughness

The area under the stress-strain curve of Fig. 2.6 is a measure of the work required to cause fracture to occur. The ability of a material to absorb energy up to fracture is also used by designers as a characteristic property of a material and is called *toughness*. This property of a material is a function of both the strength and the ductility of the material. The area under the stress-strain curve yields units of (in./in.) (lb per sq. in.) or in. lb per cu. in., i.e., energy per unit volume of material. This quantity is also known as the *modulus of toughness*. A typical value of modulus of toughness for cast iron is 500 in. lb per cu. in., while for low carbon steel it is approximately 17,000 in. lb per cu. in.

2.4 OTHER TESTS

The tensile test is the most important single test that can be conducted to evaluate the properties of a material for design purposes. Other tests are conducted on materials in order to obtain data concerning properties of materials that are needed, but cannot be obtained from the simple static tension test. The direct correlation between the results of these tests and the desired material property is often not feasible and in many cases only qualitative results are obtainable. We shall briefly indicate the more important of these tests in the following paragraphs.

1. Compression Tests

The *compression test* is used primarily to test brittle materials such as cast iron and concrete. The universal testing machine is used for this test, and data is taken in a manner similar to that discussed for the tension test. It is usual to make the test specimen in the form of a prism, whose height in the compression direction is several times larger than its lateral dimensions in order to minimize the effect of friction on the contact surfaces of the specimen. The results of the compression test are an elastic range, a proportional limit, and a yield strength. In the compression test, the cross-sectional area of the specimen increases, and this produces a continuously rising engineering stress-strain curve.

2. High Temperature Tests

The evaluation of the physical properties of materials at high temperatures is very important in modern technology. Most properties, such as the yield point and ultimate strength, are temperature dependent. In addition, when materials are kept at constant stress at high temperatures for considerable periods of time, a continuous deformation of the material, known as *creep*, occurs. An allied phenomenon, known as *relaxation*, occurs when an initial stress, such as in the bolts of a flanged joint, decreases with time at an elevated temperature, causing a decrease in the tightness of the bolted joint.

In general, the strength of metallic materials varies inversely with temperature. By definition, the *creep limit* of a material is taken to be the stress required to produce an elongation of 1% in 100,000 hours.

The creep limit decreases markedly as the temperature of the material increases. Since the creep limit decreases, the load-carrying capacity decreases as the temperature increases.

The creep test is a long-term, elaborate, expensive test. Preliminary data can be obtained by the *stress rupture test*. The stress rupture test is as the name implies. A specimen is subjected to a temperature, and a stress level to produce a failure in a given time is determined. When the data for, say, 10, 100, and 1000 hours are plotted on a semilogarithmic plot of stress versus the logarithm of time, a straight-line relation usually results. Quite often both stress-rupture data and creep data are needed for evaluation of material properties at elevated temperatures. In order to better understand the creep phenomenon, let us refer to Fig. 2.10, which shows a typical plot of strain versus time for a metal.

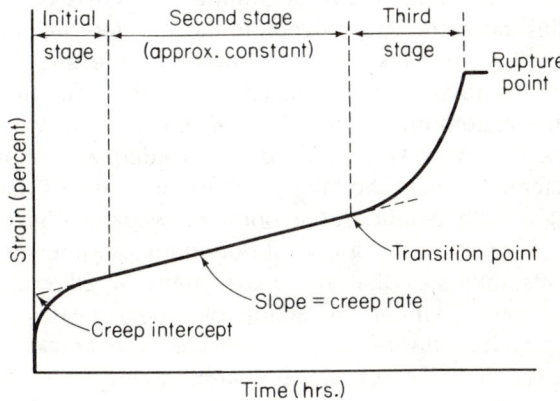

Fig 2.10 Typical creep-rupture curve

For metals tested at high values of stress or temperature, three stages in the creep-time relation are observed. The initial stage, often called the *stage of primary creep*, includes the elastic deformation and that region where the rate of creep deformation decreases rather rapidly with time. The second stage, often referred to as the *secondary creep stage*, represents a stage where the rate of strain has decreased to a constant value for a considerable time period representing the period of minimum creep rate. The third stage, often called the *tertiary creep stage*, represents the period where the reduction in cross-sectional area leads to a higher stress, a greater creep rate, and finally rupture. The inflection point between the constant creep rate of the second stage and the increasing rate of the third stage is referred to as the *transition point*. Failure generally occurs shortly after the transition point. Minimum creep rate is that indicated in the second stage, where the creep rate is practically constant.

3. Cyclic Stress (Fatigue)

It is well known that a member subjected to repeated conditions of loading or unloading, or to repeated stress reversals, will fail at a stress considerably lower than the ultimate stress obtained in a simple tension test. Failures that occur as a result of this type of repeated loading are known as *fatigue failures*. One type of fatigue test is a *rotating beam test*. In this test, a specimen is mounted in a test machine and loaded to produce a given stress. The test member is then rotated causing the stress to change sign every half-revolution (i.e., is completely reversed from compression to tension and vice versa). The number of cycles of stress is equal to the number of revolutions of the machine. The test is run until the specimen ruptures, and the number of cycles to rupture is noted. The data are usually plotted as shown in Fig. 2.11, with stress as the ordinate and the logarithm of the number of cycles to failure as the abscissa. This curve is generally known as an S-N diagram. At the beginning, the stress decreases as the number of cycles increases. After several million cycles, the curve becomes a horizontal line whose stress value is known as the *endurance limit* of the material for complete stress reversal. Thus, the endurance limit can be defined as the maximum completely reversed stress to which a material can be subjected without failure. To date no adequate theory has been developed to clearly explain the fatigue failure of materials. Fatigue failure appears to begin with a crack at a point of weakness in the material, with the crack progressing along crystal boundaries. A microscopic examination of metals indicates that there are many small cracks scattered throughout a material. Under the action of repeated stress, these small cracks open and close during the stress cycle. The cracks cause higher stress at the base of the crack as compared to the stress if there were no crack. Under this repeated concentration of stress, the cracks will

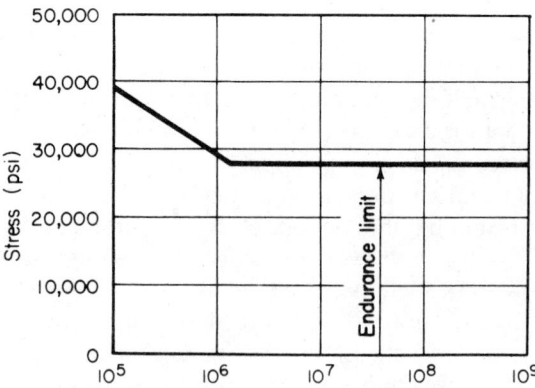

Fig 2.11 Typical S-N curve

gradually extend across the section of the member and finally cause complete failure of the member.

Although there appears to be a relation between the endurance limit and the tensile strength of a material, the influence of such items as the surface conditions of the test specimen, the speed of testing, the prior treatment of the material, etc., is such as to make generalized statements of such a relation of limited value.

4. Hardness

The *hardness test* is used as an engineering test for a quick, inexpensive method of obtaining approximate values of some of the mechanical properties of materials to be used in design. The test usually measures the ability of a material to resist penetration. The hardness test is usually designated by the instruments used to determine the hardness. These most commonly are the Brinell, Rockwell, Vickers, and Shore scleroscope. The basic test consists of applying a standard load to an indenter (a ball or pyramid) for a fixed time period and then measuring the size of the resulting indentation. Thus, in the Brinell test, the hardened steel ball indenter is 10 millimeters in diameter, and it is pressed into the test piece with a force of 3000 kilograms for hard materials (or 500 kilograms for soft materials) for 30 seconds. The *Brinell hardness number* is defined as the load in kilograms divided by the area of the impression in square millimeters. A nonpenetrating test is the Shore scleroscope test, which consists of dropping a small diamond tipped hammer from a fixed height and measuring the height of its rebound. The height of rebound corresponds to the hardness of the material. It should be noted that this test measures the resilience of a material rather than its resistance to penetration.

The results of hardness tests have been correlated to the ultimate strength of many materials. The ratio of the ultimate stress of steel to Brinell hardness is reasonably consistent and varies from approximately 470 to 530. In other words, a steel having a Brinell hardness number (BHN) of 280 will have an ultimate stress in the range of 132,000 psi to 148,000 psi. Although hardness and strength are inter-related, care should be exercised when comparing the hardness and strength of materials, especially when only limited data are available.

5. Impact Testing

When a load is applied rapidly to a part or structure, appreciable shock and vibration can occur. This type of loading, referred to as *impact loading,* therefore produces conditions in the structure that differ markedly from those produced by equal magnitude static loads. A further effect of this manner of loading is that materials behave differently than they do under static loading. There are instances when a metal is ductile when loaded statically but fractures in a brittle manner when subjected to impact loading.

There are in general two types of tests to determine the behavior of materials under impact loads. The usual impact test is referred to as the *notched bar test* and consists of subjecting notched specimens to axial, bending, and torsional loads by the Charpy or Izod impact testing machines. In both of these machines, an impact load is applied to the specimen by swinging a weight W from a certain vertical height (h) to strike and rupture the notched specimen, and then the load stops at a vertical height (h'). The energy expended in rupturing the specimen is then equal approximately to ($Wh - Wh'$). This type of test is primarily used for studying the influence of metallurgical variables.

Another type of impact testing is made on unnotched specimens and the general purpose is to obtain a stress-strain diagram of materials under impact load or a load-distortion diagram of a structural member as it is completely fractured under an impact load.

These tests are of little use in design, and the meaning of the results obtained is open to question. The tests are, at best, qualitative in nature, and the result obtained from a notched specimen is not directly applicable to design problems. The data from unnotched tests can be used directly in design problems. Tension and torsion impact testing has become the subject of increasing study in recent years. For a further discussion of this type of impact test the student is referred to general metallurgical and testing references.

2.5 CLOSURE

This chapter has as its purpose the presentation of a brief introduction to the mechanical properties of materials, and methods of obtaining data on these properties. The literature on this subject is vast and, at best, we have only touched on a small portion of it. The student will undoubtedly be exposed to more detail in materials and design courses. The importance of these studies cannot be over-emphasized since the selection of the proper material for a given application is one of the principal functions of a designer.

PROBLEMS

2.1 A rod has a uniform diameter of 0.50 in. An axial tensile force of 1000 lb causes the rod to elongate by 1% of its original length. Determine the modulus of elasticity of the material.

2.2 A rod has a uniform diameter of 50 mm. An axial tensile force of 1000 N causes the rod to elongate an amount equal to 1% of its original length. Determine the modulus of elasticity of the material.

2.3 A material is tested by applying a tensile load to a specimen that is $\frac{5}{8}$ in. in diameter. If the load is 10,000 lb, and the elongation is 0.020 in. in an 8-in. gage length, determine the modulus of elasticity of the material.

2.4 If the diameter of the specimen in Problem 2.3 is 0.6245 in. under load, determine Poisson's ratio.

2.5 A tensile test machine applies a load of 20 kN to a test specimen that is 30 mm in diameter. If the elongation is 0.5 mm in a gage length of 200 mm, determine the modulus of elasticity of the material.

2.6 If the diameter of the specimen in Problem 2.5 is 29.97 mm under load, determine Poisson's ratio.

2.7 A test specimen is originally 0.505 in. in diameter and supports an ultimate load of 16,000 lb. The specimen was found to have a length at failure of 2.8 in. and an initial gage length of 2.000 in. What is its ultimate strength and percent elongation?

2.8 A test specimen is originally 12.5 mm in diameter and supports an ultimate load of 65 kN. At failure the length is found to be 70 mm. The initial gage length was 50 mm. What is the ultimate strength of the material and its percent elongation at failure?

2.9 A test specimen is originally 0.75 in. in diameter. A load of 12,000 lb causes the stress-strain curve to become non-linear and failure is found to occur at a load of 42,000 lb. Calculate the proportional limit of the material and the stress at failure.

2.10 A test specimen is 12.5 mm in diameter and under load it is found that a load of 50 kN causes the stress strain curve to deviate from linearity. Failure in the specimen is found to occur at 170 kN. What is the stress at failure and what is the proportional limit of the material?

2.11 A $\frac{5}{8}$-in. diameter rod is tested in tension. The force on the rod at the yield point is 14,000 lb, the ultimate load is 16,000 lb, and the rod breaks at a load of 15,000 lb. Calculate the breaking stress, the ultimate stress, and the yield stress.

2.12 If the rod in Problem 2.11 necks down to a diameter of $\frac{3}{8}$ in. at fracture, determine the percent change in area and the true breaking stress.

2.13 A material is known to have a modulus of elasticity of 30×10^6 psi. If a rod is subjected to a 10,000-psi stress, what is its percent elongation?

2.14 A test specimen having an original diameter of 0.506 in. has a final diameter of 0.302 in. at failure. The original gage length was 2.000 in. and at failure it was found to be 2.75 in. Determine the percentage elongation and the percentage reduction in area of the specimen.

2.15 A test specimen has an original diameter of 12.5 mm. At failure its diameter is 8 mm. The original gage length was 50 mm and at failure it was found to be 70 mm. Determine the percent elongation and the percent reduction in area of the specimen.

2.16 The following data were obtained from a tension test:
  a) Original diameter of test specimen = 0.505 in.
  b) Minimum diameter after failure = 0.312 in.
  c) Final length at failure = 2.79 in.
  d) Initial gage length = 2.000 in.
  Determine the percent reduction of area and percent elongation.

2.17 A tensile load is applied to a $\frac{1}{2}$-in.-diameter specimen. If the measured elongation of the specimen is 0.0052 in. over an 8-in. gage length, determine the unit deformation, the stress, and the applied load if E is 30×10^6 psi.

2.18 A tensile test specimen is 12.5 mm in diameter. The specimen elongates 0.0625 mm in a gage length of 200 mm. Determine the unit deformation, the stress and the applied load if $E = 210$ GPa.

2.19 At the proportional limit it is found that a $\frac{3}{4}$-in.-diameter steel bar has elongated by 0.0030 in. in a gage length of 8 in. while its diameter was reduced by 0.000084 in. If the axial load was 5000 lb, determine the modulus of elasticity and Poisson's ratio.

2.20 A bar 20 mm in diameter elongates 0.075 mm in a gage length of 200 mm at the proportional limit where the axial load is 20 kN. If its diameter is reduced by 0.002 mm, determine E and Poisson's ratio.

2.21 A stress-strain diagram for a material has a straight line portion similar to that of steel. If a stress of 5000 psi produces a strain of 5×10^{-4} in./in. and a stress of 15,000 psi produces a strain of 15×10^{-4} in./in., determine the modulus of elasticity of the material.

2.22 A material is tested in tension and it is found that a stress of 40 MPa produces a strain of 6×10^{-4} m/m and a stress of 110 MPa produces a strain of 20×10^{-4} m/m. If these data are taken within the proportional limit of the material, determine E.

2.23 During a tension test, a stress of 2500 psi is found to cause a strain of 8.3×10^{-5} in./in., and a stress of 10,000 psi caused a strain of 33.2×10^{-5} in./in. Assuming that the proportional limit of the material is 20,000 psi, determine the modulus of elasticity of the material.

2.24 A cylindrical test specimen has an initial diameter of 0.752 in. During the tension test of this test specimen, it is found that a load of 15,000 lb is the maximum load that the specimen can carry and still maintain stress proportional to strain. Yielding occurs at a load of 20,000 lb, and failure occurs at 40,000 lb. The diameter at failure is 0.485 in. Determine the proportional limit, yield stress, breaking stress, true breaking stress, and percent change in area.

2.25 A cylindrical test specimen has an initial diameter of 20 mm. When tested in tension, it is found that the maximum load that it can carry while still maintaining stress proportional to strain is 70 kN. Yielding is found to occur at 180 kN. At failure the diameter is found to be 12.5 mm. Determine the proportional limit, yield stress, breaking stress, true breaking stress, and the percent change in area.

2.26 The following data were taken from the test of a specimen having a diameter of 0.505 in. initially and an initial gage length of 2.000 in. Plot the stress-strain curve; and determine the proportional limit, the modulus of elasticity, the percent elongation, and percent reduction of area. Also determine the ultimate strength and breaking strength based upon both the initial and final area of the piece. The final length of the piece is 2.79 in. and the final necked down diameter is 0.302 in.

Stress (psi)	Strain (in./in.)
0	0
5000	0.000167
10,000	0.00033
20,000	0.00066
30,000	0.00099
35,000	0.00120
40,000	0.0025
50,000	0.060
60,000	0.14
46,000	0.40 fracture

2.27 A tensile test is conducted on a non-ferrous alloy and the following data are obtained:

Stress MPa	Strain m/m × 10^{-4}	Stress MPa	Strain m/m × 10^{-4}
0	0	210	28.8
14	2	238	32.6
42	5.5	266	36.4
70	9.2	294	40.1
98	13.5	322	44.2
126	17.5	350	48.6
154	21.1	364	54.8
182	24.9	371	74.0
		378	94.0

Determine the proportional limit and the modulus of elasticity for this material by plotting the data.

2.28 A test is conducted on a specimen having an initial diameter of 12.5 mm and a gage length of 200 mm. Plot the data obtained as a stress-strain curve and determine the proportional limit and E. This test was stopped before failure occurred.

Load MN	Extensometer reading-mm	Load MN	Extensometer reading-mm
4.83	0.025	51.7	0.325
13.5	0.075	55.6	0.375
22.7	0.125	57.0	0.400
30.6	0.175	57.92	0.475
38.2	0.225	58.06	0.525
45.2	0.275	58.19	0.575
		58.61	0.675

REFERENCES

Arges, K. P. and A. E. Palmer. *Mechanics of Materials.* New York: McGraw-Hill Book Co., 1963.

Bruhn, E. F. *Analysis and Design of Flight Vehicle Structures.* Cincinnati: Tri-State Offset Co.

Byars, E. F. and R. D. Snyder. *Engineering Mechanics of Deformable Bodies.* 3rd ed. New York: Harper and Row, 1975.

Case, John and A. H. Chilver. *Strength of Materials.* London: Edward Arnold. Ltd.

Clark, D. S. *Engineering Materials and Processes.* International Textbook Co., Inc.

Keyser, C. A. *Materials of Engineering.* Englewood Cliffs, NJ: Prentice-Hall, Inc.

Lipson, C. and R. C. Juvinall. *Handbook of Stress and Strength.* New York: MacMillan Co.

Olsen, G. A. *Strength of Materials.* Englewood Cliffs, NJ: Prentice-Hall, Inc.

Rosenthal, E. and G. P. Bischof. *Elements of Machine Design.* New York: McGraw-Hill Book Co., 1955.

Shigley, J. E. *Machine Design.* New York: McGraw-Hill Book Co.

Slaymaker, R. R. *Mechanical Design and Analysis.* New York: John Wiley & Sons.

Timoshenko, S. and Donovan H. Young. *Elements of Strength of Materials.* 5th ed. New York: D. Van Nostrand Co., 1968.

3

Riveted and welded structures

3.1 INTRODUCTION

The connection of the members of structures to carry the loads imposed upon them and to provide the necessary structural support is usually accomplished by welding, riveting, or bolting the structures together. The principles that we shall discuss concerning riveted connections are directly applicable to bolted structures, and we will not treat bolted structures as a separate topic. A word of caution is in order at the beginning of our study. The techniques used by individual welders and riveters in making a specific joint can create situations that are not amenable to mathematical computations. A cold-worked rivet improperly set, a weld made with improper penetration of the weld material, the use of the improper weld rod, etc., can all lead to unsatisfactory structural riveted or welded attachments. In order to preclude such bad joints or at least to limit the number of them that occur, various standard codes are in effect. Of these the ASME (American Society of Mechanical Engineers) Boiler and Pressure Vessel Code and the AISC (American Institute of Steel Construction) Manual of Steel Construction are the most comprehensive and most widely used in industry. Where desirable, we shall refer to them and the rules they contain or suggest.

3.2 RIVETED CONNECTIONS

A *riveted connection* is a mechanical connection made by upsetting (or distorting) the ends of a special pin, known as a *rivet*. Once a riveted

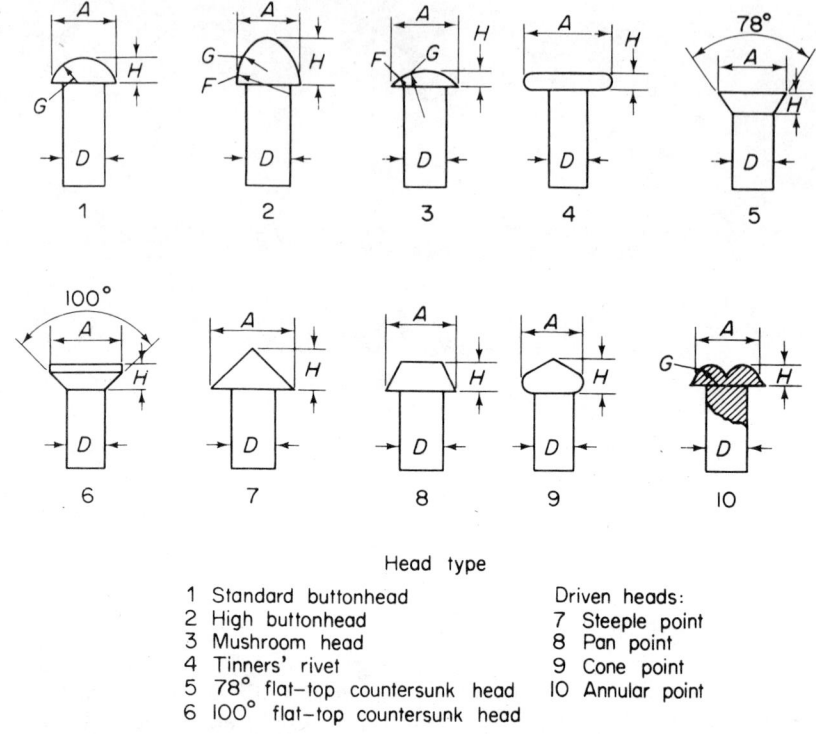

Fig 3.1 Common types of heads for rivets (Reproduced with permission from *Mechanical Design and Systems Handbook* by H. A. Rothbart, McGraw-Hill Book Co., Inc. (Figure 22-2, page 22-3))

connection is made, it is permanent and cannot be disassembled as can a bolted or screwed joint. As Fig. 3.1 shows, rivets are usually provided as a cylindrical shank on which a preformed head exists on one end. The rivet is inserted into holes drilled or punched in the plates that are to be fastened, and the rivet is "upset," i.e., formed by mechanically deforming the rivet between appropriate dies. The forming process can be carried out either hot or cold; steel rivets are usually worked hot (1600 °F to 1900 °F) while some non-ferrous rivets are kept refrigerated before being used. The steps in setting up a rivet joint are shown schematically in Fig. 3.2. The formed part of the rivet is held in place by one die known as the backing bar and the other die (the set) is positioned on the other end of the undeformed shank. The rivet is then upset by using a riveting machine (gun) operated pneumatically, mechanically, or by steam. The rivet can also be formed manually by using a hammer on the set.

Common rivets require access from both sides of the work and are installed by upsetting the extending shank to form a tail. They can be

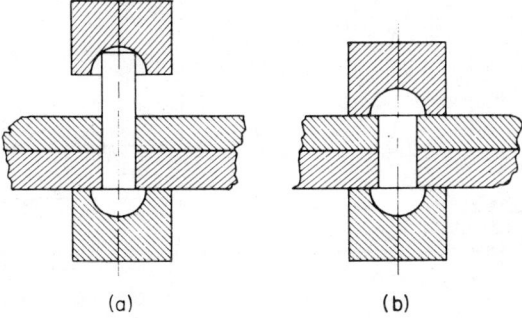

Fig 3.2   Setting-up of an ideal riveted joint

used for structural or nonstructural applications. Blind rivets require access from only one side of the work, and the tail is formed either chemically or mechanically. They can also be used for structural or nonstructural applications. Figure 3.3 gives the approximate dimensions used in selecting the length of a rivet that will yield a final rivet with the approximate tail dimensions shown. Commercially available steel structural rivets come in sizes from $\frac{3}{8}$-in. shank diameter to $1\frac{1}{4}$-in. diameter, in increments of $\frac{1}{8}$ in. SI sizes should be available in the near future.

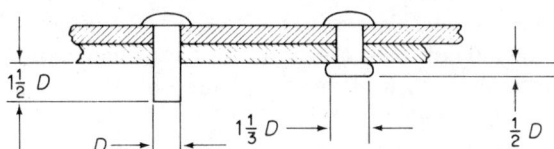

Fig 3.3   Approximate rivet sizes for structural joints

Riveted joints are usually categorized by the type and complexity of the joint. The most commonly encountered riveted joints are the *lap joint* and the *butt joint;* both are shown schematically in Fig. 3.4. In the lap joint, the two ends of the plates to be fastened are placed over one another and riveted together by placing rivets in holes made through both plates as shown on Fig. 3.4(a). Figure 3.4(b) shows the second commonly used riveted joint, the butt joint, so named because the edges of the plates to be riveted are butted against each other. The joint is made by riveting a cover plate (or plates) to each of the main plates as shown in Fig. 3.4(b).

Riveted joints are also classified by the number of rows of rivets used, i.e., a *double-riveted lap joint* is a lap joint having two rows of rivets. Thus, Fig. 3.4(b) shows a *single-riveted butt joint* since it consists of one

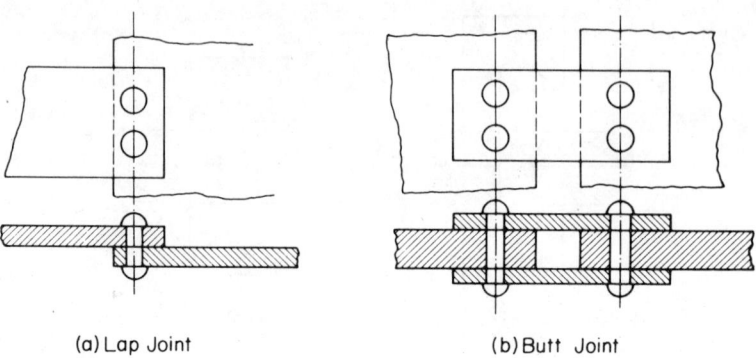

(a) Lap Joint (b) Butt Joint

Fig 3.4 Types of riveted joints

row of rivets. The student should note that it is possible to have a butt joint with differing rows of rivets on either side of the joint. In this event it is necessary to identify the joint on each side.

3.3 DESIGN STRESSES FOR RIVETED CONNECTIONS

The strengths of some steel rivet materials are given in Table 3.1, and it can be seen that there is a wide range of material properties depending upon the type of steel selected. The choice of suitable design stresses for design of riveted joints is not a simple task since consideration must be given to both the possible modes of failure of a given joint and the behavior of materials under such loading.

Additionally, the methods of fabrication of a particular joint may induce latent local stresses or physical conditions that may cause the joint to fail. Also, riveted pressure vessels operating at elevated temperatures must be designed to allow for the decrease of allowable working stress at these temperatures. Table 3.2 shows the decrease in working stress due to temperature for various grades of boiler steel. It will be noted from this table that in the temperature range of 0–700 °F, the working stress is based upon a factor of safety of 5, i.e., one-fifth of the minimum specified ultimate tensile strength of the material at room temperature.

In all riveted joints it is necessary that the holes be sufficiently larger than the rivet diameter for easy insertion and provision for sufficient clearance for the rivet in case of misalignment of the holes with plates. Such misalignment is more likely to occur in structural joints where the holes in each member are punched or drilled separately before the component parts of the joint are assembled.

The rules for the design and fabrication of riveted joints are given in the American Society of Mechanical Engineers (ASME) Boiler and Pressure

Chapter 3, Riveted and Welded Structures   81

Table 3.1 Strength of Steel Rivet Materials*

ASTM spec.	Description of material	Ultimate tensile strength psi	Ultimate tensile strength MPa	Yield point psi	Yield point MPa	Elongation in 8 in., %
A31	Boiler rivet steel and rivets, grade A	45,000–55,000	310–379	23,000	159	27
	Boiler rivet steel and rivets, grade B	58,000–68,000	400–469	29,000	200	22
A141	Structural rivet steel	52,000–62,000	359–427	28,000	193	24
A195	High-strength structural rivet steel	68,000–82,000	427–565	38,000	262	20
A406	High-strength structural-alloy rivet steel	68,000–82,000	427–565	50,000	345	20
A131	Structural steel for ships	55,000–65,000	379–448	30,000	207	23
A152	Wrought-iron rivets and rivet rounds	47,000	324	28,000	193	22–28

*Reproduced with permission from *Mechanical Design and Systems Handbook* by H. A. Rothbart, McGraw-Hill Book Co., Inc. Copyright 1964 by McGraw-Hill, Inc. SI units added.

Vessel Codes, the Steel Construction Manual of the American Institute of Steel Construction (AISC), and the American Railway Engineering Association (A.R.E.A.) Code for Railway Bridges. These codes differ due to the

Table 3.2 Values of Working Stress at Elevated Temperatures*

Maximum temperature, °F °C	Minimum of the specified range of ultimate tensile strength of the material, psi & (MPa)				
	45,000 (310)	50,000 (344)	55,000 (379)	60,000 (414)	75,000 (517)
0–700 (0–371)	9,000 (62)	10,000 (69)	11,000 (76)	12,000 (83)	15,000 (103)
750 (399)	8,220 (57)	9,110 (63)	10,000 (69)	11,200 (77)	13,000 (90)
800 (427)	6,550 (45)	7,330 (51)	8,000 (55)	9,000 (62)	10,200 (70)
850 (454)	5,440 (38)	6,050 (42)	6,750 (47)	7,400 (51)	8,300 (57)
900 (482)	4,330 (30)	4,830 (33)	5,500 (38)	5,600 (39)	6,000 (41)
950 (510)	3,200 (22)	3,600 (25)	4,000 (28)	4,000 (28)	4,000 (28)

*Reproduced with permission from *Design of Machine Members* by A. Vallance and V. L. Doughtie, McGraw-Hill Book Co., Inc., New York. 1951. SI units added.

Table 3.3 Allowable Stresses for the Design of Axially Stressed Riveted Joints*

	Shear Stress psi S_s	Shear Stress MPa S_s	Tensile Stress psi S_t	Tensile Stress MPa S_t	Bearing Stress psi S_b	Bearing Stress MPa S_b
A.S.M.E. Boiler Code	8,800	60.7	11,000†	75.8	19,000	131.0
A.R.E.A. Code for Railway Bridges						
Power-driven rivets	13,500	93.1	18,000	124.1	27,000	186.2
Hand-driven rivets	11,000	75.8	18,000	124.1	20,000	137.9
A.I.S.C. Code for Buildings	15,000	103.4	20,000	137.9	32,000** (single shear)	220.6
(Prior to 7th edition) See Table 3.4 for 7th ed. Values.					40,000 (double shear)	275.8

*Reproduced with permission from *Elements of Mechanics of Materials* by G. A. Olsen, Prentice-Hall, Inc., Englewood Cliffs, N.J. 1958. SI Units added.
†Varies with steel specified. See Table 3.1.
**A surface is in single- or double-shear bearing depending on whether one or two shear resisting surfaces bound the bearing area. If there is one, the area is in single-shear bearing. If there are two, it is in double-shear bearing.

specific applications and methods of fabrication generally used. Table 3.3 gives a summary of the stresses allowed by these codes for shear, tension, and bearing. We shall not investigate further the rules given in these codes for the fabrication of riveted joints, i.e., punching of holes, minimum plate thicknesses, etc.; and the student is cautioned at this time that these codes should be carefully studied before undertaking the design of structural or pressure vessel joints.

The 1970 AISC Manual of Steel Construction recognizes recent advances in metallurgy that have produced high-strength alloy steels. The allowable tension and shear stresses on rivets, bolts, and threaded parts as given in the 7th edition of the AISC Manual (1970) are shown in Table 3.4. In this latest AISC edition, the terms *friction-type* and *bearing-type* are used to categorize connections that transmit load by means of shear in their fasteners. Friction-type connections depend upon clamping forces sufficiently high to prevent slip of the connected parts. Bearing-type connections depend upon contact of the fasteners against the sides of their holes to transfer the load from one connected part to another.

For structural joints, the values given in Table 3.4 should be used. Also, as is noted later, *the tensile stress in the connecting plate can be taken as* $0.6 S_y$, *where* S_y *is the yield stress of the plate.*

Table 3.4 Allowable Stresses (7th ed. AISC Code)*

Description of Fastener	Tension (S_t) (MPa) 1000 psi	Shear (S_s) 1000 psi	
		Friction-Type Connections (MPa)	Bearing-Type Connections (MPa)
A502, Grade 1, hot-driven rivets	(137.9) 20.0		15.0 (103.4)
A502, Grade 2, hot-driven rivets	(186.2) 27.0		20.0 (137.9)
A307 bolts	(137.9) 20.0‡		10.0 (69.0)
Threaded parts‡‡ of steel (S_y = yield stress)	$0.60 S_y$‡		$0.30 S_y$
A325 and A449 bolts, when threading is *not* excluded from shear planes	(275.8) 40.0**	15.0 (103.4)	15.0 (103.4)
A325 and A449 bolts, when threading is excluded from shear planes	(275.8) 40.0**	15.0 (103.4)	22.0 (151.7)
A490 bolts, when threading is *not* excluded from shear planes	(372.3) 54.0**,***	20.0 (137.9)	22.5 (155.1)
A490 bolts, when threading is excluded from shear planes	(372.3) 54.0**,***	20.0 (137.9)	32.0 (220.6)

*Reproduced with permission of the American Institute of Steel Construction, New York, N.Y., from the 7th edition of the AISC Manual of Steel Construction (1970). SI units added.

‡Applied to tensile stress area equal to $0.7854(D - 0.9743/n)^2$ where D is the major thread diameter and n is the number of threads per inch.

**Applied to the nominal bolt area.

‡‡Since the nominal area of an upset rod is less than the stress area, the former area will govern.

***Static loading only.

3.4 FAILURE IN RIVETED JOINTS

Before we study the design of riveted joints and the assumptions made in conventional design procedures, let us consider the possible ways that failure can occur in such joints.

Rivet failure

(a) *Rivet Shear*   Figure 3.5 shows the condition of a joint failure due to the shear of a rivet. In Fig. 3.5(a), the mode of failure shown

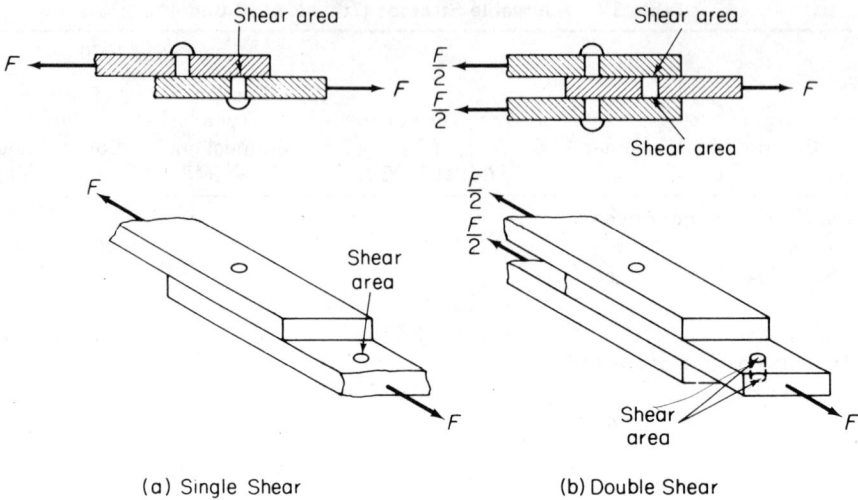

(a) Single Shear   (b) Double Shear

Fig 3.5 Failure by rivet shear

is due to the shear of the rivet at the interface planes between the plates due to the relative motion of the upper and lower plates.

Since Fig. 3.5(a) shows a single shear plane on which the rivet can fail, this type of failure is called a *single shear* of the rivet. Fig. 3.5(b) shows a failure that occurs due to the relative motion of the center plate with respect to both the upper and lower plates, and this mode of failure is called *double shear* for obvious reasons.

(b) *Rivet Bending*   Consider the conditions shown in Fig. 3.6(a) and (b). In both instances the rivet is shown to be long compared to its diameter, and this condition causes the rivet to distort excessively. In the condition shown in Fig. 3.6(c) the axial load on each plate causes a moment, *Fd*, to exist on the plate tending to bend it. This action places the rivet in

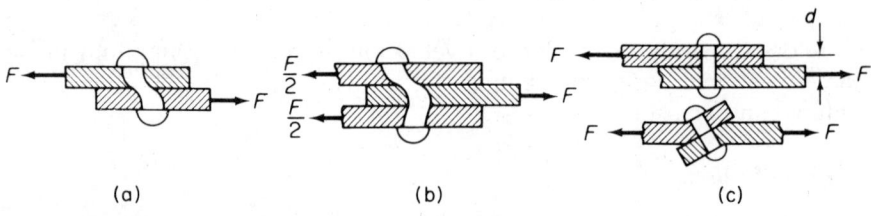

(a)   (b)   (c)

Fig 3.6 Rivets in bending

tension and subjects it to bending as well. The use of a second row of rivets decreases the tendency of the plates to bend and alleviates greatly the conditions shown in Fig. 3.6(c).

(c) *Rivet Crushing (Bearing)* The bearing of the plate against a rivet can cause the rivet to crush. The failure that occurs in this case is a compressive failure that will cause the connection to loosen and deform excessively. Figure 3.7 shows this action in a joint subject to single shear. In addition to undergoing the action shown in Fig. 3.7, rivets in single

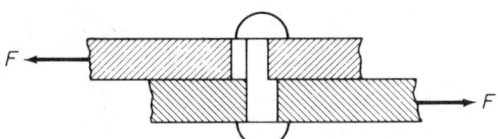

Fig 3.7 Rivet crushing

shear tend to rotate in the hole [see also Fig. 3.6(c)] causing a nonuniform distribution of bearing stresses between the plate and the rivet. When rivets are in double shear as shown in Fig. 3.6(b), the joint is loaded symmetrically reducing the tendency of the rivet to rotate and resulting in a uniform bearing stress distribution in the rivet. In order to account for these actions, the AISC decreased the allowable bearing stress for rivets in single shear to 80% of that allowed for rivets in double shear (as we can see in Table 3.3). Thus, rivets in double shear are 25% stronger in bearing and 100% stronger in shear (since there are two shear areas) than rivets in single shear, as calculated from the values in Table 3.3.

Recent tests have shown that in bearing, the critical portion of a connection is not the fastener; rather the critical portion of a connection (for bearing) is the material of the connected parts. Recognizing this, the 1970 AISC Manual permits the fastener to be checked for bearing failure using the properties of the connected part. The allowable bearing stress on the projected area of bolts and rivets in bearing-type connections as given by this code is $S_b = 1.35\ S_y$, where S_y is the yield stress of the connected part.

(d) *Rivet Set-up* After a rivet is hot formed, it cools and in this process contracts, placing the plates in compression, which causes bearing under the heads of the rivet and sets up initial tensile stresses in the rivet equal to the yield point of the material of the rivet. Due to its diametrical contraction, the rivet does not fill the hole in the plates, allowing joint slippages to occur. All of these effects leave many unknowns in the actual stress distribution in the riveted joint.

Plate failure

(a) *Tension in Net Plate Section* (*Tearing*)   A rivet decreases the area available to resist the direct load placed on a riveted joint. Figure 3.8 shows this type of plate failure. In addition to the simple joint shown in Fig. 3.8, a more complex joint, such as a butt joint, could sustain this type of failure, which occurs in the plates being joined as well as in the connecting (or *strap*) plates.

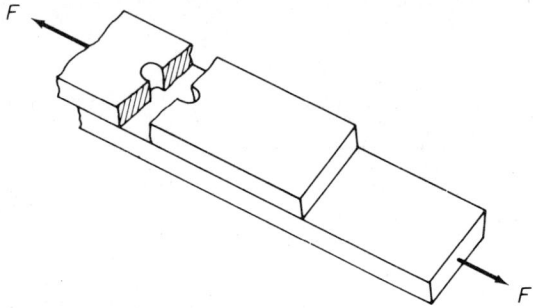

Fig 3.8   Plate tearing across net section

(b) *Bearing Failure of Plate* (*Crushing*)   In Fig. 3.7 we saw the material of a rivet being crushed by the plate. However, the plate against which the rivet bears is in compression, and the plate material can fail and wrinkle due to excessively high compressive stresses. Crushing of the plate will cause the joint to loosen due to the motion such action causes. In a butt joint the straps are also subject to this crushing action. However, by making the sum of the strap thicknesses greater than the plate thickness, it becomes unnecessary to investigate the straps in bearing.

(c) *Plate Shear* (*Edge Failure*)   If a rivet is placed too near the edge of a plate, the plate may shear as shown in Fig. 3.9. It is also possible

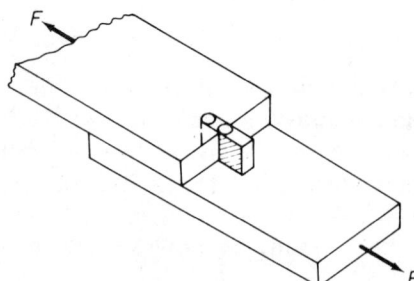

Fig 3.9   Plate edge shear failure

for the joint to fail by the opening up or tearing of this plate in the area in back of the rivet. The failures so described can be eliminated by designing the joint for the proper edge distance depending upon the rivet size.

(d) *Plate Tearing Between Rows* (*Diagonal Failure*) When a riveted connection has more than one row of rivet holes, failure can occur by a tearing of the plate between rows of holes. This type of failure (sometimes called *diagonal failure*) is due to the rivets in successive rows being too close. Figure 3.10 shows this type of failure, and it is noted that the proper proportioning of the dimensions of a riveted joint minimizes the possibility of this occurrence.

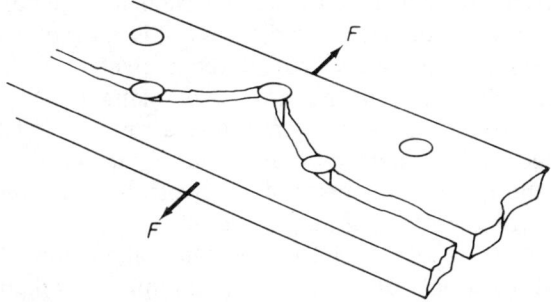

Fig 3.10   Diagonal failure in plate

3.5 DESIGN OF RIVETED JOINTS

Due to the complex interactions that occur in a riveted connection, it is a practical impossibility to design such connections in a rigorous manner. Even attempting to account for the failure modes as noted in the previous section presents a task so formidable as to be nearly impossible. Also, the various codes previously noted have design rules that differ from each other depending upon the specific application. In order to rationally approach the design of riveted joints, we will make certain simplifying assumptions, which may not be applicable to an actual situation. Due to this, the student is once again cautioned to refer to the specific applicable code before attempting to design riveted structural connections.

With the foregoing in mind, the assumptions made are:

1. Each rivet takes an equal share of the load, or the load is distributed among the rivets in proportion to the shear areas of the rivets. Actual conditions cause the outer row of a multi-row joint to take a larger than its proportional share of the load.
2. There is no failure mode due to edge tearing or shear. The design codes effectively eliminate this type of failure.

3. The effective shear area for a rivet in double shear is twice the effective shear area for a rivet in single shear. The nominal diameter of the rivet is used to calculate the shear area.
4. There are no bending, tension, or set-up stresses in the rivets.
5. The crushing (bearing) is uniformly distributed over the projected area of the rivet and plate.
6. The tensile stresses in the plate are uniform over the net area of the plate.
7. Friction between the members of the connection is assumed to be zero and does not enter into the calculation of the strength of the riveted joint. The AISC Manual should be consulted for friction-type connections.
8. The process of making the hole in the joint members does not damage the material. Actually, when holes are punched in a plate, the material adjacent to the hole is damaged, and the hole is ragged. Thus, the AISC and A.R.E.A. codes both require that punched rivet holes be considered $\frac{1}{8}$ in. (3.2 mm) larger than the nominal rivet diameter when calculating the tension stresses on the net section across a row of rivets. Even where holes are drilled and reamed, they are made $\frac{1}{16}$ in. (1.6 mm) greater in diameter than the nominal diameter of the rivets. In calculations involving pressure vessels designed according to the ASME Code, the nominal rivet diameter plus $\frac{1}{16}$ in. is to be used in all plate calculations. We shall use for our calculations a hole diameter $\frac{1}{8}$ in. (3.2 mm) larger than the nominal rivet diameter when calculating plate tension. For machined parts, the hole diameter can be taken equal to the fastener diameter.
9. After forming, the rivets completely fill the holes in the plate. For pressure vessel joints, the driven rivet is assumed to fill the hole completely; and for this application, the rivet diameter used in calculations is the diameter of the rivet hole in the plate. For structural joints, whether punched or drilled, the nominal rivet diameter will be used for rivet shear calculations and bearing calculations.

As a consequence of the previous discussions, we shall consider that a rivet joint can fail in one of three possible modes, and we express these failure modes mathematically based upon the assumptions made.

Failure by rivet shear

For single shear of a rivet subjected to the loading in Fig. 3.5(a)

$$S_s = \frac{F}{A_s} = \frac{F}{\pi d_s^2/4} \tag{3.1}$$

where S_s is the shear stress in psi, F is the shear force on the rivet, A_s is the effective shearing area of the rivet, and d_s is the nominal diameter of the rivet.

For double shear [Fig. 3.5(b)]

$$S_s = \frac{F}{2A_s} = \frac{F}{2(\pi d_s^2/4)} \tag{3.2}$$

Crushing or bearing of plate and/or rivet

Consider the condition shown in Fig. 3.11 where it is assumed that the rivet is bearing against the plate with a uniform radial pressure, which we shall denote as S_b, the bearing stress. If we draw a free-body diagram of the loading shown in Fig. 3.11(a), we obtain Fig. 3.11(b).

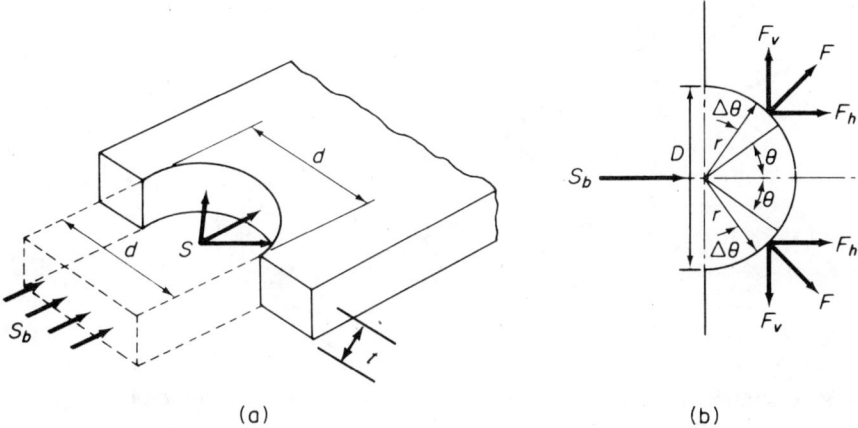

Fig 3.11 Bearing of plate

At the angle θ (both positive and negative) a small angle $\Delta\theta$ is shown. The force F on this segment is the stress (S_b) multiplied by the area of segment $t(r\Delta\theta)$. This force is directed normal to the arc, and F_h and F_v are the horizontal and vertical components of the force F. Reference to Fig. 3.11(b) shows us that the sum of all the vertical force components is zero, since there is a positive (up) and equal vertical force vector for each negative (down) vector. The horizontal vector (in each quadrant) is given by

$$F_h = S_b(t)(r\Delta\theta) \cos\theta \tag{3.3}$$

If we sum up all these horizontal vectors for angles θ from $+90°(\pi/2)$ to $-90°(-\pi/2)$, we obtain

$$F_h = S_b(t)(2r) = S_b td \text{ or } S_b = \frac{F_h}{td} \tag{3.4}$$

Thus the horizontal force is equal to the product of the bearing stress and the projected area of the hole in the plate. This product is equivalent to a uniform stress of S_b acting over the projected area of the hole in the plate as is shown in Figs. 3.11(a) and (b). The rivet bearing is obtained in an identical manner and equation (3.4) is also applicable to the calculation of rivet bearing.

Tension in the plate

Figure 3.12 shows a plate in tension. As shown, we will consider a uniform tensile stress S_t to act across the net cross-sectional area of the plate. In terms of the dimensions shown in Fig. 3.12 and denoting the plate tensile area as A_t:

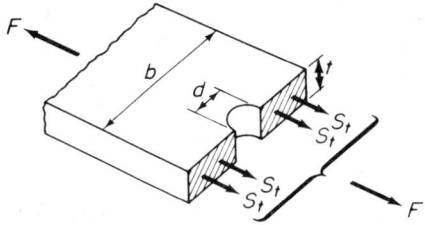

Fig 3.12 Plate in tension

$$S_t A_t = F = S_t(b - d)t \qquad (3.5)$$

If a section of a riveted connection has n areas subjected to shear, equation (3.2) can be written for the rivet shear as

$$S_s = \frac{F}{A_s n} = \frac{F}{n(\pi d_s^2/4)} \qquad (3.6)$$

for single shear. For double shear

$$S_s = \frac{F}{2A_s n} = \frac{F}{2n(\pi d_s^2/4)} \qquad (3.7)$$

For bearing of n bearing areas

$$S_b = \frac{F}{ntd} \qquad (3.8)$$

Finally, for tension across the net section of the plate with n rivets in a given row

Chapter 3, Riveted and Welded Structures   91

$$S_t = \frac{F}{(b-nd)t} \text{ or } n = \frac{b}{d} - \frac{F}{dtS_t} \quad (3.9)$$

The 7th edition of the AISC Manual gives as an allowable stress in tension a value equal to $0.6\,S_y$, where S_y is the yield stress of the material.

Illustrative problem 3.1

A riveted lap joint has to carry a load of 135 kN. (a) Using the AISC allowable stresses from Table 3.4 ($S_t = 137.9$ MPa, $S_s = 103.4$ MPa and $S_b = 220.6$ MPa for single shear bearing), design the joint. Assume each plate to be 9.5 mm thick, 12.5-mm-diameter rivets are to be used and the plate is 225 mm wide. (b) Compare this design with a connection designed using the 1970 AISC Manual, using A502 Grade 1 rivets in a connection having a plate yield stress of 344.8 MPa.

Solution:

(a)   For this problem we do not know how many rivets will be needed. The hole diameter in the plate will be taken to be 3.2 mm greater than the rivet diameter. The hole is therefore $12.5 + 3.2 = 15.7$ mm (diameter).

1. rivet shear:

$$S_s = \frac{F}{A_s n}; \; n = \frac{F}{S_s A_s} = \frac{135\,000}{103.4 \times 10^6 \times \frac{\pi}{4}(0.0125)^2} = 10.6 \text{ rivets}$$

2. rivet (plate) bearing:

$$S_b = \frac{F}{nA_b}; \; \frac{F}{S_b A_b} = n = \frac{135\,000}{220.6 \times 10^6 (0.0095 \times 0.0125)}$$
$$= 5.2 \text{ rivets}$$

3. plate tension (rivets in one row):

$$S_t = \frac{F}{(b-nd)t}; \; n = \frac{b}{d} - \frac{F}{dtS_t} = \frac{0.225}{0.0157} - \frac{135\,000}{(0.0157)(0.0095) \times 137.9 \times 10^6}$$

$= 7.8$ rivets (say 8 rivets)

Notice that at least 10.6 (say 11) rivets are needed to provide the required shear area, that at least 5.2 rivets must be provided to yield the necessary bearing area, and that the greatest number of rivets that can be used before tearing the plate is 8.0. Therefore this joint cannot be used. It is necessary either to thicken the plate or increase the rivet diameter, or both. (b) The allowable tensile stress is $0.6\,S_y = 0.6 \times 344.8 \times 10^6 = 206.9$ MPa, the allowable bearing stress is $1.35\,S_y = 1.35 \times 344.8 \times 10^6 = 465.4$ MPa, and the allowable shearing stress for A 502 Grade 1 material is 103.4 MPa. Proceeding as before,

1. rivet shear:

$$n = \frac{F}{S_s A_s} = \frac{135\,000}{103.4 \times 10^6 \times \frac{\pi}{4}(0.0125)^2} = 10.6 \text{ rivets}$$

2. rivet (plate) bearing:

$$n = \frac{F}{S_b A_b} = \frac{135\,000}{465.4 \times 10^6\,(0.0095 \times 0.0125)} = 2.4 \text{ rivets}$$

3. plate tension (rivets in one row):

$$n = \frac{b}{d} - \frac{F}{dt S_t} = \frac{0.225}{0.0157} - \frac{135\,000}{(0.0157)(0.0095)(206.9 \times 10^6)}$$
$$= 9.96 \text{ rivets}$$

For this case at least 10.6 rivets are needed for shear, at least 2.4 (say 3) rivets are needed for bearing, and the greatest number of rivets that can be used before tearing the plate is 9.96 (say 10). Therefore 10 rivets are the maximum that can be used, using these design criteria. The student will note the widely differing results obtained using these different design bases. It is for this reason that the student is cautioned to use the latest code methods.

Illustrative problem 3.2

A butt joint is used to connect two 12.5-mm thick plates. Assume the joint to be 75 mm wide. Assume also that there is a single rivet in double shear [see Fig. 3.5(b)], that each upper and lower strap plate is 9.5 mm thick, and that a 12.5-mm-diameter rivet is used (hole diameter is 12.5

+ 3.2 = 15.7 mm). What safe load can this joint (or section of a connection) carry? Use A502 Grade 1 rivets and assume the plate material has a yield stress of 275.8 MPa.

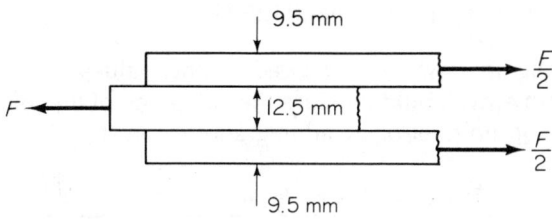

Fig 3.13 Illustrative problem 3.2

Solution:

(a) rivet shear: $F_s = A_s S_s$, $A_s = \dfrac{\pi}{4}(0.0125)^2 = 1.23 \times 10^{-4} \text{m}^2$. For double shear we have twice the area. Also $S_s = 103.4$ MPa; $F_s = 2(1.23 \times 10^{-4})(103.4 \times 10^6) = 25\ 486$ N.

(b) rivet (plate) bearing: This is a case of double shear bearing. We also note that the sum of the thicknesses of the upper and lower straps exceeds the thickness of the main plates. Referring to Fig. 3.13 we can see that half of the total force is taken in each strap, but the thickness of each strap exceeds half of the plate thickness. Therefore, it is not necessary to calculate the bearing and tensile stresses in the straps since they will not be the limiting stresses for this joint. Using $S_b = 1.35\ S_y$, we have for the main plate $F_b = S_b A_b = 275.8 \times 10^6 \times (1.35)(0.0125)(0.0125) = 58\ 177$ N.

(c) plate tension; $S_t = 0.6\ S_y = 0.6(275.8 \times 10^6) = 165.5 \times 10^6$ Pa

$$F_t = S_t A_t = 165.5 \times 10^6 (0.075 - 0.0153)(0.0125)$$
$$= 123\ 504 \text{ N}$$

The safe load is limited to the rivet shear load of 25 486 N.

A term commonly used to evaluate riveted joints is *joint efficiency*. The efficiency of a riveted joint is defined as the ratio of the load carrying capability of the joint to the tensile load that the main unperforated plate can carry. Ideally, it would be desirable to have the strength of a joint equal in bearing, shear, and tension. In practice this may not be possible, and the least load capacity is the design load for a joint. If we consider Illustrative Problem 3.2, we can determine the joint efficiency as

$$\text{Efficiency} = \eta = \left[\frac{25\,486}{(0.075)(0.0125)(165.5 \times 10^6)}\right] 100 = 16.4\%$$

which is quite low when compared to the values given in Table 3.5. For this example, we could increase the efficiency of the joint by increasing the diameter of the rivet or by adding another rivet.

Table 3.5 Typical Riveted Joint Efficiencies

Type of Joint		Efficiency, %
Lap:		
	Single row	53
	Double row	67
	Triple row	76
Butt:		
	Single row	57
	Double row	77
	Triple row	85
	Quadruple row	90

Illustrative problem 3.3

Using A502 Grade 1 rivets and a plate yield stress of 275.8 MPa, determine the safe load of the butt joint shown. What is its efficiency?

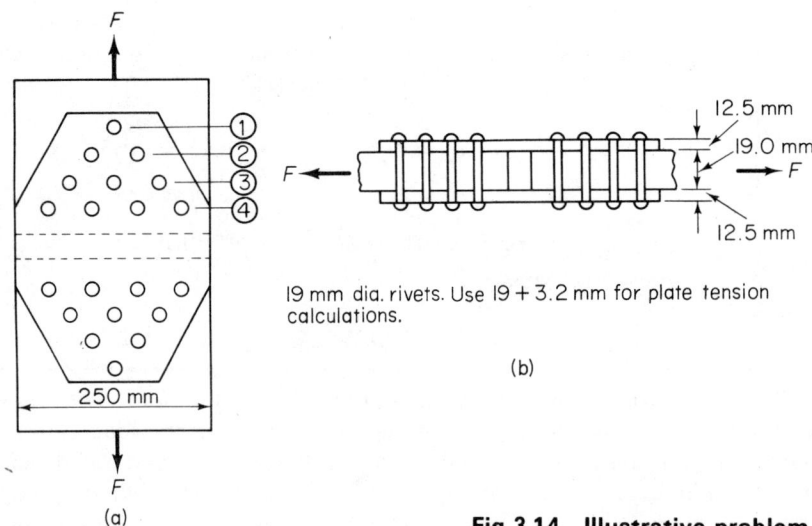

19 mm dia. rivets. Use 19 + 3.2 mm for plate tension calculations.

Fig 3.14 Illustrative problem 3.3

Solution:

The load will be assumed to be uniformly distributed over all of the rivets. The resisting load flow is shown in Fig. 3.15 where the branching flows denote the load into the straps. Note that each rivet of the 10 rivets is assumed to take $\frac{1}{10}$ of the applied load. For this problem we need not investigate the straps (except to assure ourselves that they are wide enough at any section) since their combined thickness exceeds that of the main plates.

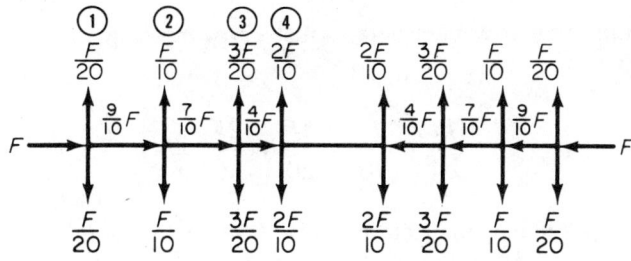

Fig 3.15 Load flow for illustrative problem 3.3

(a) Rivet shear: We have 10 double-shear rivets, giving an allowable shear load: (S_s = 103.4 MPa)

$$F_s = (10)(2)\frac{\pi}{4}(0.019)^2 \times 103.4 \times 10^6 = 586\,339 \text{ N} = 586.3 \text{ kN}$$

(b) Bearing: For (double-shear) bearing $S_b = 275.8 \times 10^6 \times 1.35 = 372.3$ MPa. Therefore

$$F_b = 372.3 \times 10^6 \times 0.019 \times 0.019 \times 10 = 1.344 \times 10^6 \text{ N}$$

(c) Tension: We shall have to investigate each section, using the information on Fig. 3.15.

(1) Section (1) must carry the full load F; therefore, we have for $S_t = 0.6 S_y = 165.5$ MPa and 1 rivet hole

$$F_t = 165.5 \times 10^6 \times 0.019(0.250 - 0.0222) = 716.3 \text{ kN}$$

(2) Section (2) carries $\frac{9}{10}F$, and there are 2 rivet holes

$$\frac{9}{10}F_t = 165.5 \times 10^6 \times 0.019\,[0.250 - 2(0.0222)]\,;$$
$$F_t = 718.3 \text{ kN}$$

(3) Section (3) carries $\frac{7}{10}F$, and there are 3 rivet holes

$$\frac{7}{10}F_t = 165.5 \times 10^6 \times 0.019 \,[0.250 - 3(0.0222)];$$
$$F_t = 823.9 \text{ kN}$$

(4) Section (4) carries $\frac{4}{10}F$, and there are 4 rivet holes

$$\frac{4}{10}F_t = 165.5 \times 10^6 \times 0.019 \,[0.250 - 4(0.0222)];$$
$$F_t = 1.267 \times 10^6 \text{ N}$$

(5) At the inner row each strap must carry the sum of the strap load, i.e.,

$$\frac{F}{20} + \frac{F}{10} + \frac{3F}{20} + \frac{2F}{10} = \frac{F}{2}$$

[This is a check since by inspection of Fig. 3.5(b) the inner force is obviously $F/2$.]

The strap has four holes at the inner row and can carry in tension

$$\frac{F_t}{2} = 0.0125 \,[0.250 - 4(0.0222)] \, 165.5 \times 10^6; \, F_t = 667.0 \text{ kN}$$

The joint can therefore carry only 586.3 kN; the limiting factor being rivet shear. The joint efficiency is;

$$\eta = \frac{586.3 \times 10^3}{0.250 \times 0.019 \times 165.5 \times 10^6} \times 100 = 74.6\%$$

which compares well with the efficiency for the quadruple row butt joint given in Table 3.5.

3.6 ECCENTRICALLY LOADED RIVETED JOINTS

Thus far we have considered each rivet to be loaded equally. In order for this condition to exist, it is necessary for the resultant transmitted force to pass through the center of gravity of the rivet group. Figure 3.16(a) shows the resultant force passing through the center of gravity of the rivet group. In this case all of the resisting forces are equal, parallel to, and opposed to the applied force, F. The situation shown in Fig. 3.16(b) differs from the case shown in Fig. 3.16(a) in that the former joint is being twisted

Chapter 3, Riveted and Welded Structures   97

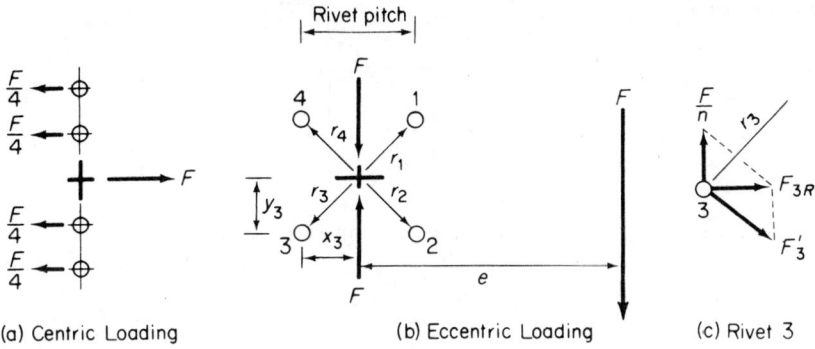

(a) Centric Loading   (b) Eccentric Loading   (c) Rivet 3

Fig 3.16

by the eccentric loading. We can consider this joint to be loaded by the force F through the center of gravity of the rivet group and the couple Fe which is shown. The direct shear of each of the rivets due to the load F through the center of gravity of the rivet group is simply the load divided by the total rivet shearing area. In addition, the couple rotates the bracket about the center of gravity of the rivet group. Denoting the distances of the rivets from the center of gravity as r_1, r_2, r_3, and r_4 respectively, we will assume each rivet to be distorted (and therefore stressed) an amount proportional to its distance from the center of gravity. Therefore we write as a general equation

$$\frac{F'_1}{F'_n} = \frac{r_1}{r_n} \quad \text{or} \quad F'_n = \left(\frac{F'_1}{r_1}\right) r_n \tag{3.10}$$

where F'_n denotes the force on rivet n due to twisting. The moment of this force about the center of gravity of the rivet group is

$$F'_n r_n = \left(\frac{F'_1}{r_1}\right) r_n^2 \tag{3.11}$$

But the sum of all of the moments on the rivets about the center of gravity of the rivet group must equal the applied moment Fe. Denoting this summation for n rivets as $\sum_{1}^{n}$ we have

$$Fe = \frac{F'_1}{r_1} \sum_{1}^{n} r_n^2 \tag{3.12}$$

Solving for F'_1

$$F'_1 = r_1 \left[\frac{Fe}{\sum_{1}^{n} r_n^2} \right] \quad (3.13)$$

If we once again refer to Fig. 3.16(b), we can write r_n^2 as

$$r_n^2 = x_n^2 + y_n^2$$

x_n and y_n are coordinate distances respectively from the center of gravity. Using this and rearranging equation (3.13), we arrive at the twisting force on any rivet F'_n

$$F'_n = r_n \left[\frac{Fe}{\sum_{1}^{n} x_n^2 + \sum_{1}^{n} y_n^2} \right] \quad (3.14)$$

This force must be added vectorially to the direct force to obtain the total force on the rivet. For rivet 3 in Fig. 3.16(c) we have the direct resisting force F/n and the force due to twisting F'_3 acting as shown with the ensuing resultant F_{3R}. The foregoing is best illustrated by a specific example.

Illustrative problem 3.4

If the rivet group shown in Fig. 3.16(b) has a square pattern in which the rivet pitch in either the x- or y-directions is equal to 100 mm and a load F of 45 000 N is applied at a distance e of 250 mm, determine the total force acting on rivet 1.

Solution:

The direct load on each rivet is $F/4 = 45\ 000/4 = 11\ 250$ N. The twisting moment is $45\ 000 \times 0.250 = 11\ 250$ N·m. For each rivet, $x = y = 50$ mm and $r = \sqrt{(0.050)^2 + (0.050)^2} = 0.0707$ m. The sum

$$\sum_{1}^{4} x_n^2 = (0.050)^2 + (0.050)^2 + (0.050)^2 + (0.050)^2 = 0.01$$

and

$$\sum_{1}^{4} y_n^2 = (0.050)^2 + (0.050)^2 + (0.050)^2 + (0.050)^2 = 0.01$$

Applying equation (3.14)

$$F'_1 = \frac{0.0707 \times 11\,250}{0.01 + 0.01} = 39\,768 \text{ N}$$

A vector diagram of the forces on rivet 1 is shown in Fig. 3.17. The resultant of the two vectors shown will have its x and y components equal to the vector sum of the x and y components of these vectors. For the 39 768 N force, the x component is 39 768 cos 45° = 28 120 N, and the y component is 39 768 sin 45° = 28 120 N. Thus the resultant will have a left-directed x component of 28 120 N and an upward directed component of 28 120 + 11 250 = 39 370 N. The resultant will therefore be $R_1 = \sqrt{(28\,190)^2 + (39\,370)^2} = 48\,422$ N directed as shown.

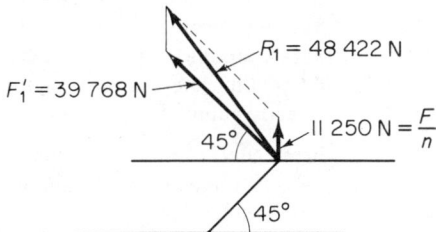

Fig 3.17   Illustrative problem 3.4

3.7 WELDED STRUCTURES—GENERAL

Welding is a metal joining process characterized by the melting of the parts being joined or by the placing of these parts at a temperature approaching their melting point. The welding process, when properly executed, offers the advantage of being more economical than riveted or bolted connections, does not require the drilling or punching of holes in the plates that the latter process does, yields a lighter joint, and yields a more uniform load distribution than riveted joints.

The three common types of welding are *resistance welding, gas welding,* and *arc welding.* Resistance welding obtains the required heat from the electrical resistance at the interface of the parts being joined; gas welding provides the required heat from the combustion of the gas (acetylene, etc.) with air; arc welding generates heat from the resistance of an electric current across a gas gap (filler metal may or may not be added to the joint), and an inert gas may be used to envelope the area being welded. Table 3.6 compares the advantages and disadvantages of these various welding processes.

Table 3.6 Welding Process Comparison*

Process	Advantages	Disadvantages
Resistance welding	Very flexible Economical Joins three or more pieces with one weld Properties of joint independent of process Can join any combination of metals that alloy together	Poor in fatigue Requires expensive equipment Must be able to reach both sides of the joint
Gas welding	Good control of weld metal Heat input anneals weld metal and heat-affected zone	High cost Low production rate Higher heat input; steep temperature gradient More warpage, distortion, etc. Material thicknesses must be nearly equal
Arc welding	High production rate Minimum distortion Low annealing of base metal	Stress concentrations at edge of welds, undercutting, overlaps, etc., likely

*Table 3.6, and Figures 3.19, 3.20, and 3.21 are taken from *Mechanical Design and Systems Handbook* by H. A. Rothbart, McGraw-Hill Book Co., Inc., 1964, with permission.

As in any technology, a certain amount of empirical background has evolved as well as a nomenclature specific to the topic. Figure 3.18 shows the nomenclature used for certain basic weld joints and Fig. 3.19 shows the standard weld proportions used to prepare a joint for welding. Experience has shown that these proportions are necessary to produce good welds. A specific type of joint is not suitable for all applications, and Fig. 3.20 shows the applicability of different types of welds to basic weld types. Figure 3.21 gives a comparison of the usage, efficiency, and application of many specific weld types. Once again the student should be aware that a considerable body of technology has evolved concerning welding, and that the information presented in this and subsequent sections of the

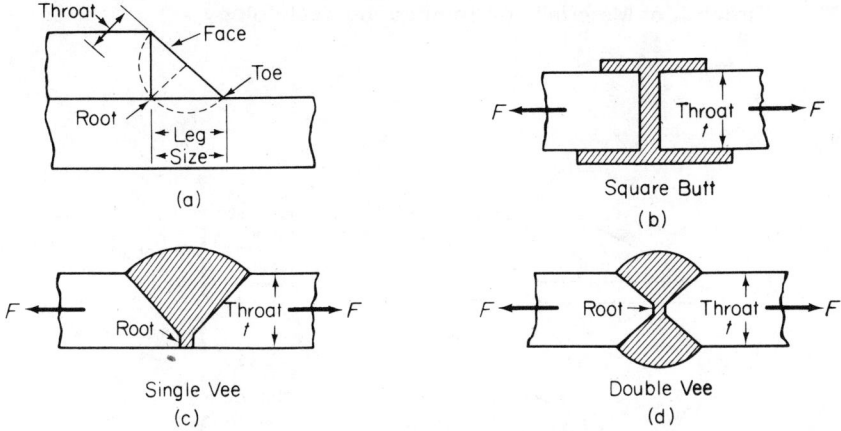

Fig 3.18 Basic weld nomenclature

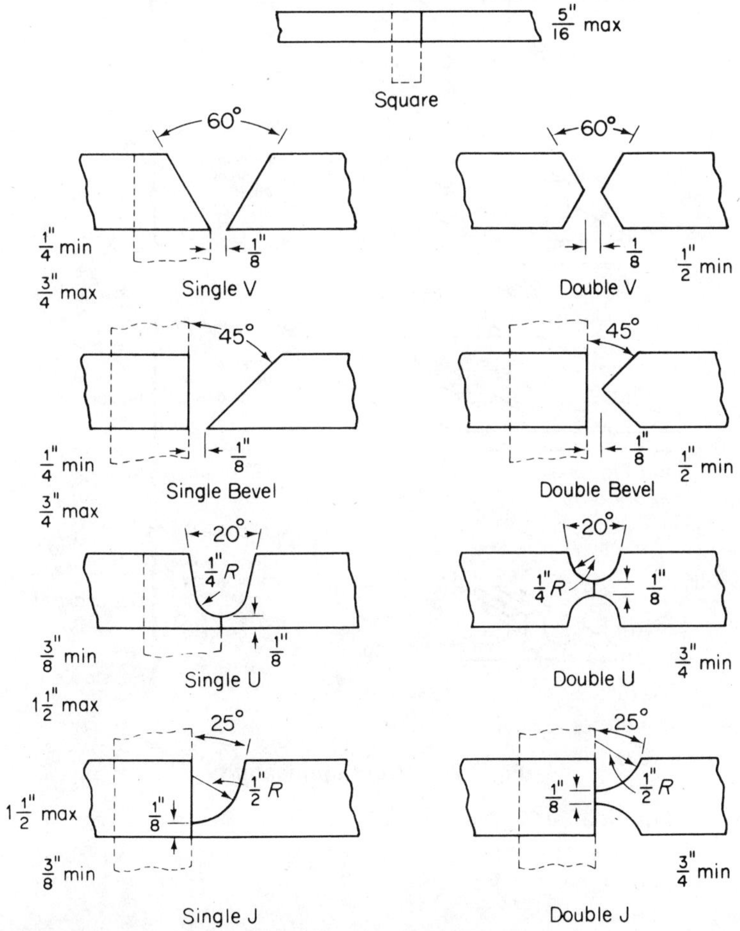

Fig 3.19 Proportions for weld joint preparation

102 Strength of Materials for Engineering Technology

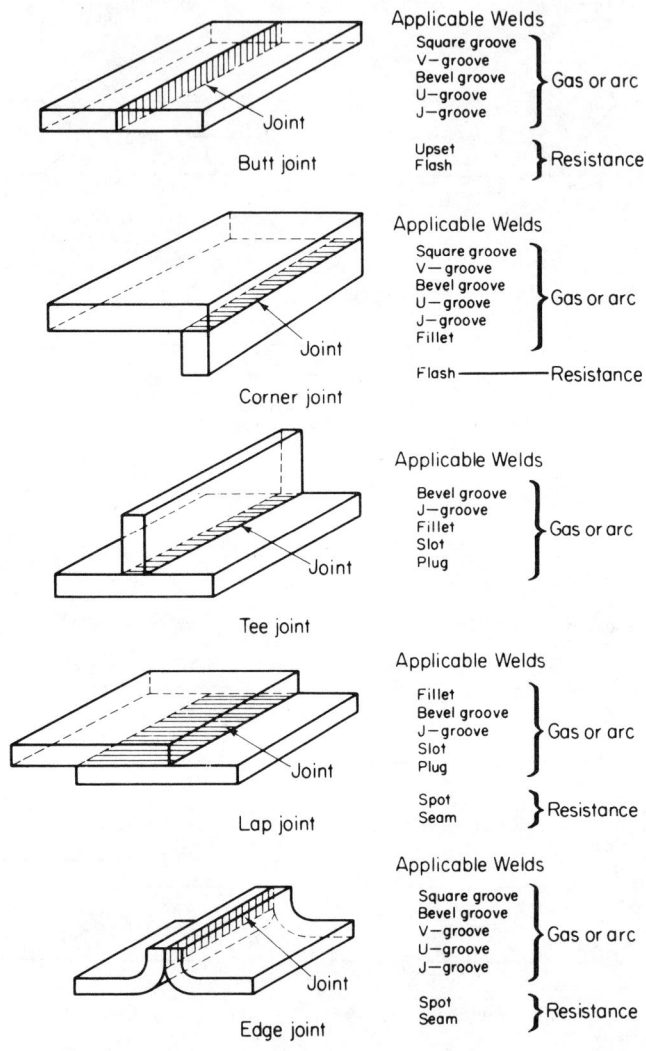

Fig 3.20 Weld applicability

Chapter 3, Riveted and Welded Structures

Type of weld	Illustration	Sheet gage	Efficiency in Shear	Efficiency in Tension	Efficiency in Compression	Fatigue resistance	Application
Square butt welded one side	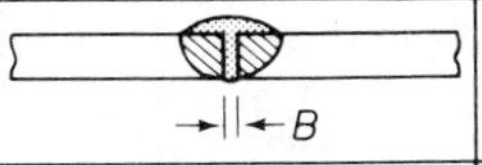 →‖← B	Up to 0.124	High	High	High	Fair	
Single V welded one side	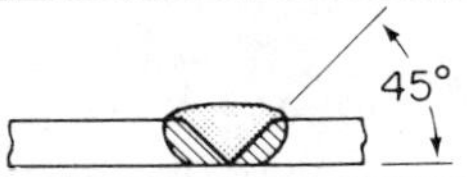 45°	0.125 and over	High	High	High	Good	All general applications. The choice depends upon the material, gage, loading, fatigue requirements, etc.
Single bevel welded one side	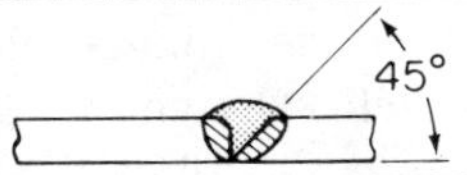 45°	0.125 and over	High	High	High	Good	
Double V welded both sides	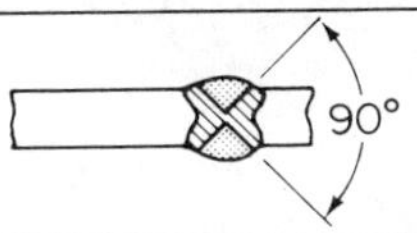 90°	0.375 and over	High	High	High	Good	
Open-square groove corner	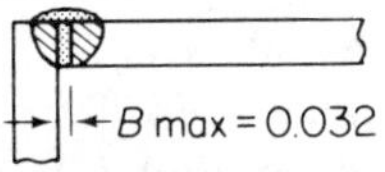 ‖← B max = 0.032	Up to 0.065	Medium	Low	High	Poor	
Single V corner fillet	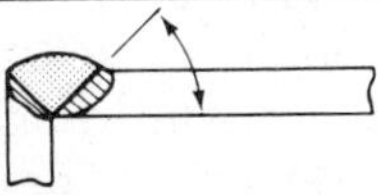	0.065 and over	Medium	Low	....	Poor	Closed structures and heavy-gage tanks
Outside or inside single fillet	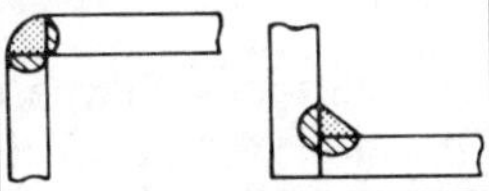	Up to 0.094	High	Low	....	Poor	
Single fillet tee	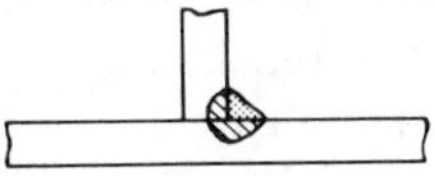	Any gage	Medium	Low	Low	Poor	General purposes
Double fillet tee	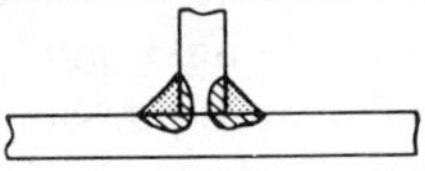	Any gage	High	High	High	Good	
Edge fillet weld	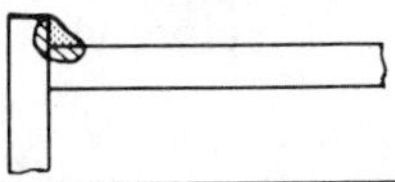	Any gage	Medium	Low	....	Fair	Closed structure and heavy-gage tanks
Double-lap fillet	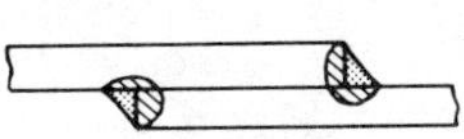	Up to 0.250	High	High	....	Fair	
Flange weld		Up to 0.081	Medium	Low	....	Poor	Tanks and closed non-structural parts

Fig 3.21 Weld type comparison

present chapter is introductory material. Also standard codes, such as the AISC Code, govern in many cases and should be consulted for specific applications

3.8 WELD DESIGN

Reference to Fig. 3.21 shows that the *butt joint* and the *fillet joint* are the two basic joint types in welding. In the design of butt joints, the cross-sectional area used for tension is that of the thinner plate being joined away from the weld. Thus for the butt welds shown in Fig. 3.18(b), (c), and (d), the minimum throat dimension is used. Denoting the minimum throat dimension as t

$$F_t = (S_t)(t)L \qquad (3.15)$$

where L is the length of the weld and S_t is the allowable tensile stress (see Table 3.7). Due to the possibility of heat damage in the parent metal adjacent to the weld, equation (3.15) is sometimes modified by introducing an additional efficiency term on the right hand side to decrease the design load carrying capacity of the joint. We will assume this joint efficiency to be 100% and use equation (3.15) as it stands since Code design stresses already include this factor.

Fillet welds are used to make joints having surfaces that are usually at right angles to each other. The basic assumptions made in the design of a fillet weld are that failure will occur due to shearing of the weld across the minimum section of the weld [the throat in Fig. 3.18(a)] and that the load is taken up uniformly over the length of the weld. For the weld shown in Fig. 3.22, the side welds, if they are the only ones present, will resist the shearing force F throughout their length and the failure section is assumed to be throat of the weld. For an equal-sided triangle, $t = \left(\dfrac{\sqrt{2}}{2}\right)a = 0.707a$. The joint design criteria therefore yield the follow-

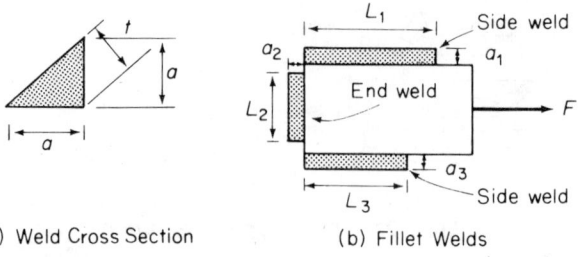

(a) Weld Cross Section (b) Fillet Welds

Fig 3.22 Fillet weld analysis

Table 3.7 Permissible Weld Stresses*

Kind of Stress	Permissible Stress	Required Electrode	"Matching" Base Metal
Tension and Compression parallel to axis of any complete penetration groove weld	Same as for base metal		
Tension normal to effective throat of complete-penetration groove weld	Same as allowable tensile stress for base metal		
Compression normal to effective throat of complete or partial-penetration groove weld	Same as allowable compressive stress for base metal		
Shear on effective throat of complete-penetration groove weld and partial-penetration groove weld	Same as allowable shear stress for base metal		
Shear stress on effective throat of fillet weld regardless of direction of application of load; tension normal to the axis on the effective throat of a partial-penetration groove weld; and shear stress on effective area of a plug or slot weld. The given stresses shall also apply to such welds made with the specified electrode on steel having a yield stress greater than that of the "matching" base metal. The permissible stress, regardless of electrode classification used, shall not exceed that given	18,000 psi (124.1 MPa)	AWS A5.1, E60XX electrodes AWS A5.17, F6X-EXXX fluxelectrode combination AWS A5.20, E60T-X electrodes	A500 Grade A A570 Grade D
	21,000 psi (144.8 MPa)	AWS A5.1 or A5.5 E70XX electrodes AWS A5.17, F7X-EXXX fluxelectrode combination AWS A5.18, E70S-X or E70U-1 electrodes AWS A5.20, E70T-X electrodes	A36 A53 Grade B A242 A375 A441 A500 Grade B

Table 3.7 *(continued)*

Kind of Stress	Permissible Stress	Required Electrode	"Matching" Base Metal
in the table for the weaker "matching" base metal being joined.			A501 A529 A570 Grade E A572 Grades 42 to 60 A588
	24,000 psi (165.5 MPa)	AWS A5.5, E80XX electrodes Grade 80 Submerged Arc, Gas Metal-Arc or Flux Cored Arc Weld Metal	A572 Grade 65
	27,000 psi (186.2 MPa)	AWS A5.5, E90XX electrodes Grade 90 Submerged Arc, Gas Metal-Arc or Flux Cored Arc Weld Metal	A514 over $2\frac{1}{2}$ in. thick (63.5 mm)
	30,000 psi (206.7 MPa)	AWS A5.5, E100XX electrodes Grade 100 Submerged Arc, Gas Metal-Arc or Flux Cored Arc Weld Metal	A514 over $2\frac{1}{2}$ in. thick (63.5 mm)
	33,000 psi (227.5 MPa)	AWS A5.5, E110XX electrodes Grade 110 Submerged Arc, Gas Metal-Arc or Flux Cored Arc Weld Metal	A514, $2\frac{1}{2}$ in. and less in thickness (63.5 mm)

*From the seventh edition *AISC Manual of Steel Construction*, 1970, with permission of the American Institute of Steel Construction, New York, N.Y.

ing equation for each side weld

$$F_1 = 0.707 a_1 L_1 S_s; \quad F_3 = 0.707 a_3 L_3 S_s \qquad (3.16)$$

Table 3.7 gives the values of permissible stress for welds from the 1970 edition of the AISC Manual of Steel Construction. It will be noted that for tension or compression, the allowable stress is that of the base metal (as is also the case for shear on groove welds). For fillet welds, the allowable shear stress is determined by the combination of electrode and base metal. The allowable shear stress in fillet welds varies from 124.1 MPa to 227.5 MPa.

If we use a lower value of S_s, of 127.3 MPa, equation (3.16) becomes

$$F = 90.0 \times 10^6 \, a_1 L_1 \qquad (3.17)$$

End fillet welds are probably stronger than side fillet welds, but this is not usually considered in design calculations. An end weld is considered to behave like a side weld, and the design is based on the throat section of the weld. Thus for only the end fillet weld [Fig. 3.22(b)]

$$F = 0.707 \, a_2 L_2 S_s \qquad (3.18)$$

And if S_s is 127.3 MPa,

$$F = 90 \times 10^6 \, a_2 L_2 \qquad (3.19)$$

If both end and side fillets are used

$$F = 90 \times 10^6 \, (a_1 L_1 + a_2 L_2 + a_3 L_3) \qquad (3.20)$$

Illustrative problem 3.5

If both side welds shown in Fig. 3.22(b) are equal length 12.5-mm welds, F is 225 kN and L_1 is 50 mm, determine the L_2 required. Assume $S_s = 127.3$ MPa.

Solution:

Applying equation (3.20) and noting $a_1 = a_2 = 0.0125$ m

$$225\,000 = 90 \times 10^6 \times 0.0125 (2 \times 0.050 + L_2)$$

and $L_2 = 0.1$ m $= 100$ mm

108   Strength of Materials for Engineering Technology

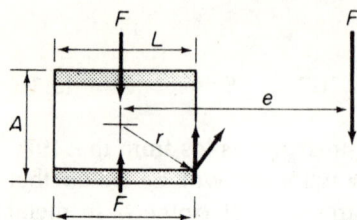

Fig 3.23   An eccentrically loaded welded joint

In the foregoing discussion we had assumed the welds to be subjected to shear only. This type of shear stress, produced in the direction of the applied load, is known as *primary shear*. In a manner analogous to that considered for eccentrically loaded riveted joints, we find that non-concentric loads on welds tend to rotate the welded joint giving rise to secondary shear stresses. Consider the condition shown in Fig. 3.23, where two fillet welds of equal length are subjected to a twisting moment Fe about their center of gravity.

Table 3.8   Polar Moment of Inertia For Weld Calculations

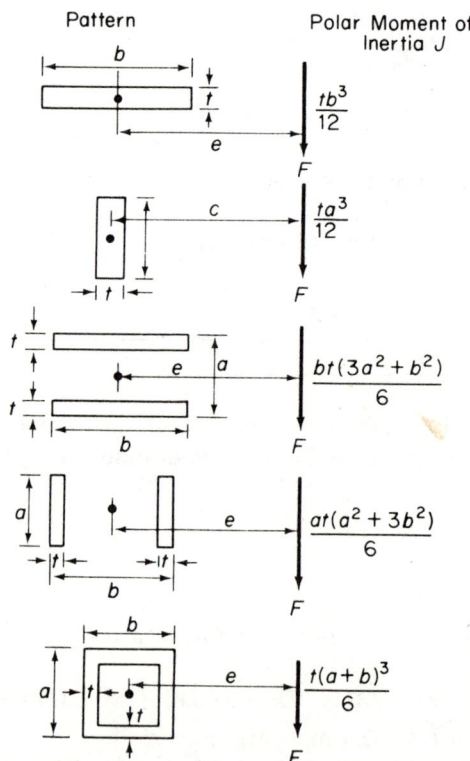

Pattern	Polar Moment of Inertia J
	$\dfrac{tb^3}{12}$
	$\dfrac{ta^3}{12}$
	$\dfrac{bt(3a^2 + b^2)}{6}$
	$\dfrac{at(a^2 + 3b^2)}{6}$
	$\dfrac{t(a+b)^3}{6}$

Chapter 3, Riveted and Welded Structures   109

In addition to the primary direct shear load on the welds, the structure will tend to rotate about the center of gravity of the welds inducing a secondary shear stress. The considerations involved in the calculation of the secondary stress in the weld are similar to those for riveted joints and also involve considerations taken up in Chapters 4 and 5. Rather than deriving these items at this point, we shall merely cite them. The student should return and re-study this section after completion of these next two chapters. The secondary shear stress is

$$S'_s = \frac{Fer}{J_o} \qquad (3.21)$$

where J_o is known as the *polar moment of inertia* of the weld group. Table 3.8 gives the polar moment of inertia of common weld configurations. The total stress in the weld at a given point will be the vector sum of the direct and secondary stresses.

Illustrative problem 3.6

If the value of F is 45 kN, $e = 100$ mm, $L = 125$ mm, and $A = 100$ mm and all welds in Fig. 3.23 are 12.5-mm welds, determine the stress on the lower right-hand corner of the weld.

Solution:

The direct stress is the force divided by the shear area of the weld. Therefore

$$S = \frac{45\,000}{0.707 \times (0.0125 \times 0.125) \times 2} = 20.37 \text{ MPa}$$

From Table 3.8,

$$J_o = \frac{bt(3a^2 + b^2)}{6} = \frac{0.125(0.0125)(3 \times 0.100^2 + 0.125^2)}{6}$$
$$= 1.188 \times 10^{-5} \text{ m}^4$$

and

$$S' = \frac{Fer}{J} = \frac{45\,000 \times 0.100 \times \sqrt{0.050^2 + (0.125/2)^2}}{1.188 \times 10^{-5}} = 30.32 \text{ MPa}$$

A vector diagram at the point in question yields

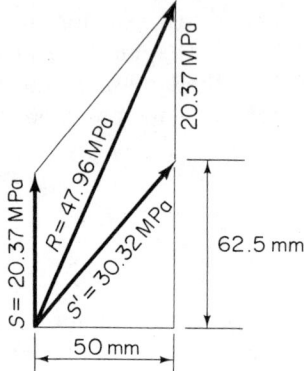

Fig 3.24 Illustrative problem 3.6

The x and y components of the 30.32 MPa are: $x = (0.050/0.08) \times 30.32 = 18.95$ MPa, $y = (0.0625/0.08) \times 30.32 = 23.69$ MPa. The resultant stress $R = \sqrt{(18.95)^2 + (23.69 + 20.37)^2} = 47.96$ MPa

3.9 CLOSURE

The methods of material joining discussed in this chapter, riveting and welding, are widely accepted and used throughout industry. The designs of these connections, however, are quite complex; and, to an extent that will not be found elsewhere in this book, it has been necessary to make a multitude of simplifications in our analysis in order to obtain solutions. Also, the connections in question are the subject of extensive empirical studies that have been incorporated into specific, detailed design codes. While the author is quite aware of the frustration of the student when cautioned to consult other works before doing the design of a particular structural connection, it cannot be helped. Quite often the rationale for a given rule or design criteria is not evident—the best that can be said for these instances is that *these rules work*. These codes are continuously reviewed by cognizant code committees to insure that their adequacy and applicability are in line with current state-of-the-art knowledge and practices. Revisions of these design rules are incorporated into these codes continually, and even the most experienced structural designer must spend time with the code to insure that his data are current. No other satisfactory approach has yet evolved, and the student should heed all of the precautionary notes scattered throughout this chapter.

PROBLEMS

In all of the following problems, except as specifically noted, use $S_t = 137.9$ MPa, $S_s = 103.4$ MPa, $S_b = 465.4$ MPa. (These values correspond to A502 Grade 1 rivets.) Use 127.3 MPa as the allowable shear stress in fillet welds. *Use $0.6S_y$ (Plate) for S_t in riveted plates, in accordance with the latest AISC code, with S_y (Plate) = 344.8 MPa.*

3.1 A single-riveted lap joint is made as shown in Fig. P3.1. Using rivet and hole diameters equal to the nominal rivet diameter, determine the dimensions of the joint for equal bearing, and shear and tensile strength. (*HINT:* Solve for t from the condition of equal bearing and shear strength. Then calculate W from the tension criteria.)

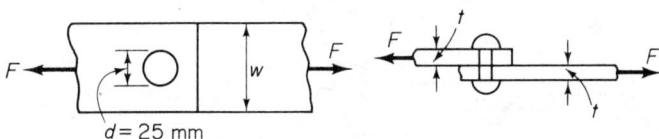

Fig P3.1

3.2 Determine the efficiency of the joint designed in Problem 3.1.
3.3 If three 16-mm-diameter rivets are used to carry the load in the joint shown in Fig. P3.3, determine the shear stress in the rivets.

```
           o
Fig P3.3   o ———————→ 80 kN
           o
```

3.4 Four 12.5-mm rivets carry the load equally in the arrangement shown. Determine the shear stress in each of the rivets if each rivet is in double shear.

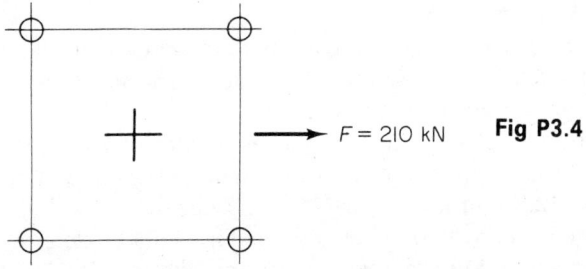

3.5 Two 12.5-mm-diameter rivets are used to carry the load in a lap joint as shown in Fig. P3.5. Determine the maximum load the joint can carry.

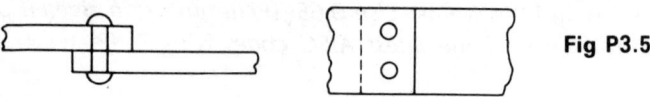

Fig P3.5

3.6 A 19-mm-diameter rivet is used to fasten two plates as shown in Fig. P3.6. Determine the load the joint can sustain.

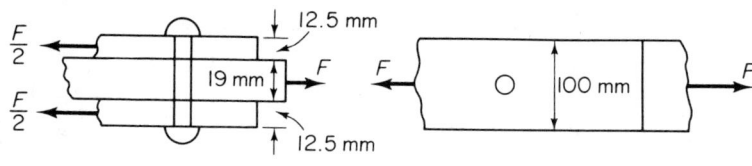

Fig P3.6

3.7 Assuming that three rivets are used in Problem 3.6, determine the load that the joint can carry.

3.8 A butt joint is made as shown in Fig. P3.8 using 16-mm-diameter rivets. Determine the load that this joint can carry.

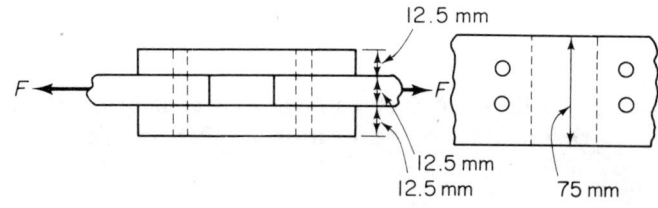

Fig P3.8

3.9 A double-row, riveted lap joint has an efficiency of 67%. If there are four rivets in each row and the plate is 25 mm thick and 200 mm wide, what size rivet is required?

3.10 A double-row, riveted lap joint has an efficiency of 65%. Determine the size of rivet required if there are two rivets in each row and the plate is 12.5 mm thick and 100 mm wide.

3.11 A lap joint is fabricated as shown in Fig. P3.11. Determine the joint efficiency and the maximum load that the joint can sustain.

Chapter 3, Riveted and Welded Structures

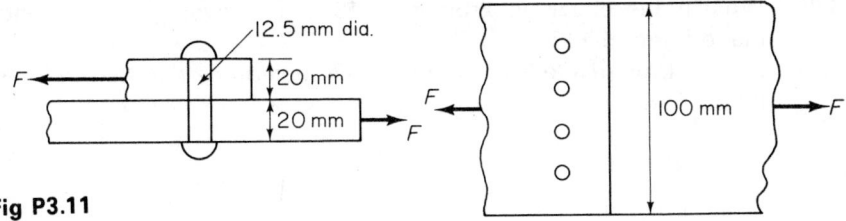

Fig P3.11

3.12 A lap joint is made as shown in Fig. P3.12. Determine the maximum load the joint can sustain. What is its efficiency?

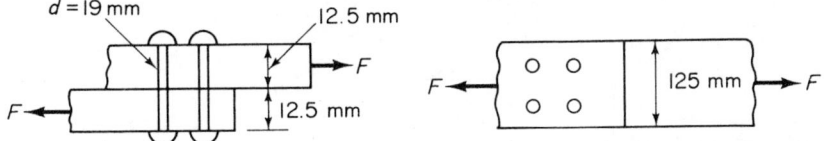

Fig P3.12

3.13 Four 19-mm diameter rivets are used to make the joint shown in Fig. P3.13. Determine the load that this joint can carry.

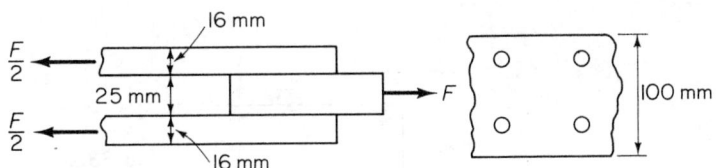

Fig P3.13

3.14 If the lower plate in Problem 3.12 is made 9.5 mm thick, determine the maximum load the joint can sustain and its efficiency.

3.15 A butt joint has two rows of rivets as shown in Fig. P3.15. If the nominal rivet diameter is 12.5 mm, determine the allowable strength of the joint.

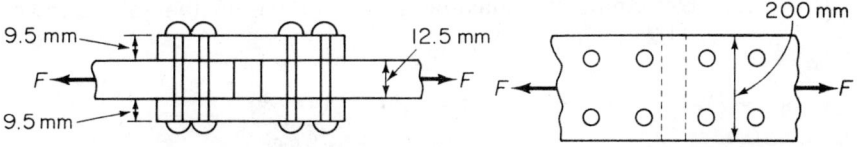

Fig P3.15

3.16 What is the effect on Problem 3.15 if the cover plates are each made 8 mm thick?

3.17 For the triple-riveted butt joint shown, determine the load that can be carried.

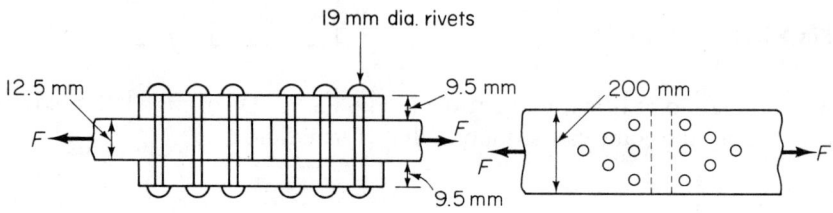

Fig P3.17

3.18 Solve Problem 3.17 if the main plate is made 16 mm thick.

3.19 If a quadruple-riveted butt is made as shown in Fig. 3.14, but the upper and lower plates are made 6.5 mm thick, determine the strength of this joint. Is the 6.5-mm thickness of cover plates desirable? If not, why not?

3.20 Determine the maximum shear force in the joint shown in Fig. P3.20.

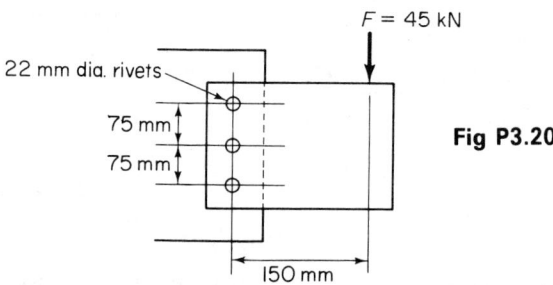

Fig P3.20

3.21 If the distance from the load to the centerline of the rivet group in Problem 3.20 is made 100 mm, what will the maximum shear force be in the joint shown?

3.22 Determine the maximum shear force in the joint shown in Fig. P3.22.

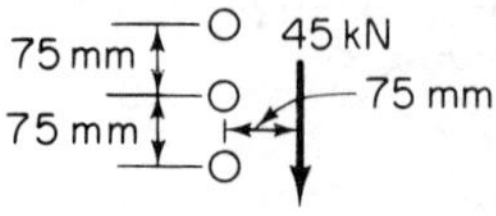

Fig P3.22

Chapter 3, Riveted and Welded Structures   115

3.23 Determine the maximum shear force in the joint shown in Fig. P3.23.

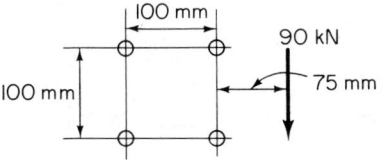

Fig P3.23

3.24 What is the maximum shearing force in the connection shown in Fig. P3.24?

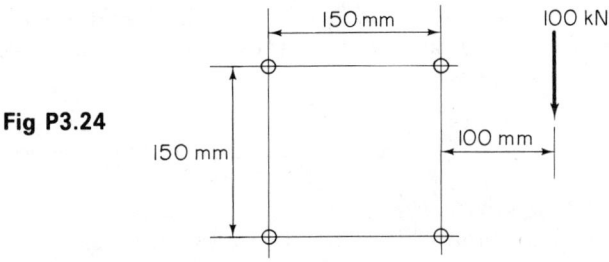

Fig P3.24

3.25 Determine the maximum shear force in the joint shown in Fig. P3.25.

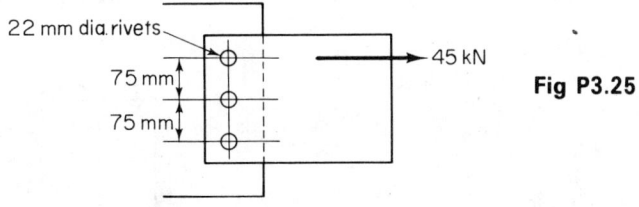

Fig P3.25

3.26 Which of the joints shown in Fig. P3.26 is stronger?

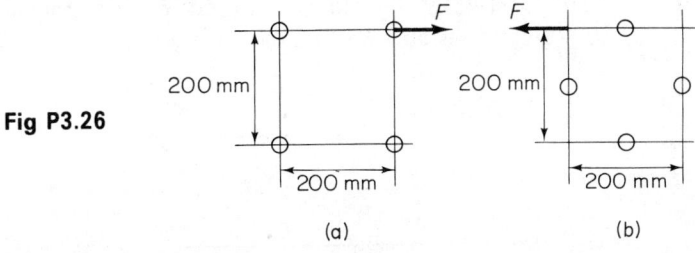

Fig P3.26

3.27 What is the maximum efficiency that a single fillet weld can have?
3.28 Determine the maximum load that the joint in Fig. P3.28 can carry. Assume A36 steel and AWS 5.1 electrodes.

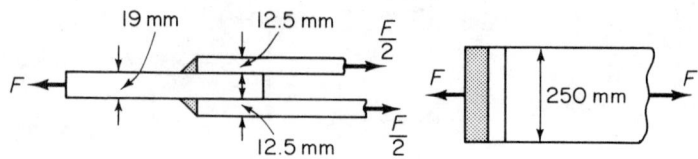

Fig P3.28

3.29 Determine the efficiency of the welded joint shown in Fig. P3.28; $S_t = 0.6 \times 248.2$ MPa $= 148.9$ MPa for A36 steel.
3.30 For the arrangement shown, what is the maximum force that can be applied?

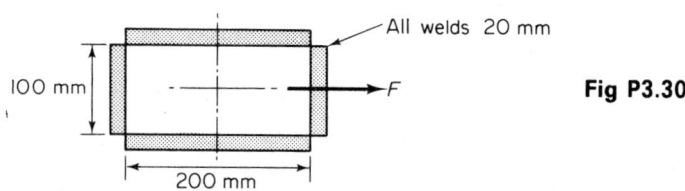

Fig P3.30

3.31 A 125-mm square, 12.5-mm-thick plate is welded to a large base plate. Determine the maximum centric force the welds can resist.

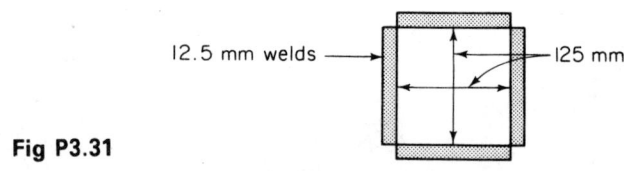

Fig P3.31

3.32 A plate is welded to a base plate as shown in Fig. P3.32. What length of 12.5-mm weld is required to carry the load? Use AWS A5.1 electrodes and A500 Grade A steel.

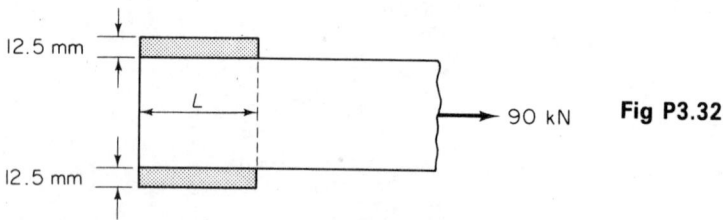

Fig P3.32

Chapter 3, Riveted and Welded Structures   117

3.33 What weld size is required to carry a load of 150 kN for the arrangement shown in Fig. P3.33 if $L = 50$ mm?

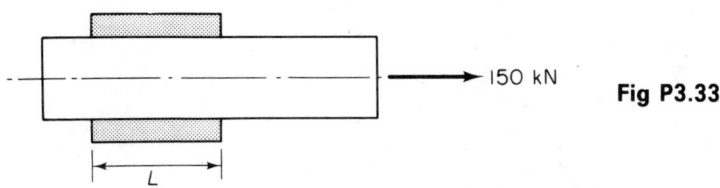

Fig P3.33

3.34 A plate is welded to a large base plate by two fillet welds as shown in Fig. P3.34. If the applied load is 450 kN, determine L.

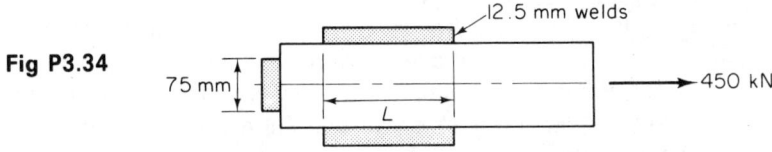

Fig P3.34

3.35 Two plates are to be welded as shown in Fig. P3.35. Which joint is stronger?

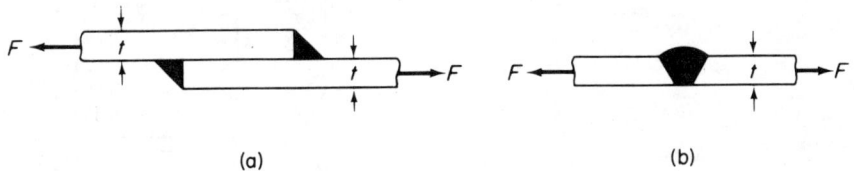

Fig P3.35

3.36 Which arrangement in Fig. P3.36 yields a lower shear stress in the welds?

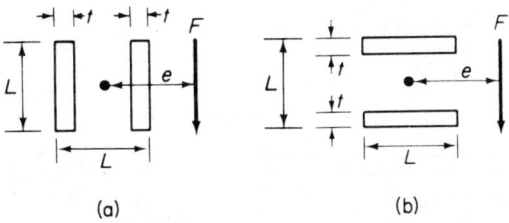

Fig P3.36

3.37 Determine the lengths of weld required in the joint shown in Fig. P3.37. Assume the side and end fillet welds are equally long. S_s = 126.9 MPa allowable.

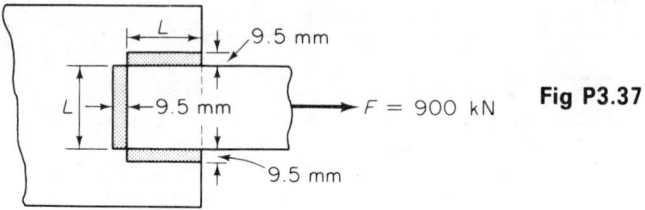

Fig P3.37

3.38 Refer to Fig. P3.37. If L = 500 mm, what size weld is required to resist a load of 1.5 MN?

3.39 Two 19-mm plates are to be butt-welded together. If the plates are 200 mm wide and tests show that the weld yields at 1.16×10^6 N, what is the joint efficiency? Assume the plate yield stress is 344.8 MPa. Comment on the definition of joint efficiency as applied to this problem.

3.40 In order to increase the strength of a welded joint, a designer suggests that the design in Fig. P3.40 be used. Comment on it.

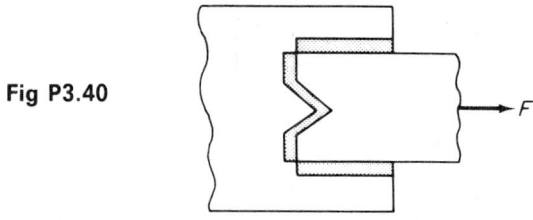

Fig P3.40

3.41 Determine the condition for the placement of F if each length of each side weld is to be subjected to the same stress. Let L equal the total length, $L_A + L_B$. The joint is not to be subject to twisting.

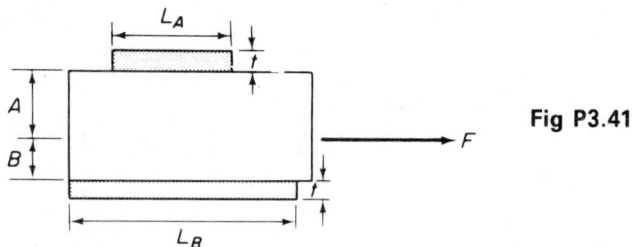

Fig P3.41

3.42 If $L_A = 100$ mm and $L_B = 200$ mm, determine the location of the safe force F that can be applied without twisting the connection, shown in Fig. P3.41. Assume $A + B = 300$ mm.

3.43 If an end weld is added to the joint shown in Fig. P3.41, determine the location of F that would have zero net moment on the weldment. Assume the end weld to have a length equal to $A + B$.

3.44 There is an option to either rivet or spot weld a joint as shown in Fig. P3.44. Neglecting all considerations other than strength, which joint is more desirable? Assume the rivet diameter equals the spot weld diameter.

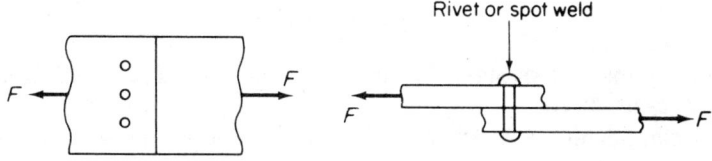

Fig P3.44

3.45 A 100-mm-square bar is to be welded to a plate, and the shear force is in the plane of the weld. Determine the maximum shear load that this joint can sustain if the load passes through the center of gravity of the welds. The welds are fillet welds with 12.5-mm legs, and $S_s = 126.9$ MPa.

3.46 If the weld described in Problem 3.45 is subjected to a 178 kN load located 250 mm from the center of gravity of the weld, determine the stress due only to the moment. Compare it to the value that would be obtained if this load were passed through the center of gravity of the weld as in Problem 3.45.

3.47 Which corner of the square in Fig. P3.47 will have the highest stress if the periphery of the square is welded completely around?

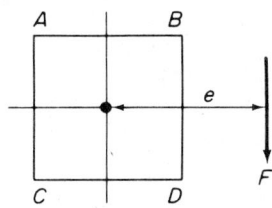

Fig P3.47

3.48 A welded joint is loaded as shown in Fig. P3.48. Determine the maximum shear stress.

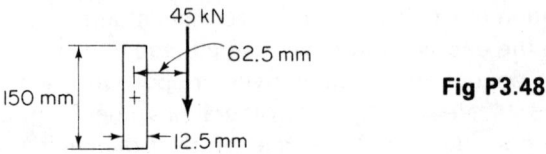

Fig P3.48

REFERENCES

American Institute of Steel Construction. *The Manual of Steel Construction.* 7th ed., 1970.

American Society of Mechanical Engineers. *The Boiler and Pressure Vessel Code.*

Arges, K. P. and A. E. Palmer. *Mechanics of Materials.* New York: McGraw-Hill Book Co., 1963.

Black, P. H. and O. E. Adams, Jr. *Machine Design.* 3rd ed. New York: McGraw-Hill Book Co., 1968.

Doughtie, V. L. and A. Vallance. *Design of Machine Members.* 4th ed. New York: McGraw-Hill Book Co., 1964.

Jensen, A. and H. H. Chenoweth. *Applied Strength of Materials.* 3rd ed. New York: McGraw-Hill Book Co., 1971.

Olsen, G. A. *Elements of Mechanics of Materials.* 3rd ed. Englewood Cliffs, NJ: Prentice-Hall, Inc., 1974.

Rosenthal, E. and G. P. Bischof. *Elements of Machine Design.* New York: McGraw-Hill Book Co., 1955.

Rothbart, H. A. *Mechanical Design and Systems Handbook.* New York: McGraw-Hill Book Co., 1965.

Sheiry, E. S. *Elements of Structural Engineering.* International Textbook Co.

Timoshenko, S. and Donovan H. Young. *Elements of Strength of Materials.* 5th ed. New York: D. Van Nostrand Co., 1968.

4

Center of gravity and moment of inertia— the first and second moment of area

4.1 INTRODUCTION

In subsequent chapters of our study it will become necessary to calculate two properties of an area, namely the first and second moments of the area with respect to a specified axis. Because these terms bear a resemblance to similar expressions that are developed in mechanics for solid bodies, the terms "center of gravity" (or centroid) and "moment of inertia," respectively, are commonly given them. Although an area cannot be said to have a center of gravity or inertia (since it has no mass), these terms are in such universal use that we will subsequently use them in the development of this chapter. As the expressions for center of gravity and moment of inertia are studied, the student should note that use is again made of principles developed in statics for parallel force systems.

4.2 CENTER OF GRAVITY

Recall that the weight of a body is the resultant force of attraction on each particle of the body with respect to the earth. Since each unit of mass of the body experiences the same force, and each force is parallel to the forces acting on other particles, the weight of the body is simply the resultant of a parallel force system. Weight, therefore, is a force and can be represented and treated as a vector. Like all vectors, it must have three characteristic attributes: magnitude, direction, and line of application. If the body is irregularly shaped and oriented differently in space, its weight will remain constant and the direction of its resultant weight vector will be toward the center of the earth, but the line of action of the resultant will be in a different position with respect to the various parts of the body. Most important, however, is the fact that regardless of the orientation of the body, the resultant weight vector (extended if necessary) will pass through a point in the body (or outside of the body) known as the *center of gravity* (or *centroid*) of the body. This point can be considered to be the point of application of the weight of the body.

Consider the system shown in Fig. 4.1, which consists of three weights,

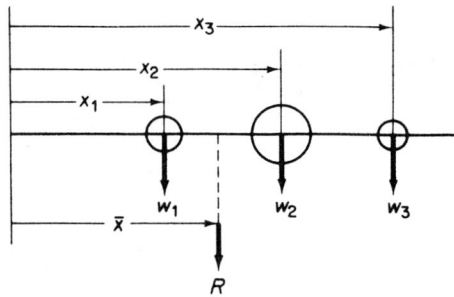

Fig 4.1 The center of gravity of discrete weights

w_1, w_2, and w_3, mounted on a rigid, weightless rod. We wish to determine the resultant of this system, which requires us to evaluate both its magnitude and location along the rod. It will be noted that the arrangement shown in Fig. 4.1 consists of three parallel forces due to the effect of gravity on each of the bodies. From statics we may state that the magnitude of the resultant of a parallel force system is equal to the algebraic sum of the forces on the system. Thus the resultant R for the system shown in Fig. 4.1 can be written as

$$R = w_1 + w_2 + w_3 \qquad (4.1)$$

The location of the resultant is found from the condition that the moment of the resultant about an arbitrary point must equal the sum of the moments of the individual weights about this same point. For the situation being considered, a moment summation about the left end of the bar yields

$$R\bar{x} = w_1 x_1 + w_2 x_2 + w_3 x_3 \qquad (4.2)$$

Chapter 4, Center of Gravity and Moment of Inertia

or

$$\bar{x} = \frac{w_1 x_1 + w_2 x_2 + w_3 x_3}{R} \qquad (4.2a)$$

where $\bar{x}$ is the location of the resultant from the left end of the bar. Combining equations (4.1) and (4.2) yields

$$\bar{x} = \frac{w_1 x_1 + w_2 x_2 + w_3 x_3}{w_1 + w_2 + w_3} \qquad (4.2b)$$

The location of the resultant $\bar{x}$ is the location of the center of gravity of the arrangement shown in Fig. 4.1. If we now introduce the Σ notation so that

$$\sum_{1}^{3}(wx) = w_1 x_1 + w_2 x_2 + w_3 x_3 \qquad (4.3)$$

and

$$\sum_{1}^{3}(w) = w_1 + w_2 + w_3 \qquad (4.4)$$

where the symbol Σ indicates a summation of the wx terms and w terms respectively, we can rewrite equation (4.2b)

$$\bar{x} = \frac{\sum_{1}^{3}(wx)}{\sum_{1}^{3}(w)} \qquad (4.5)$$

Illustrative problem 4.1

If w_1 is 50 N, w_2 is 100 N, and w_3 is 25 N with corresponding distances $x_1 = 5$ m, $x_2 = 10$ m, and $x_3 = 14$ m for the arrangement shown in Fig. 4.1, determine the location of the center of gravity.

Solution:

The magnitude of the resultant is equal to the algebraic sum of the weights, namely

$$R = 50 + 100 + 25 = 175 \text{ N}$$

The moment of each weight about the left edge of the bar is, $w_1 x_1 = 50(5)$, $w_2 x_2 = 100(10)$ and $w_3 x_3 = 25(14)$. The total moment is therefore $250 + 1000 + 350 = 1600$ N·m. The location of the center of gravity, $\bar{x}$, is given by equation (4.5) as

$$\bar{x} = \frac{1600}{175} = 9.14 \text{ m}$$

Now consider a thin body, like the one shown in Fig. 4.2, having a uniform thickness. If we freely suspend the body from point A, a plumb bob dropped from A will give us the line of action of the weight of the

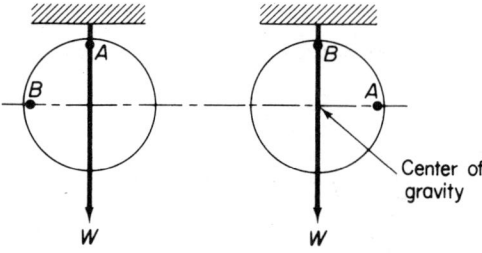

Fig 4.2 Center of gravity experimentally

body passing through point A, which we can scribe on the face of the plate. If we now suspend the plate from B and drop a plumb bob as before, we can again scribe the line of action of the weight, this time passing through point B. The intersection of these two lines gives us a point O on the face of the body that locates the center of gravity in the face plane.

We can summarize the foregoing and generalize it to any plane area in the following manner: first let the thickness decrease until we have a plane area and then subdivide the area into n small areas, each having an area ΔA and each having a center of gravity located distances x_1, $x_2, \ldots, x_n$ from the y axis as shown in Fig. 4.3. Each of the small areas can be considered to have a force acting on it proportional to its area. Applying the same reasoning that we used for the situation shown in Fig. 4.1, we have

$$\bar{x}(\Delta A_1 + \Delta A_2 + \Delta A_3 + \ldots + \Delta A_n)$$
$$= (\Delta A_1)x_1 + (\Delta A_2)x_2 + (\Delta A_3)x_3 + \ldots + (\Delta A_n)\bar{x}_n \quad (4.6)$$

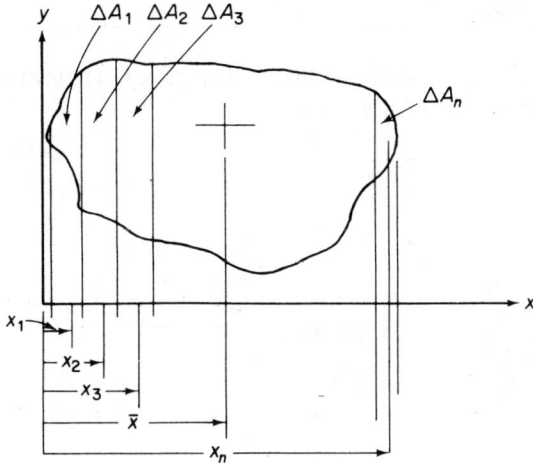

Fig 4.3 The center of gravity of an irregular area

Using the symbolic summation notation

$$\bar{x} = \frac{\sum_{1}^{n} (\Delta A) x}{\sum_{1}^{n} \Delta A} \qquad (4.7)$$

and by similarity

$$\bar{y} = \frac{\sum_{1}^{n} (\Delta A) y}{\sum_{1}^{n} \Delta A} \qquad (4.8)$$

Equations (4.7) and (4.8) are quite general; and in each of these expressions we note that the denominator is simply the area of the plate, and the numerator is the sum of the first moment of each area with respect to either the x or y axis. It is the numerator that gives us the name of first moment of area for the center of gravity of an area. Also from equations (4.7) and (4.8), we can conclude that the first moment with respect to an axis through the centroid must be zero since the summation in the numerator will be zero. This is due to the fact that there must be equal positive and negative moments on either side of the centroid. If the area has an axis of symmetry, the centroid must lie on this axis (or axes).

Illustrative problem 4.2

A beam has the cross section shown in Fig. 4.4. Determine the location of the center of gravity of this area.

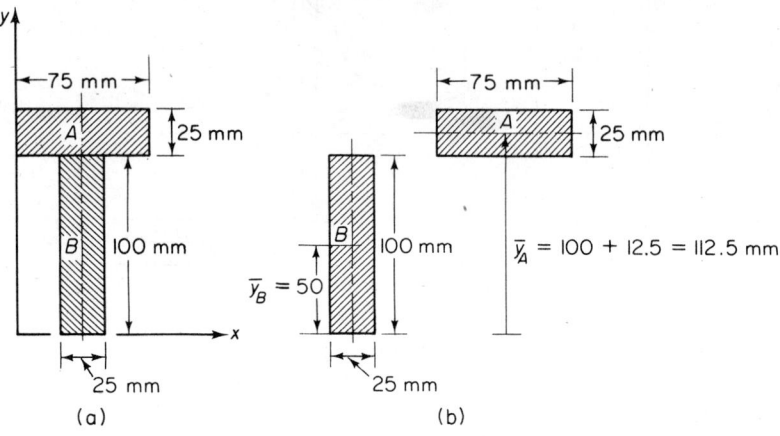

Fig 4.4 Illustrative problem 4.2

Solution:

Figure 4.4(a) shows the physical body divided into two rectangular parts, A and B. Since the figure is symmetrical, $\bar{x}$ is located on the axis of symmetry. In Fig. 4.4(b) each part is shown separately, and the locations of the centers of gravity for both part are shown as $\bar{y}_A$ and $\bar{y}_B$, respectively. In order to apply equation (4.8) we must calculate each area and its moment with respect to the y axis.

A_A = Area A = 75 × 25 = 1875 mm² $\bar{y}_A$ = 100 + 12.5 = 112.5 mm
A_B = Area B = 100 × 25 = 2500 mm² $\bar{y}_B$ = 50 mm
Σ Area = 4375 mm²

Applying equation (4.8)

$$\bar{y} = \frac{A_A \bar{y}_A + A_B \bar{y}_B}{A_A + A_B}$$

$$\bar{y} = \frac{1875 \times 112.5 + 2500 \times 50}{1875 + 2500} = 76.79 \text{ mm}$$

Illustrative problem 4.3

Determine the center of gravity of the angle shown in Fig. 4.5.

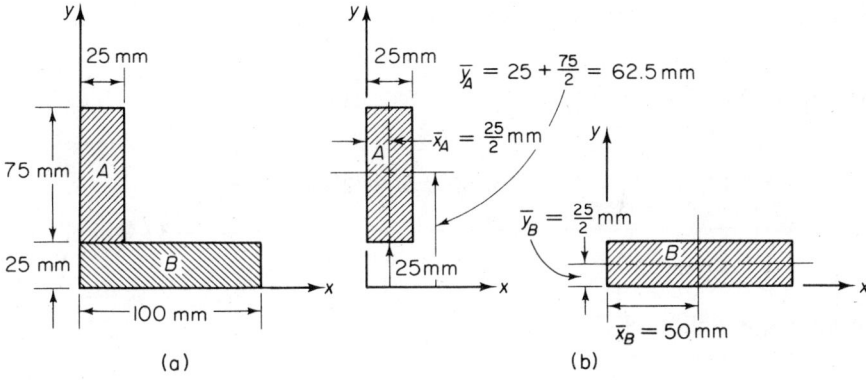

Fig 4.5 Illustrative problem 4.3

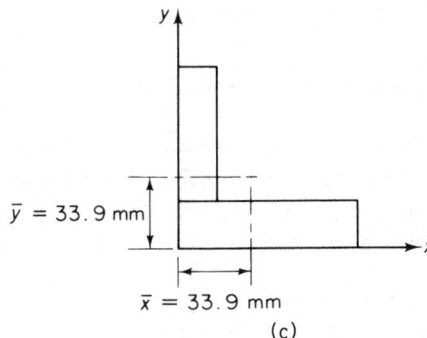

Fig 4.5(c) Illustrative problem 4.3

Solution:

Since there is no axis of symmetry, it will be necessary to determine both $\bar{x}$ and $\bar{y}$. For convenience we select the x and y reference axes as shown in Fig. 4.5(a) to coincide with the left and bottom edges of the figure. Also, we subdivide the angle into two parts, A and B as shown. In Fig. 4.5(b), each part of the figure is shown separately with the location of its center of gravity also shown, i.e., $\bar{x}_A$, $\bar{y}_A$, $\bar{x}_B$, $\bar{y}_B$.

Let us proceed as was done in Illustrative Problem 4.2.

Area $A = A_A = 75 \times 25 = 1875$ mm^2 $\bar{y}_A = 25 + \dfrac{75}{2} = 62.5$ mm

Area $B = A_B = 25 \times 100 = 2500$ mm^2 $\bar{y}_B = \dfrac{25}{2}$ mm

Applying equation (4.8),

$$\bar{y} = \frac{A_A \bar{y}_A + A_B \bar{y}_B}{A_A + A_B} = \frac{1875(62.5) + 2500 \times \dfrac{25}{2}}{1875 + 2500} = 33.9 \text{ mm}$$

Applying equation (4.7),

$$(4.7), \bar{x} = \frac{A_A \bar{x}_A + A_B \bar{x}_B}{A_A + A_B} = \frac{1875 \times \dfrac{25}{2} + 2500 \times 50}{1875 + 2500} = 33.9 \text{ mm}$$

where $\bar{x}_A = \dfrac{25}{2}$ mm and $\bar{x}_B = 50$ mm.

Thus far we have considered simple figures that could be built up as the sum of several rectangles. Let us now consider the case of an area with a portion cut out of it as shown in Fig. 4.6. This figure can be

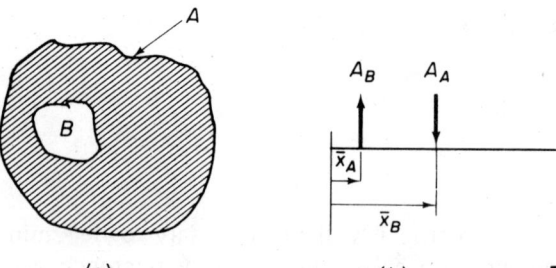

Fig 4.6 Area with cut-out

thought of as consisting of the entire area A, minus the area B. For this case we can consider the area B to be negative and portray it as a vector in a direction opposite to that of the vector for the entire area. The moment of B, $A_B x_B$ is therefore the negative of the moment of A, $A_A x_A$. Thus the moment effect of a cut-out is opposite to the moment effect of the solid area of the plate.

Illustrative problem 4.4

Determine the center of gravity of the rectangle with the cut-out shown in Fig. 4.7.

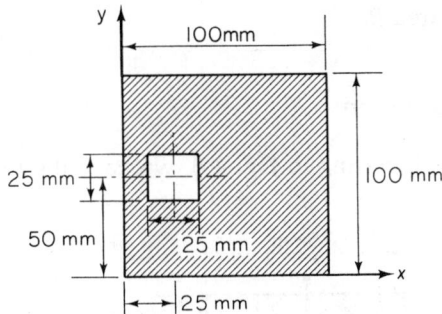

Fig 4.7 Illustrative problem 4.4

Solution:

Proceed as before, denoting the area of the entire plate as A_A and the area of the cut-out as A_B.

$$A_A = 100 \times 100 = 10\,000 \text{ mm}^2 \quad \bar{x}_A = 50 \text{ mm} \quad \bar{y}_A = 50 \text{ mm}$$
$$A_B = -(25 \times 25) = -625 \text{ mm}^2 \quad \bar{x}_B = 25 \text{ mm} \quad \bar{y}_B = 50 \text{ mm}$$

Applying equations (4.7) and (4.8) respectively

$$\bar{y} = \frac{(A_A)\bar{y}_A + A_B \bar{y}_B}{A_A + A_B} = \frac{10\,000 \times 50 + (-625)(50)}{10\,000 - 625} = 50 \text{ mm}$$

Note that the cut-out is on the y axis of symmetry and that, consequently, $\bar{y}$ for the composite figure (50 mm) falls along this axis. Proceeding

$$\bar{x} = \frac{(A_A)\bar{x}_A + (A_B)\bar{x}_B}{A_A + A_B} = \frac{10\,000 \times 50 + (-625)(25)}{10\,000 - 625} = 51.7 \text{ mm}$$

The effect of the cut-out is to shift the center of gravity away from the cut-out toward the uncut portion of the area.

Illustrative problem 4.5

Solve Illustrative Problem 4.2 using as reference axes the centroid of Area B.

Solution:

Referring to Fig. 4.8, we have the following

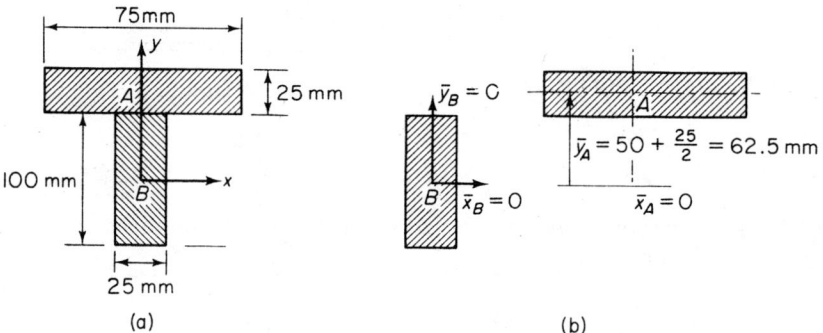

Fig 4.8 Illustrative problem 4.5

$A_A = 75 \times 25 = 1875 \text{ mm}^2 \quad \bar{y}_A = 50 + \dfrac{25}{2} = 62.5 \text{ mm} \quad \bar{x}_A = 0$

$A_B = 25 \times 100 = 2500 \text{ mm}^2 \quad \bar{y}_B = 0, \quad \bar{x}_B = 0$

From equation (4.8)

$$\bar{y} = \frac{A_A \bar{y}_A + A_B \bar{y}_B}{A_A + A_B} = \frac{1875 \times 62.5 + 2500(0)}{1875 + 2500} = 26.79 \text{ mm}$$

and for $\bar{x}$, $\bar{x} = 0$ by symmetry or from equation (4.7)

$$\bar{x} = \frac{A_A \bar{x}_A + A_B \bar{x}_B}{A_A + A_B} = \frac{1875(0) + 2500(0)}{1875 + 2500} = 0$$

The center of gravity is therefore located 26.79 mm above the center of gravity of area B, or 26.79 + 50 = 76.79 mm from the bottom of the figure. This result is the same as that obtained from Illustrative Problem 4.2. Obviously, the choice of axis does not change the result; it can however simplify the calculations for a particular problem if some preliminary thought is given to the selection of the reference axis before starting the problem.

Thus far we have treated simple rectangular shapes. In principle, it is possible to divide the most complex figure into a number of small areas sufficient to obtain the center of gravity of the figure. This procedure is most readily carried out using the methods of calculus, but can be demonstrated as follows in Illustrative Problem 4.6.

Illustrative problem 4.6

Determine the center of gravity for the right triangle shown in Fig. 4.9.

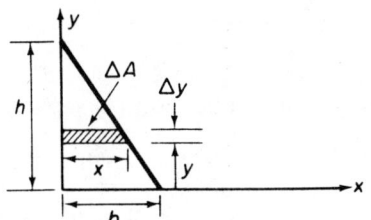

Fig 4.9 Illustrative problem 4.6

Solution:

Divide the triangular area into small rectangular areas ΔA where $\Delta A = x \Delta y$. The centroid of this area with respect to the base of the triangle is located at y and the moment of the area with respect to the base of the triangle is $(y) \Delta A = y(x \Delta y)$. From similar triangles we also have

$$\frac{x}{b} = \frac{h - y}{h}$$

or

$$x = \frac{b}{h}(h - y)$$

132   Strength of Materials for Engineering Technology

We can now express the moment of the area in terms of the variable y and the constants b and h

$$M = y(x\Delta y) = x(y\Delta y) = \frac{b}{h}(h - y)y\Delta y$$

expanding

$$M = by\Delta y - \frac{b}{h}y^2\Delta y$$

If we sum all such terms as y goes from zero at the base to h at the top and denote this summation as $\sum_{y=0}^{y=h}$, we have

$$\sum M = \sum_{y=0}^{y=h} by\Delta y - \sum_{y=0}^{y=h} \frac{b}{h}y^2\Delta y$$

Since b and h are constants, we can consider them to be constant multipliers and remove them from the summation.

$$\sum M = b\sum_{y=0}^{y=h} y\Delta y - \frac{b}{h}\sum_{y=0}^{y=h} y^2\Delta y \qquad (A)$$

Let us consider each term on the right-hand side of equation (A) separately. If we plot the following curve, we note that the $\sum_{y=0}^{y=h} y\Delta y$ is the shaded area under the line. Since the shaded area is the area of a triangle of base y and altitude y, the area is $(\frac{1}{2}y)y$ or between 0 and h, $(\frac{1}{2}h)h$. Thus the first term on the right side of equation (A) is given as

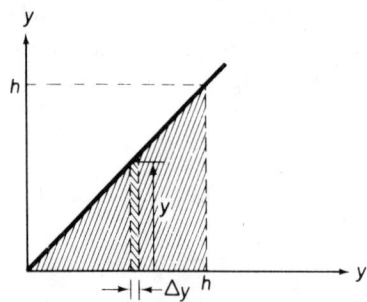

Fig 4.10   Evaluation of $y\Delta y$

Fig 4.11 Evaluation of $y^2 \Delta y$

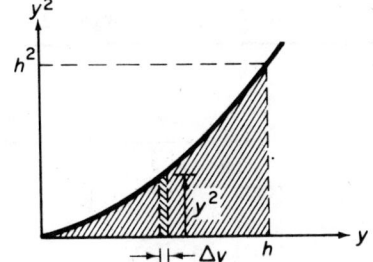

$$b \sum_{y=0}^{y=h} y \Delta y = \frac{bh^2}{2}$$

The second term can be evaluated in a similar manner by plotting y^2 against y. The area under this curve can be evaluated numerically by the methods of analytic geometry or calculus and is found to be $y^3/3$, which for the case under study is $h^3/3$. The second term on the right hand side of equation (A) is therefore

$$-\frac{b}{h} \sum_{y=0}^{y=h} y^2 \Delta y = \left(-\frac{b}{h}\right)\frac{h^3}{3} = -\frac{bh^2}{3}$$

and equation (A) is therefore

$$\sum M = \frac{bh^2}{2} - \frac{bh^2}{3} = \frac{bh^2}{6}$$

$\bar{y}$ is $\Sigma M/\Sigma A$, where ΣA is the area of the triangle. Substituting $bh/2$ for the area, we obtain

$$\bar{y} = \frac{bh^2/6}{bh/2} = \frac{h}{3}$$

and if we proceed in an analogous manner we obtain

$$\bar{x} = \frac{b}{3}$$

Using similar methods, we can obtain the location of the center of gravity for various areas. Table 4.1 gives the properties of an area for many of the areas commonly encountered in engineering.

Table 4.1 Properties of an Area

Section	Area	$\bar{x}$	$\bar{y}$
Square	b^2	$\dfrac{b}{2}$	$\dfrac{b}{2}$
Rectangle	bh	$\dfrac{b}{2}$	$\dfrac{h}{2}$
Triangle	$\tfrac{1}{2} bh$	$\dfrac{b}{3}$	$\dfrac{h}{3}$
Semi-circle	$\dfrac{\pi R^2}{2}$	0	$\dfrac{4R}{3\pi}$ or $0.4244R$
Quarter-circle	$\dfrac{\pi R^2}{4}$	$\dfrac{4R}{3\pi}$ or $0.4244R$	$\dfrac{4R}{3\pi}$ or $0.4244R$
Quadrant of ellipse	$\dfrac{\pi ab}{4}$	$\dfrac{4a}{3\pi}$ or $0.4244a$	$\dfrac{4b}{3\pi}$ or $0.4244b$
Spandrel $y = kx^n$, Vertex	$\dfrac{bh}{n+1}$	$\dfrac{b}{n+2}$	$\left\{\dfrac{n+1}{4n+2}\right\}h$

Illustrative problem 4.7

Determine the location of the centroid of a quarter circle, with a 150-mm radius with respect to the x and y axes, as shown in Fig. 4.12.

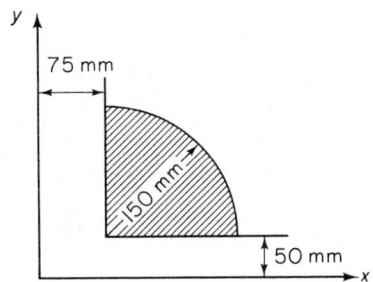

Fig 4.12 Illustrative problem 4.7

Solution:

From Table 4.1, $\bar{x} = \bar{y} = 0.4244R$ for the quarter circle. Thus, for the present problem, $\bar{x} = \bar{y} = 0.4244(150) = 63.66$ mm. From the x reference axis, $\bar{y} = 50 + 63.66 = 113.66$ mm, and from the y reference axis, $\bar{x} = 75 + 63.66 = 138.66$ mm.

Illustrative problem 4.8

A triangle is located with respect to the x and y axes as shown in Fig. 4.13. Determine the centroid of the triangle with respect to these axes.

Fig 4.13 Illustrative problem 4.8

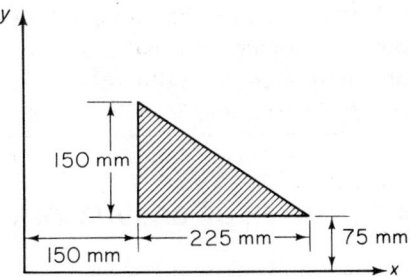

Solution:

From Table 4.1 we have $\bar{x} = b/3 = \dfrac{225}{3} = 75$ mm. For $\bar{y}$ we have $\bar{y} = h/3 = \dfrac{150}{3} = 50$ mm. With respect to the reference axes these become $\bar{x} = 150 + 75 = 225$ mm and $\bar{y} = 75 + 50 = 125$ mm.

Illustrative problem 4.9

Determine the centroid of the semicircle shown in Fig. 4.14 with respect to the axes shown.

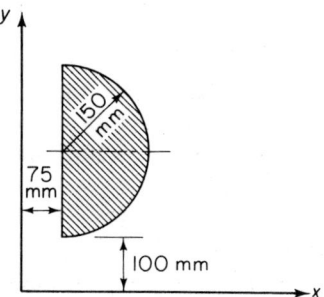

Fig 4.14 Illustrative problem 4.9

Solution:

Using the data of Table 4.1, $\bar{x} = 0.4244R = 0.4244(150) = 63.66$ mm, $\bar{y}$ is located on the axis of symmetry, 75 mm above the lowest point of the semicircle. With respect to the reference axes, $\bar{x} = 75 + 63.66 = 138.66$ mm and $\bar{y} = 75 + 100 = 175$ mm.

4.3 CENTER OF GRAVITY OF COMPOSITE AREAS

Using the methods of the preceding section, we have seen how we can evaluate the center of gravity of several simple areas. The use of

Table 4.1 enables us to extend these concepts to more complex plane areas. The procedure in such calculations consists of dividing the figure into simple areas, which are given in Table 4.1, and then selecting a convenient set of reference axes. Most calculations of this type are best handled in a tabular manner as will be shown in Illustrative Problem 4.10.

Illustrative problem 4.10

Locate the center of gravity of Fig. 4.15.

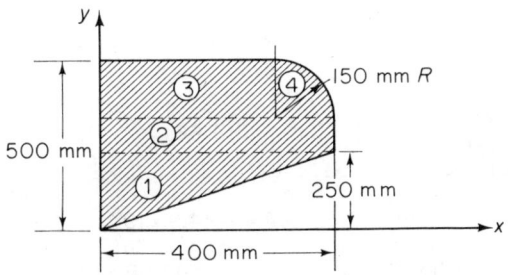

Fig 4.15   Illustrative problem 4.10

Solution:

Let us subdivide the figure into 4 parts, a triangle 400 mm × 250 mm, a rectangle 100 mm × 400 mm, a rectangle 150 mm × 250 mm, and a quarter circle with a 150-mm radius. We will also select the x and y axes as shown. For each subdivision (numbered ①, ②, ③, and ④) we will tabulate the areas, $\bar{x}$, $\bar{y}$, $A\bar{x}$, and $A\bar{y}$, as is done in the following table. Note that the area of the quarter circle is $\pi R^2/4$, and the center of gravity is $0.4244R$ plus the distance of the base of the figure from the axis. Also note the utility of using the tabular method as was done in this problem.

Table 4.2 Solution of Illustrative Problem 4.10

Section	Description	Area	$\bar{x}$	$\bar{y}$	$\dfrac{A\bar{x}}{10^6}$	$\dfrac{A\bar{y}}{10^6}$
①	triangle, base 400, height 250	50 000	$\dfrac{400}{3}$	$\dfrac{2}{3}(250)$	6.67	8.33
②	rectangle 400 × 100, at y=250	40 000	200	300	8.00	12.00
③	rectangle 250 × 150, at y=350	37 500	125	425	4.69	15.94
④	quarter circle R=150, at (250, 350)	17 671	313.66	413.66	5.54	7.31

$$\Sigma A = 145\,171 \qquad \Sigma A\bar{x} = 24.9 \times 10^6 \qquad \Sigma A\bar{y} = 43.58 \times 10^6$$

$$\bar{x} = \frac{\Sigma A\bar{x}}{\Sigma A} = \frac{24.9 \times 10^6}{145\,171} = 171.5\text{ mm} \qquad \bar{y} = \frac{43.58 \times 10^6}{145\,171} = 300.2\text{ mm}$$

Illustrative problem 4.11

Locate the center of gravity of the figure used in Illustrative Problem 4.10 with the subdivisions shown in Fig. 4.16.

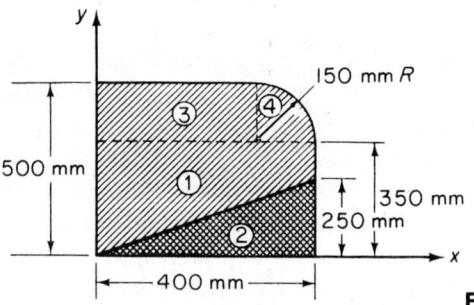

Fig 4.16 Illustrative problem 4.11

Solution:

In this alternative solution, we have chosen the following four subdivisions: a 350-mm × 400-mm rectangle, a triangle cut-out of 250 mm × 400 mm, a 400-mm × 250-mm rectangle and, a quarter circle with a 150-mm radius. In this solution the cut-out is treated as a negative area. As is to be expected, both approaches yield the same solution for $\bar{x}$ and $\bar{y}$. The choice of method is up to the student and is based upon the procedure that is most comfortable to him.

Table 4.3   Solution of Illustrative Problem 4.11

Section	Description	Area	$\bar{x}$	$\bar{y}$	$\dfrac{A\bar{x}}{10^6}$	$\dfrac{A\bar{y}}{10^6}$
①	400 × 350 rectangle	140 000	200	175	28.0	24.5
②	400 × 250 triangle	−50 000	266.67	833	−13.33	−4.17
③	250 × 150 rectangle at y=350	37 500	125	425	4.69	15.94
④	Quarter circle R = 150 at (250, 350)	17 671	313.66	413.66	5.54	7.31

$\Sigma A = 145\ 171 \qquad \Sigma A\bar{x} = 24.9 \times 10^6 \qquad \Sigma A\bar{y} = 43.58 \times 10^6$

$$\bar{x} = \frac{\Sigma A \bar{x}}{\Sigma A} = \frac{24.9 \times 10^6}{145\ 171} = 171.5\ \text{mm} \qquad \bar{y} = \frac{\Sigma A \bar{y}}{\Sigma A} = \frac{43.58 \times 10^6}{145\ 171} = 300.2\ \text{mm}$$

4.4 THE SECOND MOMENT OF AREA—MOMENT OF INERTIA

When Newton's Second Law of Motion is applied to the angular motion of rigid bodies, one of the terms that is developed is known as the *moment of inertia of the body*. As we will see later in our study of the stresses and deflections in beams and shafts, an analogous term appears for the property of an area that is also called the *moment of inertia of the area*. Since area has no mass, this term too is a misnomer; and, since there is no physical analogy for this property of an area, we will simply resort to evaluating it based upon its mathematical definition. The moment of inertia, or more properly, the second moment of area, is defined as

$$I_x = \Sigma y^2 \Delta A \tag{4.9}$$

where I, the moment of inertia, has the physical units of (length)^4. In every case it is necessary to refer the moment of inertia to an axis. We can show this very clearly by referring to Fig. 4.17, where $I_x = y^2 \Delta A$

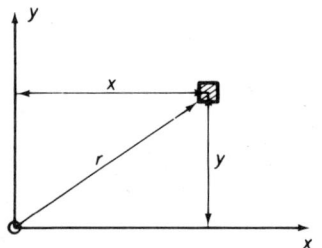

$I_x = y^2 \Delta A$
$I_y = x^2 \Delta A$

Fig 4.17 Definition of moment of inertia

and $I_y = x^2 \Delta A$. In this case, the axes in question are in the plane of the area (the page), and the moment of inertia as defined by equation (4.9) is known as the *rectangular moment of inertia*. Where the axis is perpendicular to the plane of the area, the resulting moment of inertia with respect to this axis is known as the *polar moment of inertia*. For example, let us consider an axis passing through O and perpendicular to the plane of the paper. (In mathematics this axis is the z axis.) The polar moment of inertia J_o is defined as

$$J_o = r^2 \Delta A \tag{4.10}$$

Note however that $r^2 = x^2 + y^2$. Therefore

$$J_o = x^2 \Delta A + y^2 \Delta A = I_x + I_y \tag{4.11}$$

We can express the results of equation (4.11) as follows: the polar moment of inertia of an area about an axis perpendicular to the plane of the area is equal to the sum of the rectangular moments of inertia of any two mutually perpendicular axes in the plane of the area whose intersection lies on the polar axis.

The rectangular moment of inertia defined by equation (4.9) will appear in those chapters of this book dealing with the stresses and deflections of beams as well as in the chapters dealing with stresses in columns subjected to eccentric loads. In addition, we have already used the polar moment of inertia in Chapter 3 when dealing with eccentrically loaded welds and shall also find it again when we study torsion in Chapter 5.

Illustrative problem 4.12

Determine the moment of inertia of a rectangle about an axis coinciding with its base.

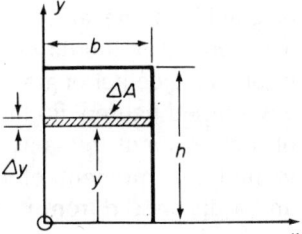

Fig 4.18 Illustrative problem 4.12

Solution:

By definition, the moment of inertia of the small area ΔA about the x axis is $y^2 \Delta A$ where $\Delta A = b\Delta y$. Using this and indicating the summation of all such areas, we have

$$I_x = \sum_{y=0}^{y=h} y^2 b \Delta y$$

but since b is a constant, we can simplify the foregoing to read

$$I_x = b \sum_{y=0}^{y=h} y^2 \Delta y$$

It will be recalled from Illustrative Problem 4.6 that the term $\Sigma y^2 \Delta y$ is the second term on the right side of equation (A). The result for this summation was noted to be $y^3/3$ and for y equal h it is $h^3/3$. Therefore

$$I_x = \frac{bh^3}{3}$$

For this figure $I_y = hb^3/3$. The polar moment of inertia about an axis perpendicular to the plane of the area and through "O" can be obtained by applying equation (4.11) to this situation; that is

$$J_o = I_x + I_y = \frac{bh^3}{3} + \frac{hb^3}{3}$$

4.5 THE PARALLEL AXIS TRANSFER THEOREM

The most useful reference axis for moment of inertia is an axis through the center of gravity of the area. The symbol $\bar{I}_x$ is used to denote the moment of inertia of the area about an axis passing through the center of gravity of the area and parallel to the x axis. It is possible to obtain the moment of inertia of an area about any axis parallel to an axis passing through the center of gravity by considering the following situation. Consider any shaped figure with the $\bar{x}$ axis as shown. Let us assume that the moment of inertia about this axis is known, i.e., $\bar{I}_x = \Sigma y^2 \Delta A$, and that we desire to find the moment of inertia about the axis x, parallel to the $\bar{x}$ axis, and a distance d from it. The expression we seek is

$$I_x = \Sigma(y + d)^2 \Delta A \qquad (4.12)$$

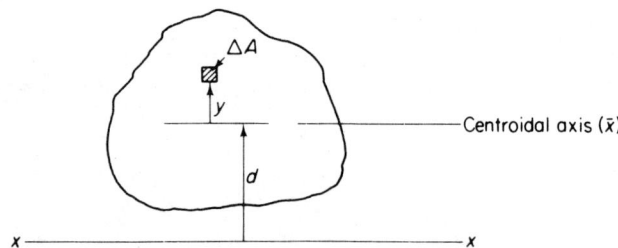

Fig 4.19 The parallel axis transfer theorem

Expanding the terms in equation (4.12)

$$I_x = \Sigma y^2 \Delta A + \Sigma 2dy \Delta A + \Sigma d^2 \Delta A \qquad (4.13)$$

In the second term we may remove d from the summation since a constant distance separates the two axes and we can also remove d^2 from the summation of the third term. Thus

$$I_x = \Sigma y^2 \Delta A + 2d\Sigma y \Delta A + d^2 \Sigma \Delta A \qquad (4.14)$$

Let us now examine each term on the right hand side of equation (4.14). The first term is the definition of $\bar{I}_{x'}$; the second term contains the first moment of area about the center of gravity, and by definition it must be zero; the last summation simply yields the total area. Therefore, we have

$$I_x = \bar{I}_x + Ad^2 \qquad (4.15)$$

By analogous reasoning

$$J_o = \bar{J}_o + Ad^2 \qquad (4.15a)$$

Equation (4.15) is known as the *parallel axis transfer theorem* and it expresses the fact that the moment of inertia of an area with respect to any rectangular axis is equal to the moment of inertia of the area about a parallel centroidal axis plus the product of the area multiplied by the square of the distance separating the axes. The student will note that $I_x > \bar{I}_{x'}$, which indicates that the minimum moment of inertia of an area with respect to any axis in a given direction occurs when the axis passes through the center of gravity of the area.

Illustrative problem 4.13

Determine the rectangular moment of inertia of a rectangle about its center of gravity: (a) using the results of Illustrative Problem 4.12; and (b) using the parallel axis transfer theorem.

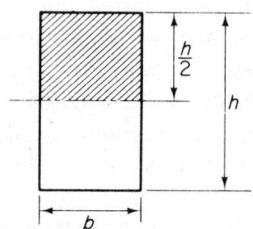

Fig 4.20   Illustrative problem 4.13

Solution:

(a) Using the results of Illustrative Problem 4.12: Consider the shaded area above the center of gravity axis. The moment of inertia of this area about its base is $b(h/2)^3/3$. If we now consider the unshaded portion, we note that the definition of moment of inertia is $y^2 \Delta A$ and that, unless the area is negative (as is the case for a cut-out), the moment of inertia is always positive. Thus we find that the moment of inertia of the unshaded area about its base is $b(h/2)^3/3$. The total moment of inertia of the area about the x axis through the center of gravity is therefore the sum of the two terms just evaluated and we have

$$\bar{I}_x = \frac{b(h/2)^3}{3} + \frac{b(h/2)^3}{3} = \frac{bh^3}{12}$$

(b) Using the parallel axis transfer theorem:

The moment of inertia about the center of gravity is equal to the moment of inertia about the base minus Ad^2

$$\bar{I}_x = I - Ad^2$$

Therefore

$$\bar{I}_x = \frac{bh^3}{3} - bh\left(\frac{h}{2}\right)^2 = \frac{bh^3}{12}$$

which is the same result the obtained in part (a) of this problem.

4.6 THE MOMENT OF INERTIA OF COMPOSITE AREAS

By the process of subdividing a body into sufficiently small areas, it is possible to calculate the moment of inertia of an area about any axis. However, most areas can be subdivided into squares, rectangles, circles, etc., and for these shapes analytical expressions exist for the rectangular moment of inertia about axes through the center of gravity of the area. Table 4.4 gives the moment of inertia for various geometric shapes about both the centroidal axis and a convenient parallel axis.

In dealing with composite bodies, we find it convenient to use Table 4.4 with the caution that, before the moment of inertia is determined as the sum of the various shapes that comprise the figure, it is necessary that the moments of inertia of all shapes must be found with respect to the same axis.

Table 4.4*

Shape	Moment of Inertia	Radius of Gyration
Rectangle	$\bar{I}_x = \frac{bh^3}{12}$ *about centre* $I_x = \frac{bh^3}{3}$ *about base*	$\bar{k}_x = \frac{h}{\sqrt{12}}$ $k_x = \frac{h}{\sqrt{3}}$
Any triangle	$\bar{I}_x = \frac{bh^3}{36}$ $I_x = \frac{bh^3}{12}$	$\bar{k}_x = \frac{h}{\sqrt{18}}$ $k_x = \frac{h}{\sqrt{6}}$
Circle	$\bar{I}_x = \frac{\pi r^4}{4}$ $\bar{J}_o = \frac{\pi r^4}{2}$	$\bar{k}_x = \frac{r}{2}$ $\bar{k}_z = \frac{r}{\sqrt{2}}$
Semicircle	$I_x = \bar{I}_y = \frac{\pi r^4}{8}$ $\bar{I}_x = 0.11 r^4$	$k_x = \bar{k}_y = \frac{r}{2}$ $\bar{k}_x = 0.264r$
Quarter circle	$I_x = I_y = \frac{\pi r^4}{16}$ $\bar{I}_x = \bar{I}_y = 0.055 r^4$	$k_x = k_y = \frac{r}{2}$ $\bar{k}_x = \bar{k}_y = 0.264r$
Ellipse	$\bar{I}_x = \frac{\pi a b^3}{4}$ $\bar{I}_y = \frac{\pi b a^3}{4}$	$\bar{k}_x = \frac{b}{2}$ $\bar{k}_y = \frac{a}{2}$

*Reproduced with permission from *Strength of Materials* by F. L. Singer, 2nd Ed., 1962, Harper and Row Publishers, Inc., New York (Table A-1, p. 545, Moment of Intertia Geometric Shapes).

Illustrative problem 4.14

Determine the moment of inertia of the "I" section shown with respect to the x centroidal axis.

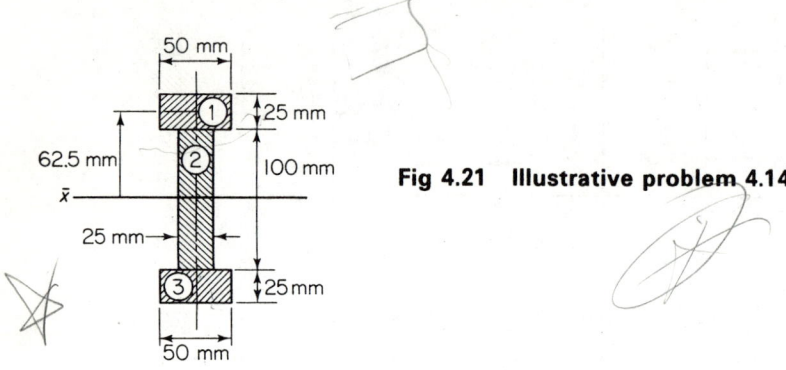

Fig 4.21   Illustrative problem 4.14

Solution:

Consider area 1; $\bar{I}_x = bh^3/12 = \dfrac{50(25)^3}{12} = 65\,104.2$ mm^4. Transferred to the $\bar{x}$ axis of the figure,

$$I_{1\bar{x}} = \bar{I} + Ad^2 = 65\,104.2 + 50(25)(62.5)^2 = 4.948 \times 10^6$$

By symmetry, area 3 contributes an equal amount to the moment of inertia of the composite figure about the $\bar{x}$ axis of the figure, namely 4.948×10^6. For area 2, its center of gravity coincides with the center of gravity of the composite figure. Thus its moment of inertia about this axis is $bh^3/12 = \dfrac{25(100)^3}{12} = 2.083 \times 10^6$. Summing,

$$\bar{I}_x = 4.948 \times 10^6 + 4.948 \times 10^6 + 2.083 \times 10^6 = 11.979 \times 10^6 \text{ mm}^4$$

Illustrative problem 4.15

Determine the moment of inertia of the box section shown in Fig. 4.22 about its x centroidal axis.

Chapter 4, Center of Gravity and Moment of Inertia 147

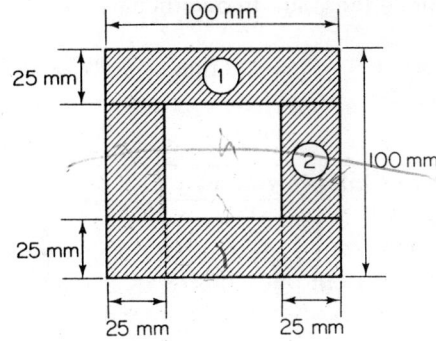

Fig 4.22 Illustrative problem 4.15

Solution:

Let us consider this problem to consist of a square with a square cut-out. Thus, for the 100-mm × 100-mm square,

$$\bar{I}_x = \frac{bh^3}{12} = \frac{100(100)^3}{12} = 8.333 \times 10^6 \text{ mm}^4$$

For the 50-mm × 50-mm cut-out,

$$\bar{I}_x = \frac{-bh^3}{12} = \frac{-50(50)^3}{12} = -0.513 \times 10^6 \text{ mm}^4$$

(negative due to negative area)

$$\bar{I}_{x \text{ entire figure}} = 7.812 \times 10^6 \text{ mm}^4$$

As an alternate solution, let us calculate the moment of inertia as the sum of each of its four rectangular parts. For 1,

$$I = \frac{bh^3}{12} + Ad^2 = \frac{100 \times (25)^3}{12} + 100(25)\left(25 + \frac{25}{2}\right)^2 = 3.646 \times 10^6 \text{ mm}^4$$

For 2,

$$I = \frac{bh^3}{12} = \frac{25(50)^3}{12} = 0.260 \times 10^6 \text{ mm}^4$$

Since there are two vertical rectangles and two horizontal rectangles

$$\bar{I}_x = 2(3.646 \times 10^6) + 2(0.260) = 7.812 \times 10^6 \text{ mm}^4$$

Illustrative problem 4.16

Determine the moment of inertia of the composite figure with a semicircular cut-out with respect to the x axis shown in Fig. 4.23.

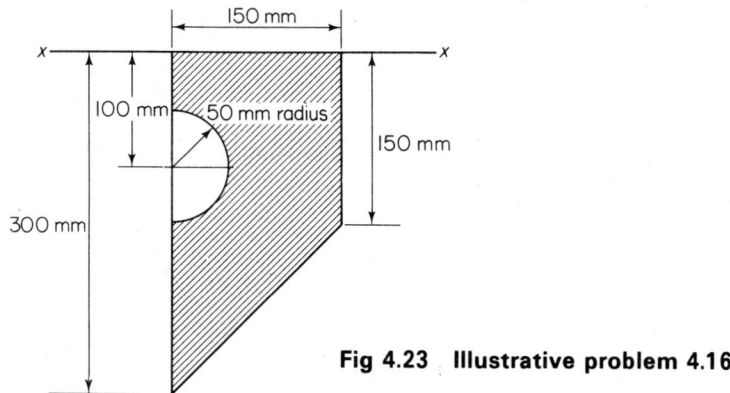

Fig 4.23 Illustrative problem 4.16

Solution:

Let us divide the area into three shapes: a 150-mm × 150-mm square, a 150-mm × 150-mm triangle, and a semicircular cut out with a 50-mm radius. For the square:

$$I_x = \frac{bh^3}{3} = \frac{150(150)^3}{3} = 168.75 \times 10^6 \text{ mm}^4$$

For the triangle:

$$I_x = \bar{I}_x + Ad^2 = \frac{bh^3}{36} + (\tfrac{1}{2})(bh)(200)^2$$

$$= \frac{150 \times (150)^3}{36} + \tfrac{1}{2}(150)(150)(200)^2$$

$$= I_x = 464.063 \times 10^6 \text{ mm}^4$$

For the semicircle:

$$I_x = \bar{I} + Ad^2 = \frac{\pi r^4}{8} + \frac{\pi r^2}{2}(100)^2 = \frac{\pi (50)^4}{8} + \frac{\pi (50)^2}{2}(100)^2$$
$$= -41.724 \times 10^6 \text{ mm}^4 \quad \text{(negative)}$$

For the composite figure:

$$\bar{I}_x = 168.75 \times 10^6 + 464.063 \times 10^6 - 41.724 \times 10^6$$
$$= 591.089 \times 10^6 \text{ mm}^4.$$

4.7 THE RADIUS OF GYRATION

Later, in our study of columns, we shall use the term *radius of gyration*. The radius of gyration is defined as that length that, when squared and then multiplied by the area, will give us the moment of inertia of the area with respect to a given axis. Thus

$$Ak^2 = I \quad (4.16)$$
$$Ak_x^2 = I_x \quad (4.16a)$$
$$Ak_y^2 = I_y \quad (4.16b)$$
$$A\bar{k}_x^2 = \bar{I}_x \quad (4.16c)$$

For a parallel axis we can write

$$I_x' = \bar{I}_x + Ad^2 = A\bar{k}_x^2 + Ad^2 = Ak_x^2$$

or

$$k_x^2 = \bar{k}_x^2 + d^2 \quad (4.17)$$

For a polar axis we can similarly obtain

$$J_o = Ak_o^2 = \bar{J}_o + Ad^2 = A\bar{k}_o^2 + Ad^2$$

and

$$k_o^2 = \bar{k}_o^2 + d^2 \quad (4.18)$$

Illustrative problem 4.17

Determine the radius of gyration about the x axis for Fig. 4.23 (Illustrative Problem 4.16).

Solution:

The area of the figure is

$$A = 150 \times 150 + \frac{150 \times 150}{2} - \frac{\pi}{2}(50)^2 = 29\,823 \text{ mm}^2$$

Since $I_x = 591.089 \times 10^6$, $Ak_x^2 = 591.089 \times 10^6 \text{ mm}^4$

$$k_x = \sqrt{\frac{591.089 \times 10^6}{29\,823}} = 140.78 \text{ mm}$$

If we wish to give the radius of gyration a physical meaning, it may be useful for us to think of it as that distance from an axis at which all of an area is concentrated, resulting in the same moment of inertia about the axis as the actual area. With this in mind, the student should note that such a concentrated area will not have the same first moment about the given axis as does the actual area.

Equation (4.18) also leads us to the conclusion that the radius of gyration is always greater than the distance from the reference axis to the center of gravity of the area. Also, the symbol k has been used for radius of gyration rather than the AISC symbol r to avoid confusion with the common use of r for radius.

Illustrative problem 4.18

Determine the location of the center of gravity of the section in Fig. 4.23. Compare this result with that obtained for the radius of gyration in Illustrative Problem 4.17.

Solution:

Using the same three shapes and the x axis shown as the reference axis

$$\bar{x} = \frac{150 \times 150 \times 75 + \frac{1}{2}(150 \times 150)(200) - \frac{\pi}{2}(50)^2(100)}{150 \times 150 + 150 \times 150/2 - \pi(50)^2/2}$$

$= 118.86$ mm

In Illustrative Problem 4.17, we found $k_x = 140.78$ mm. Comparison with the results above shows k_x to be greater than $\bar{x}$, as it must be.

4.8 CLOSURE

At this point in our study of "Strength of Materials," it may be difficult for the student to place a physical interpretation on the material covered in this chapter. If this material is thought of as a tool that will be required at a later time and background material that must be understood before further study can be undertaken, it may provide the student the rationale for mastering it. The mathematical definitions given in this chapter for the various properties discussed can be utilized without resorting to a physical model for these properties. However, where possible, an attempt has been made to relate the mathematical definitions to physical quantities. For the present, the student will have to accept the fact that we shall make use of all of these items at some later time in our studies and should make every attempt to understand them and to be able to apply them when they are needed. In this chapter the student will note that the solution of a particular problem often involves a great deal of arithmetical computation. An orderly, systematic approach to each solution will help to minimize errors in calculations.

PROBLEMS

4.1 Determine the location of the center of gravity with respect to the x axis for Fig. P4.1.

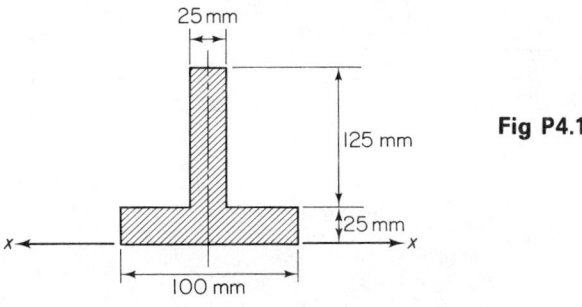

Fig P4.1

4.2 Locate the center of gravity of the section shown in Fig. P4.2.

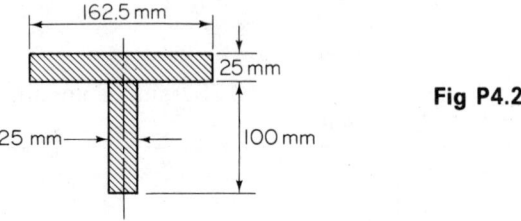

Fig P4.2

4.3 If the section shown in Fig. P4.2 is inverted, determine the location of its center of gravity.

4.4 Locate the center of gravity of the section shown in Fig. P4.4.

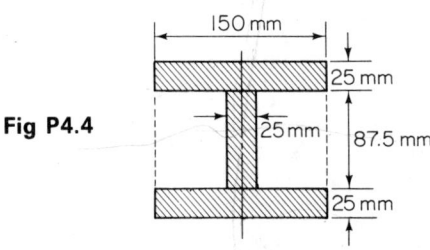

Fig P4.4

4.5 Determine the location of the center of gravity with respect to the x and y axes for the sections shown in Fig. P4.5.

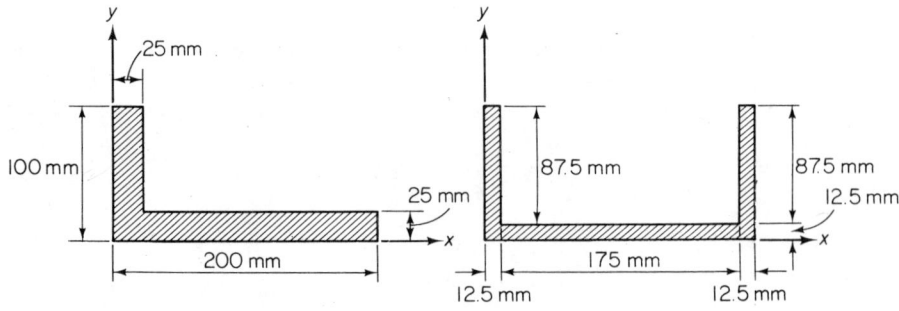

Fig P4.5

Chapter 4, Center of Gravity and Moment of Inertia   153

4.6 Locate the center of gravity of the area shown in Fig. P4.6 with respect to an axis parallel to the base.

Fig P4.6

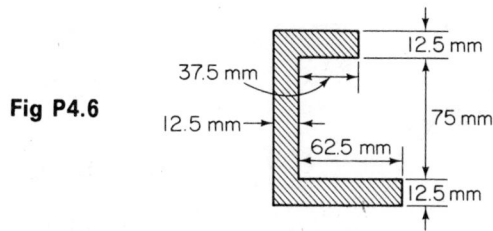

4.7 Locate the center of gravity of the figure shown in Problem 4.6 with respect to an axis coinciding with the left side of the figure.

4.8 Determine the center of gravity with respect to an axis coinciding with the base of the figure shown.

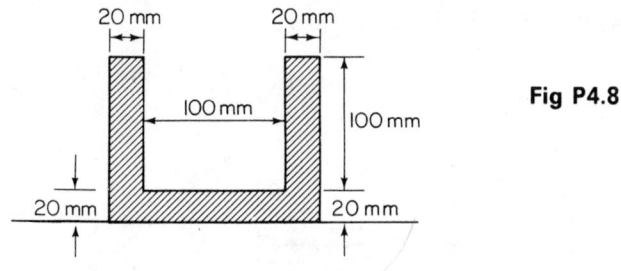

Fig P4.8

4.9 Locate the center of gravity of the area shown in Fig. P4.9.

Fig P4.9

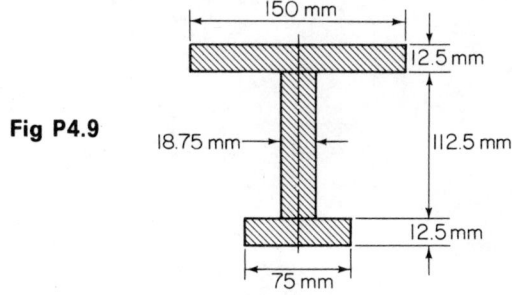

4.10 Determine the location of the center of gravity with respect to both the x and y axes for Fig. P4.10.

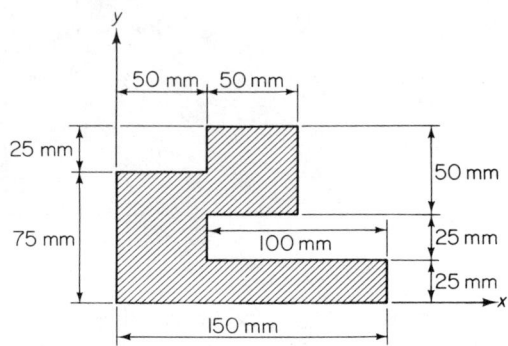

Fig P4.10

4.11 Determine the location of the center of gravity with respect to both the x and y axes for Fig. P4.11.

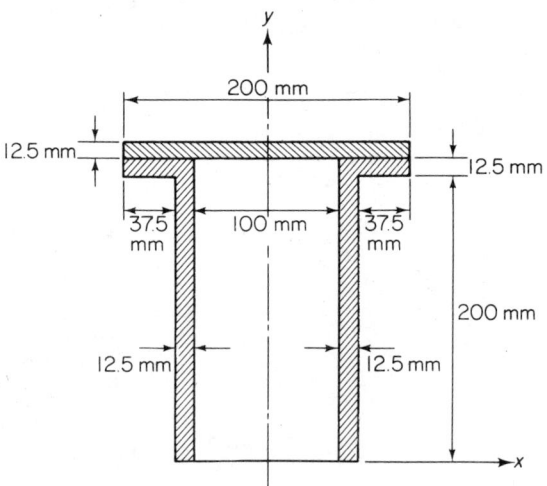

Fig P4.11

4.12 Determine the location of the center of gravity of the figure shown.

Chapter 4, Center of Gravity and Moment of Inertia   155

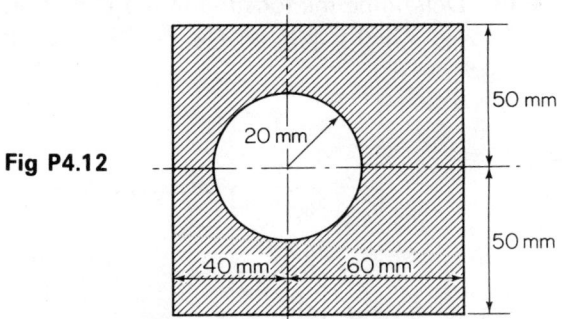

Fig P4.12

4.13 Determine the location of the center of gravity with respect to both the x and y axes for Fig. P4.13.

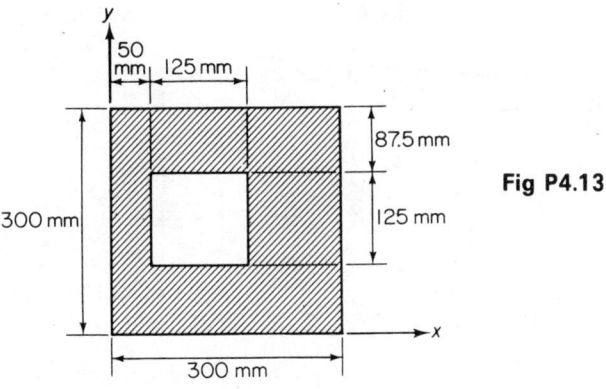

Fig P4.13

4.14 Locate the center of gravity of the area shown in Fig. P4.14

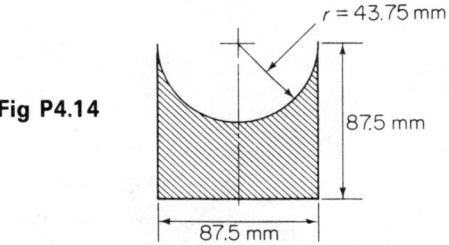

Fig P4.14

4.15 Determine the location of the center of gravity with respect to both the x and y axes for Fig. P4.15.

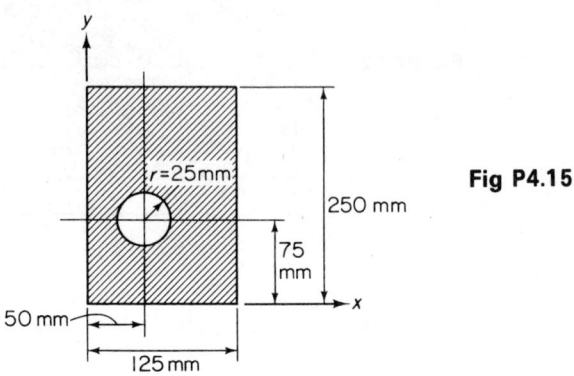

Fig P4.15

4.16 Locate the center of gravity of the figure shown in Fig. P4.16.

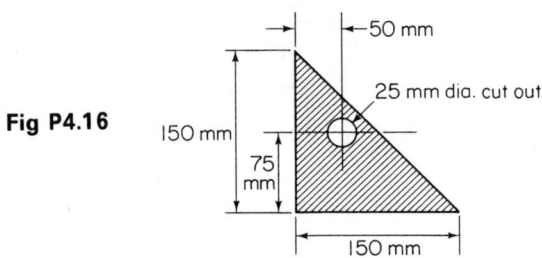

Fig P4.16

4.17 Determine the location of the center of gravity with respect to both the x and y axes for Fig. P4.17.

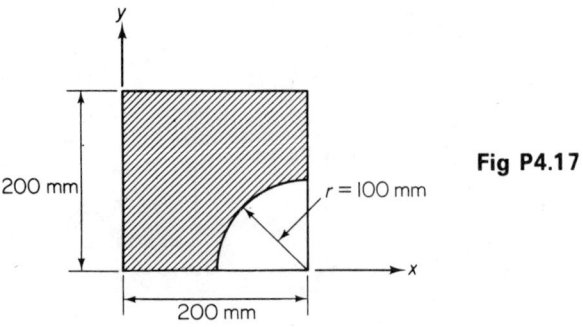

Fig P4.17

Chapter 4, Center of Gravity and Moment of Inertia 157

4.18 Locate the center of gravity of the figure shown with respect to the x and y axes shown.

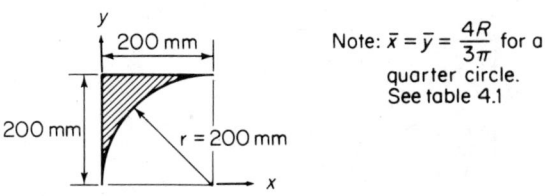

Note: $\bar{x} = \bar{y} = \dfrac{4R}{3\pi}$ for a quarter circle. See table 4.1

Fig P4.18

4.19 Determine the location of the center of gravity with respect to the x axis for Fig. P4.19.

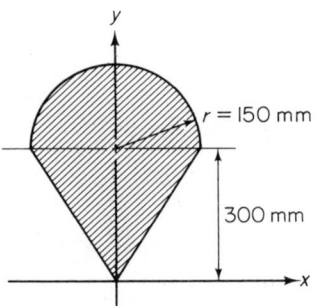

Fig P4.19

4.20 Determine the radius of the semicircle that will cause the center of gravity of the section in Fig. P4.20 to lie on the y axis.

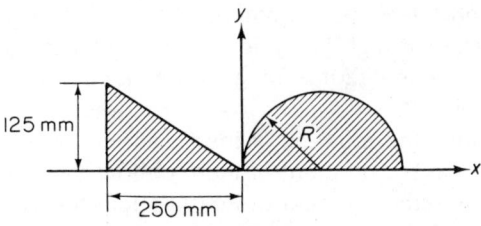

Fig P4.20

4.21 Determine the location of the center of gravity with respect to the x axis for Fig. P4.21.

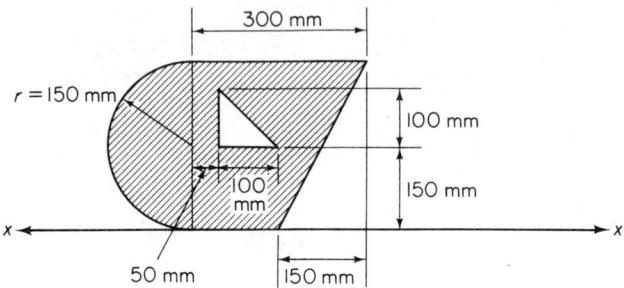

Fig P4.21

4.22 Determine the moment of inertia for the section in Fig. P4.22 to the x axis and the y axis.

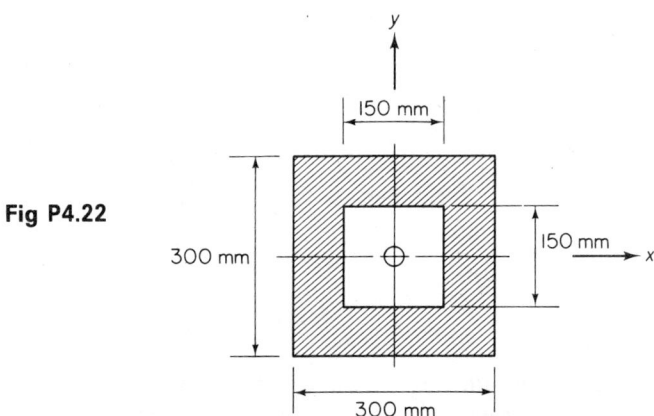

Fig P4.22

4.23 Determine the polar moment of inertia for Fig. P4.22 with respect to an axis through "O" and perpendicular to the plane of the figure.
4.24 Determine the radius of gyration about the x axis and the base of Fig. P4.22.
4.25 Determine the moment of inertia about the base of Fig. P4.1.
4.26 Determine the moment of inertia about the base of Fig. P4.8.
4.27 Determine the moment of inertia about the base of Fig. P4.12.
4.28 Determine the moment of inertia about the base of Fig. P4.13.
4.29 Determine the moment of inertia about the base of Fig. P4.15.
4.30 Determine the moment of inertia about the base of Fig. P4.17.

4.31 A solid circular member is used as a beam. If it replaces a square 150-mm section, and we desire to maintain the same moment of inertia, calculate the required diameter of the circular section.

4.32 Calculate the moment of inertia about a horizontal axis coinciding with the base of Fig. P4.2.

4.33 Calculate the moment of inertia about a horizontal axis coinciding with the base of Fig. P4.4.

4.34 Determine the moment of inertia of the "T" section shown in Fig. P4.34 with respect to its x centroidal axis.

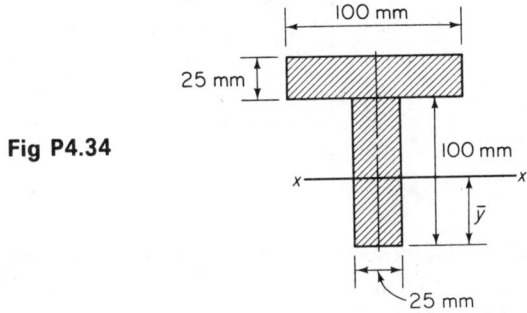

Fig P4.34

4.35 Calculate the moment of inertia about a horizontal axis coinciding with the base of Fig. P4.6.

4.36 Calculate the moment of inertia about a horizontal axis coinciding with the base of Fig. P4.9.

4.37 Calculate the moment of inertia about a horizontal axis coinciding with the base of Fig. P4.14.

4.38 Calculate the moment of inertia about a horizontal axis coinciding with the base of Fig. P4.16.

4.39 Determine the moment of inertia and radius of gyration for Fig. P4.39 about the centroidal axes.

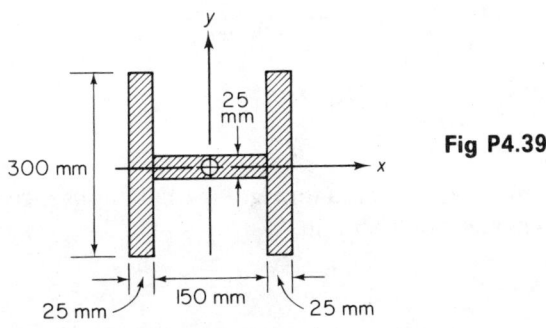

Fig P4.39

4.40 Using the result of Problem 4.39, determine the polar moment of inertia about the "O" axis, perpendicular to the plane of the figure. Also determine the polar radius of gyration about this axis.

4.41 Determine the polar moment of inertia for Fig. P4.41 with respect to an axis through "O" and perpendicular to the plane of the figure.

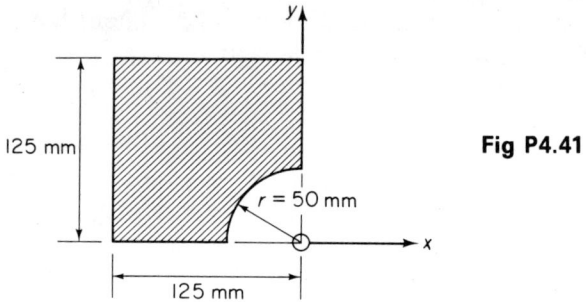

Fig P4.41

4.42 Determine the polar radius of gyration for Fig. P4.41 about the "O" axis.

4.43 Determine the moment of inertia and radius of gyration of the Z section shown in Fig. P4.43 with respect to both the x- and y-centroidal axes.

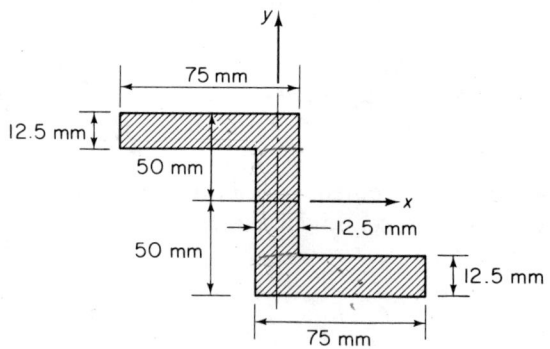

Fig P4.43

4.44 Determine the moment of inertia for Fig. P4.44 about its x-centroidal axis if all thicknesses are 12.5 mm.

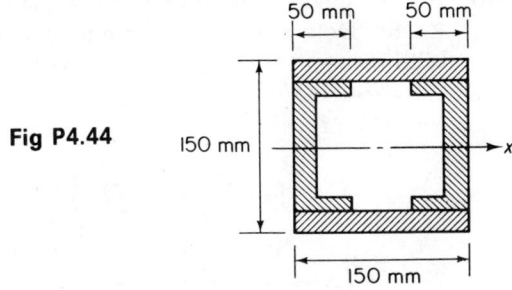

Fig P4.44

4.45 Determine the radius of gyration about the x-centroidal axis for Fig. P4.44.
4.46 If it is desired that $\bar{I}_x$ and $\bar{I}_y$ be equal, determine the width of each rectangle for Fig. P4.46.

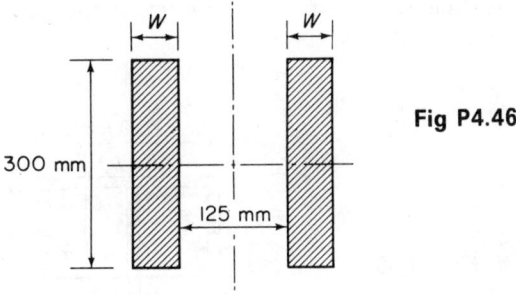

Fig P4.46

4.47 Determine the moment of inertia about the x- and y-centroidal axes for Fig. P4.47 consisting of two equal channels.

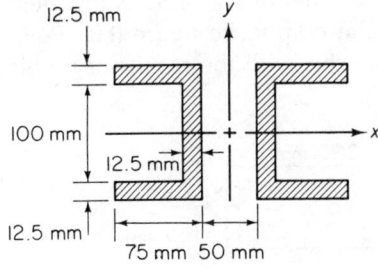

Fig P4.47

4.48 An "I" beam is reinforced by a plate at the top as shown in Fig. P4.48. Determine the moment of inertia of this composite beam with respect to its x-centroidal axis.

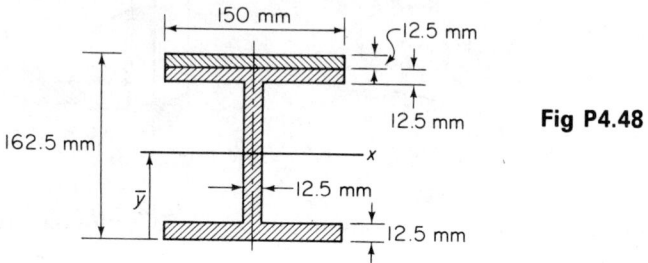

Fig P4.48

4.49 Determine the radius of gyration for the x-centroidal axis for Fig. P4.48.

4.50 Two equal channels are welded together to form the section shown in Fig. P4.50. Calculate the moment of inertia of the composite figure about its x-centroidal axis.

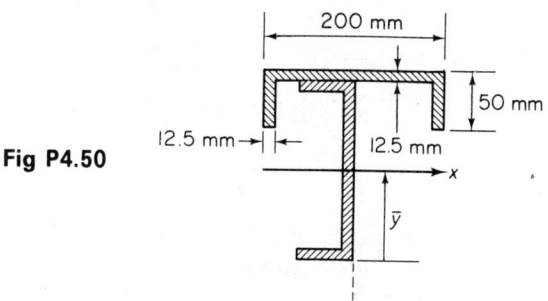

Fig P4.50

4.51 If the upper channel of Fig. 4.50 is inverted, determine the moment of inertia of the composite figure (Fig. P4.51) about its x-centroidal axis. Compare this with the results of Problem 4.50.

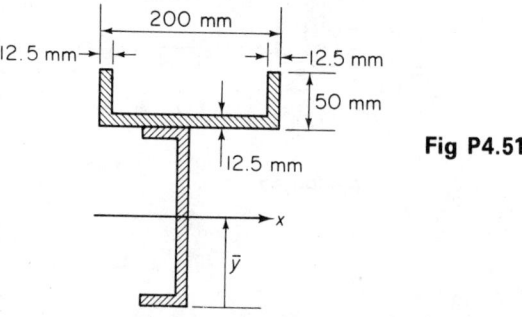

Fig P4.51

Chapter 4, Center of Gravity and Moment of Inertia 163

4.52 Determine the moment of inertia and radius of gyration about the base of the figure for Fig. P4.52.

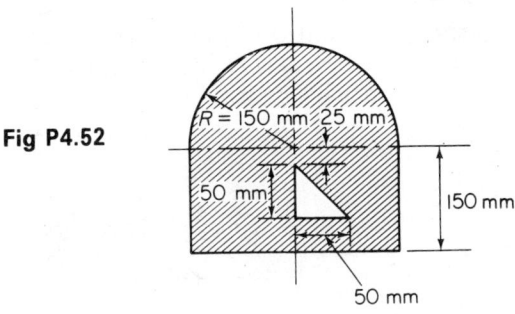

Fig P4.52

4.53 Determine the moment of inertia of Fig. P4.53 about its center of gravity ($\bar{I}_x$ and $\bar{I}_y$). Also determine the moment of inertia about the x' and y' axes shown.

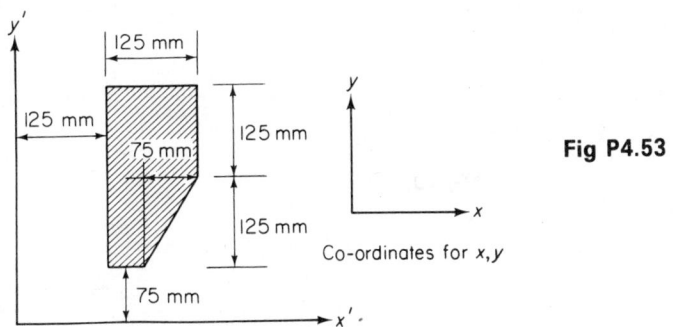

Fig P4.53

REFERENCES

Bassin, M. E. and S. M. Brodsky. *Statics and Strength of Materials*. 2nd ed. New York: McGraw-Hill Book Co., 1969.

Higdon, A. E. and W. B. Stiles. *Engineering Mechanics*. 3rd ed. Englewood Cliffs, NJ: Prentice-Hall, Inc., 1968.

Higdon, A., E. E. Ohlsen, and W. B. Stiles. *Mechanics of Materials*. 3rd ed. New York: John Wiley & Sons, 1976.

Levinson, I. J. *Introduction to Mechanics*. 2nd ed. Englewood Cliffs, NJ: Prentice-Hall, Inc., 1968.

Meriam, J. L. *Mechanics*. New York: John Wiley & Sons.

Olsen, G. A. *Strength of Materials*. Englewood Cliffs, NJ: Prentice-Hall, Inc.

Shames, I. *Engineering Mechanics.* 2 vols. 2nd ed. Englewood Cliffs, NJ: Prentice-Hall, Inc., 1966-67.

Singer, F. L. *Engineering Mechanics.* 2 vols. 3rd ed. New York: Harper and Row, 1975.

Timoshenko, S. and Donovan H. Young. *Elements of Strength of Materials.* 5th ed. New York: D. Van Nostrand Co., 1968.

5

Torsion

5.1 INTRODUCTION

When a shaft transmits power, or a spring absorbs energy, or a torsion bar is used to provide a restoring force, the elastic properties of the member resist the applied twisting force. This twisting action is resisted by the material principally by internal shearing forces, and this twist is known as *torsion*; a member subjected to this action is said to be *in torsion*. The term *torque* will also be used, and it will denote a couple acting in a plane at right angles to the longitudinal axis of the member in question. Since the conditions for equilibrium require that the member have equal and opposite couples lying in parallel planes, we note that the portion of the member between these couples is in torsion. Torsion always produces rotation, and we shall be concerned with determining both the stresses and strains caused by these rotations. The student should note that the theory developed in this chapter is primarily limited to cases of pure twisting in circular cross sections. We shall consider some instances where torsion is combined with a direct stress (in the case of springs), and we shall also briefly consider those cases of torsion in noncircular shafts that can be treated by methods within the scope of this text.

5.2 THE HORSEPOWER-TORQUE EQUATION

The transmission of power, the development of torque, and the performance of work at a specified rate are requirements imposed upon shafts by common industrial usage. It is therefore of interest to us to develop

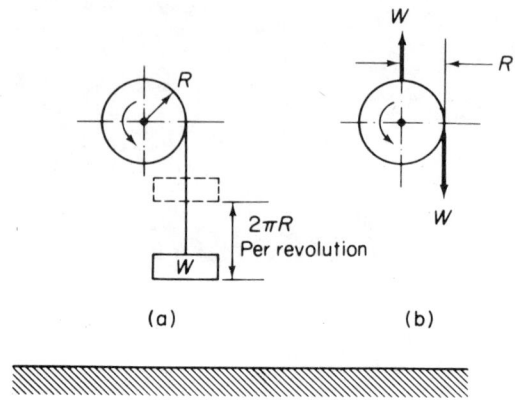

Fig 5.1 Diagram for horsepower-torque equation

a relationship between horsepower, torque, and speed to be used in designing and rating shafts.

Consider the situation shown in Fig. 5.1 where a weight is being lifted at a constant rate as the cable is wrapped around the shaft due to the steady rotation of the shaft. As shown by the free-body diagram of Fig. 5.1(b), dynamic equilibrium can exist if it is considered that the shaft is supported by a force equal and opposite to the weight W (and directed through the center) to satisfy the condition that the sum of the forces in the vertical direction must be zero.

As a result of this condition, i.e., that the sum of the moments must be zero, it is necessary to supply a moment (torque) T equal and opposite to the applied moment of WR, which yields the relation

$$T = WR \quad (5.1)$$

Now let us look at the weight W and its motion. If the weight is moving upward at a steady speed, we can evaluate the rate at which work is being done using the following reasoning. *Work* is defined as the product of a force multiplied by the displacement in the direction of the force. Each time the shaft completes one revolution, the weight is lifted by a distance equal to the circumference of the shaft, $2\pi R$ m, if R is expressed in metres. Thus the work done per shaft revolution is $W(2\pi R)$ N·m. However, the shaft is turning at the rate of N revolutions per minute so that per minute the work done is $(2\pi RWN)$ N·m/min.

At this point we note that *power* is defined as the rate at which work is being done per unit time, and that one horsepower is defined in common English units as a rate of work done at 33,000 ft lb per minute. In terms of watts, one horsepower is 746 W. Utilizing this definition with the rate

Chapter 5, Torsion   167

of doing work by the shaft in raising the weight W and the results of equation (5.1), we have

$$\text{horsepower (H.P.)} = \frac{2\pi RWN}{60 \times 746} = \frac{2\pi TN}{60 \times 746} \tag{5.2}$$

or, rearranging,

$$T = \frac{7124 \times \text{H.P.}}{N} \text{ (N·m)} \tag{5.3}$$

where T is N·m and N is in revolutions per minute.

Illustrative problem 5.1

A shaft is used to transmit 100 H.P. while rotating at 150 R.P.M. Calculate: (a) The torque being transmitted by the shaft; (b) the work done by the shaft per revolution.

Solution:

The torque being transmitted is obtained by direct substitution in equation (5.3). Thus

(a) $\quad T = \dfrac{7124 \times \text{H.P.}}{N} = \dfrac{7124 \times 100}{150} = 4749 \text{ N·m}$

or

$$T = 4749 \times 10^3 \text{ N·mm}$$

(b) 100 H.P. is a rate of doing work equal to 746×100 N·m per second. The work per revolution is found as follows:

$$\text{work per revolution} = \frac{\text{work/time}}{\text{revolution/time}}$$

$$= \frac{\text{H.P.} \times 746 \times 60}{N} \text{ N·m/revolution}$$

Therefore, work per revolution $= \dfrac{100 \times 746 \times 60}{150}$ N·m/revolution

$$= 29\ 840 \text{ N·m/revolution}$$

As an alternate solution we can consider that the torque arises from a weight being lifted by a cable attached to the shaft. Since

$$T = WR; \quad W = \frac{T}{R} = \frac{4749}{R} = \frac{4749 \text{ N}}{R}$$

Per revolution, the weight rises an amount equal to the circumference of the shaft. Thus per revolution the lift is $\pi D = 2\pi R = 6.283R$, and the work done is $6.283\,R \times \dfrac{4749}{R} = 29\,839$ N·m per revolution.

Illustrative problem 5.2

A shaft has 70 H.P. taken off it [as shown in Fig. 5.2(a)] by a motor supplying this power at 150 R.P.M. Determine the torque transmitted in each portion of the shaft.

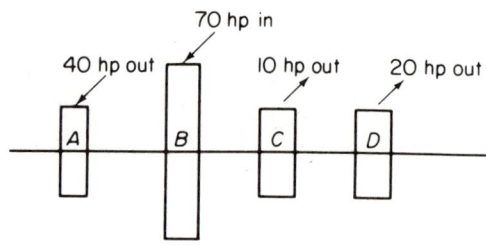

Fig 5.2(a) Illustrative problem 5.2

Solution:

We can consider the power flow from pulley *B* to the other pulleys just as we would consider the flow of a fluid from a reservoir. Thus, 70 flow units enter at *B* and branch off toward *A*, *C*, and *D*. This is shown diagramatically in Fig. 5.2(b).

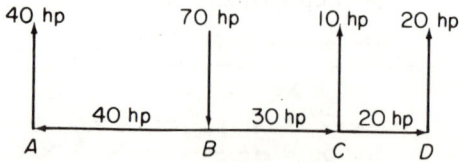

Fig 5.2(b) Illustrative problem 5.2

and we have

$$H.P._{A-B} = 40; \quad T = \frac{7124 \text{ H.P.}}{N} = \frac{7124 \times 40}{150} = 1900 \text{ N·m}$$

$$H.P._{B-C} = 30; \quad T = \frac{7124 \text{ H.P.}}{N} = \frac{7124 \times 30}{150} = 1425 \text{ N·m}$$

$$H.P._{C-D} = 20; \quad T = \frac{7124 \text{ H.P.}}{N} = \frac{7124 \times 20}{150} = 950 \text{ N·m}$$

5.3 TORSION OF CIRCULAR SHAFTS

In order to study the torsion of circular shafts, we shall have to make several assumptions concerning the nature of the imposed loads and the internal resistance of the shaft to these loads. Consider the situation shown in Fig. 5.3(a), where a circular shaft is fixed at its left end and is subjected

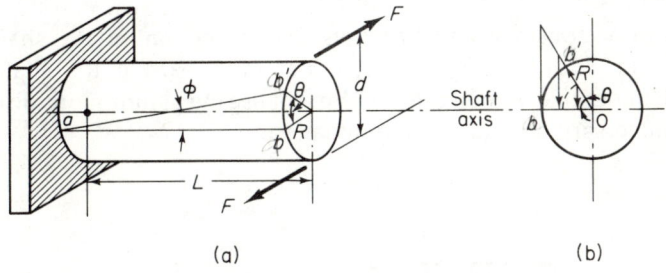

Fig 5.3 A shaft subjected to twisting only

to a torque at its right end by the applied couple $F \times d$. If the longitudinal line ab is scribed on the shaft prior to the application of this couple, it will be located as shown by the line ab' after the torque is applied. The radius originally shown as Ob will rotate through the angle θ and will be located at Ob' after equilibrium is established. At this point we will make two initial assumptions: namely,

1. Every diameter of any cross section through the shaft remains straight, and all rotate through the same angle. Thus, a diameter before twisting remains a diameter after twisting.
2. As a consequence of assumption 1, all cross sections of the shaft remain plane and rotate as if they were absolutely rigid.

170 Strength of Materials for Engineering Technology

Figure 5.3(b) shows the consequence of assumption 2. The deformation of the shaft is proportional to the distance from the center of the shaft, and therefore the strain varies directly as the distance from the center of the shaft. Since we will assume a linear stress-strain relation for the material, the stress in the shaft [as shown superimposed in Figure 5.3(b)] must vary linearly with the distance from the shaft center and is a pure shearing stress.

The third assumption we will make is a consequence of the first two and is self-explanatory.

3. The motion of any point in the shaft lies in a plane transverse to the shaft, and its direction is in a circle whose center is on the longitudinal axis of the shaft. Also, the resultant of the forces on any transverse plane must be a couple.

The final explicit assumption that we shall make concerns the line *ab* that we originally scribed on the outside of the shaft. This line will be assumed to be undergoing the following action.

4. A longitudinal straight line parallel to the axis of the shaft will be twisted to yield a helical line after twisting, having a constant rise (lead) per unit length of the shaft.

We can now consider any transverse cross section of the shaft as a free body and study the forces on it that must exist if the shaft is in equilibrium under the applied torque. We will use the assumptions discussed above, as necessary, to analyze this situation:

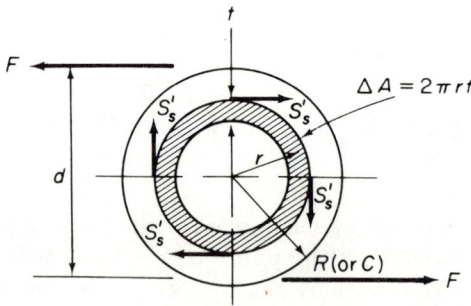

Fig 5.4 Development of the torsion equation

Figure 5.4 shows a cross section of the shaft subjected to the couple $F \times d$. Consider the circular element of area shown which we shall take to be equal to ΔA. The thickness of annulus (t) will be taken to be small enough to consider all the elements of the area to be located at a distance r from the center.

Since all elements are located at the same distance from the center, the shear stress on this area will be constant, and the shear forces will be tangent to the annulus as shown. Summing up the internal moments and equating them to the external moment we have

1. The external moment (torque) is $F \times d = T$.
2. S'_s at any radius is related to the maximum shearing stress at the outside S_s linearly as the distance from the center. Therefore, $S'_s = S_s(\frac{r}{R})$.
3. The resisting moment of the area ΔA is $(S'_s \Delta A) r$ or $(S_s \Delta A \frac{r}{R}) r$. Thus

$$F \times d = T = \sum \frac{S_s \Delta A r^2}{R} = \frac{S_s}{R} \sum \Delta A r^2 \qquad (5.4)$$

But we have already defined the summation of the $\Delta A r^2$ terms to be the polar moment of inertia (J_o). Using this definition, denoting $R = C$ (by convention and also by analogy to the terms that will be used when considering the bending of beams), we can rearrange equation (5.4) to yield

$$S_{s\,max} = \frac{TR}{J} = \frac{TC}{J} \qquad (5.5)$$

Equation (5.5) is known as the *torsion equation* and it has been developed for circular shafts. From Chapter 4 we have the polar moment of inertia of a solid circular shaft about a longitudinal central axis

$$\bar{J}_o = \frac{\pi d^4}{32} \qquad (5.6)$$

which, when combined with equation (5.5), yields the torsion equation for a solid shaft, namely

$$S_{s\,max} = \frac{16T}{\pi d^3} \qquad (5.7)$$

Since all of the assumptions that we have made concerning solid circular shafts can be applied to hollow concentric circular shafts, we can also evaluate the torsion equation for hollow circular shafts by noting $\bar{J}_o$ for this case to be

$$\bar{J}_o = \frac{\pi}{32} (d^4 - d_i^4) \qquad (5.8)$$

where d_i is the inside diameter of the shaft. Therefore for a hollow circular shaft

$$S_{Smax} = \frac{16Td}{\pi(d^4 - d_i^4)} \qquad (5.9)$$

Illustrative problem 5.3

Compare the stresses in a solid shaft having a diameter d to a hollow shaft whose mean annulus diameter is d and whose weight per metre is equal to the weight per metre of the solid shaft.

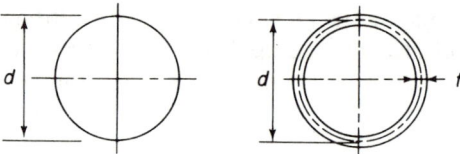

Fig 5.5 Illustrative problem 5.3

Solution:

If the weight per foot of each shaft is the same, then cross-sectional areas must be equal. Thus

$$\pi \frac{d^2}{4} = \pi d t$$

and

$$t = \frac{d}{4}$$

The O.D. of the hollow shaft is $d + \frac{d}{4}$ or $\frac{5}{4}d$ and the I.D. is $d - \frac{d}{4} = \frac{3}{4}d$. Using these values and equation (5.9)

$$S_{Smax} = \frac{16T(\frac{5}{4})d}{\pi[(\frac{5}{4}d)^4 - (\frac{3}{4}d)^4]} = \frac{16T(\frac{5}{4})d}{\pi d^4(\frac{625}{256} - \frac{81}{256})}$$

and

$$S_{Smax} = \frac{16T\frac{5}{4}d \times 256}{\pi d^4 \times 544} = \frac{16T}{\pi d^3} \times 0.588 = \frac{16T}{\pi d^3}\left(\frac{1}{1.7}\right)$$

But the stress in the solid shaft is $S_{s\,max} = 16T/\pi d^3$. Therefore $\dfrac{S_{s\,max} \text{ solid}}{S_{s\,max} \text{ hollow}}$ = $\dfrac{1.7}{1}$, and we can conclude that a hollow shaft of a given weight can carry a torque 70% greater than a solid shaft of the same weight whose outside diameter equals the mean diameter of the hollow shaft.

Illustrative problem 5.4

A hollow circular shaft has an outside diameter of 50 mm and an inside diameter of 25 mm. If the maximum allowable shear stress is 140 MPa, determine: (a) The torque that it can transmit; (b) the shear stress at the inside and at a radius of 37 mm.

Solution:

Applying equation (5.9)

(a) $\quad T = \dfrac{S_{s\,max} \pi (d^4 - d_i^4)}{16 d} = \dfrac{140 \times 10^6 (\pi)(0.050^4 - 0.025^4)}{16 (0.050)}$

$\quad\quad = 3221.4 \text{ N·m}$

(b) Since the stress is proportional to the distance from the center of the shaft

$$S_{S\,inside} = 140 \times 10^6 \times \dfrac{0.025}{0.050} = 70 \times 10^6 \text{ Pa} = 70 \text{ MPa}$$

and at $r = 37$ mm

$$S_s = 140 \times 10^6 \times \dfrac{0.037}{0.050} = 103.6 \times 10^6 \text{ Pa} = 103.6 \text{ MPa}$$

In addition to our being concerned with the stresses in the shaft, we are interested in the deflection (angular) that occurs when the shaft is subjected to torsion. Figure 5.3 (repeated) shows point b, an initial location in the right face and on the outside of the shaft. Twisting of the shaft moves b to b'. The amount of this motion ($b\ b'$) can be determined

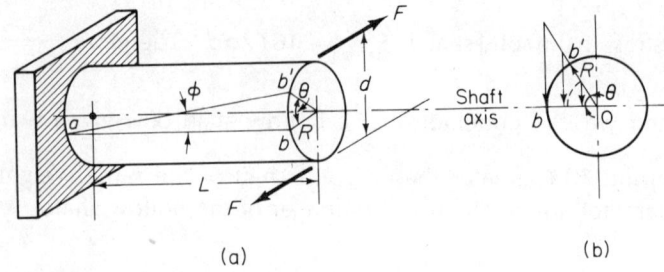

(a)　　　　　　　　　　　　(b)

Fig 5.3 (repeated) A shaft subjected to twisting only

from considering both the motion in the plane and also along the shaft. Thus,

$$b\,b' = R\theta$$

But (5.10)

$$b\,b' = L\phi$$

Therefore, we can express θ in terms of ϕ

$$\theta = \frac{L\phi}{R} \tag{5.11}$$

By definition, G, *the shear modulus,* is defined as the shearing stress divided by the shearing strain, giving us

$$G = \frac{S_s}{\phi} \tag{5.12}$$

Placing equation (5.12) into equation (5.11) yields

$$\theta = \frac{S_s L}{GR} \tag{5.13}$$

But

$$S_s = \frac{TR}{J} \tag{5.5}$$

thus, the angular twist of the shaft at a given cross section is

$$\theta = \frac{TL}{JG} \tag{5.14}$$

The student should note that θ is given in radians where 1 radian is approximately = 57.3 degrees, and that dimensionally this requires the right hand side of equation (5.14) to be dimensionless. Thus if T is in N·m and L is in metres, J must be in m^4 and G must have the dimension of N/m^2, i.e., Pa. Any self-consistent set of units will give the correct solution, and care should be exercised on this point.

Illustrative problem 5.5

A solid shaft is to transmit 30 H.P. at 1000 R.P.M. If the allowable shear stress is 70 MPa and $G = 84$ GPa, determine the shaft diameter required if: (a) it is based upon the stress only; (b) the allowable angular deflection is $1\frac{1}{2}$ degrees per metre of shaft.

Solution:

From equation (5.3)

$$T = \frac{7124 \text{ H.P.}}{N} = \frac{7124 \times 30}{1000} = 213.72 \text{ N·m}$$

(a) Applying equation (5.7)

$$d^3 = \frac{16T}{\pi S_{S\max}} = \frac{16 \times 213.72}{\pi (70 \times 10^6)} = 1.555 \times 10^{-5} \text{ m}^3$$

and $d = 2.496 \times 10^{-2}$ m $= 24.96$ mm, say 25 mm.

(b) $1\frac{1}{2}$ degrees per metre $= \dfrac{1.5}{57.3} = 2.618 \times 10^{-2}$ radians per metre

$$\theta = \frac{TL}{JG} = 2.618(10)^{-2} = \frac{T(1)}{(\pi d^4/32)(84 \times 10^9)} = \frac{T(1)(32)}{\pi d^4 (84 \times 10^9)}$$

but $T = 213.72$ N·m. Therefore

$$d^4 = \frac{213.72 \times (1)(32)}{2.618 \times 10^{-2} \times \pi \times 84 \times 10^9} = 9.9 \times 10^{-7} \text{ m}^4$$

and

$d = 3.154 \times 10^{-2}$ m $= 31.54$ mm

For this example, the required shaft diameter is determined by the limiting angular deflection.

Illustrative problem 5.6

A hollow shaft having an outside diameter of 50 mm and an inside diameter of 25 mm is transmitting power at 2000 R.P.M. If the allowable shear stress is 140 MPa, and the allowable angular deflection is 3 degrees per metre of shaft, determine the horsepower capacity of the shaft. Use $G = 84 \times 10^9$ Pa.

Solution:

For a hollow shaft

$$J = \frac{\pi(d^4 - d_i^4)}{32} = \frac{\pi}{32}(0.050^4 - 0.025^4) = 5.752 \times 10^{-7} \text{ m}^4$$

Since

$$S_s = \frac{TC}{J}; \quad T = \frac{S_s J}{C} = \frac{140 \times 10^6 \times 5.752 \times 10^{-7}}{0.025} = 3221 \text{ N·m}$$

Therefore

$$\text{H.P.} = \frac{TN}{7124} = \frac{3221 \times 2000}{7124} = 904 \text{ H.P.}$$

If the allowable angular deflection is 3 degrees per metre of shaft, θ $= \frac{3}{57.3} = 5.236 \times 10^{-2}$ radians/metre. Thus

$$T = \frac{\theta J G}{L} = \frac{5.236 \times 10^{-2} \times 5.752 \times 10^{-7} \times 84 \times 10^9}{1} = 2530 \text{ N·m}$$

$$\text{H.P.} = \frac{TN}{7124} = \frac{2530 \times 2000}{7124} = 710 \text{ H.P.}$$

Illustrative problem 5.7

The shaft shown is rigidly supported at each end. If a torque of T in. lb is applied at the intersection of the two shafts, determine the distribution of the torque T in each shaft.

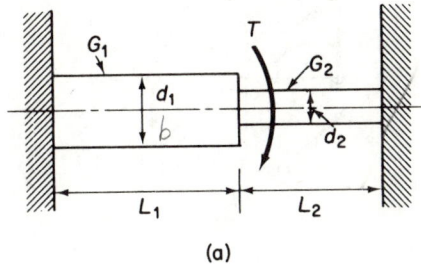
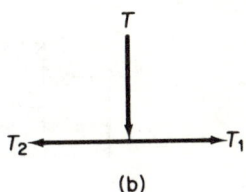

Fig 5.6 Illustrative problem 5.7

Solution:

Consider that the torque T distributes as shown in Fig. 5.6(b). Thus

$$T = T_1 + T_2$$

At the intersection of the two shafts, $\theta_1 = \theta_2$, yielding the condition that

$$\frac{T_1 L_1}{J_1 G_1} = \frac{T_2 L_2}{J_2 G_2}$$

We can solve these two equations to obtain

$$T_1 = \frac{T}{1 + (L_1/L_2)(J_2/J_1)(G_2/G_1)}$$

And

$$T_2 = \frac{T}{1 + (L_2/L_1)(J_1/J_2)(G_1/G_2)}$$

5.4 SHAFT COUPLINGS

Shaft couplings are commonly used in machinery to provide for misalignment of shafts, to connect differing machines, to join lengths of shafts, and to absorb shock loads. Of the many types of shaft couplings that are used, we shall be concerned here with the *flange coupling,* an essentially rigid coupling used to connect shafts that are in relatively good alignment. A coupling of this type is shown diagrammatically in Fig. 5.7. The two

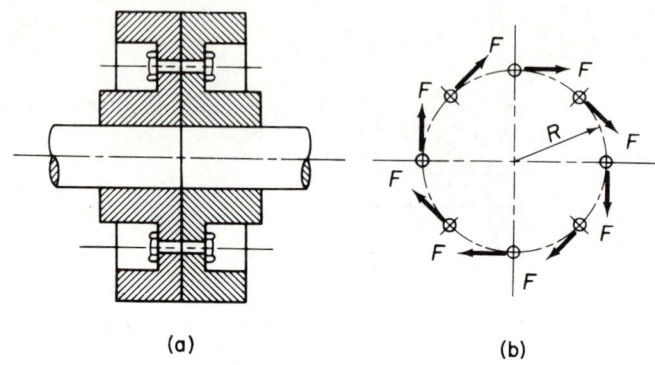

Fig 5.7 A flange coupling

parts of this coupling are held together by a single circle of bolts. In order to properly distribute the loads uniformly in the bolts, the bolts should be ground and fitted into close tolerance with reamed holes in the radial part of the flange.

Let us assume that we have such a coupling with a single circle of n bolts, each having a diameter d and arranged in a circle of radius R. The shear area of each bolt (A_s) is $\frac{\pi}{4}d^2$, the shear force is $\frac{\pi}{4}d^2 S_s$, and the torque exerted by each bolt on the driven shaft is $F \times R = \frac{\pi}{4}d^2 S_s R$. For n bolts the torque is

$$T = \frac{\pi d^2}{4} S_s R n \qquad (5.15)$$

Illustrative problem 5.8

A 25-mm diameter shaft is used to transmit 600 H.P. at 1500 R.P.M. from a motor. Using a material with an allowable shear stress of 70 MPa, find how many 10-mm-diameter bolts would be needed if they were to be placed on a 75-mm-diameter circle in a flange coupling.

Solution:

Calculating the torque, we have

$$T = \frac{7124 \text{ H.P.}}{N} = \frac{7124 \times 600}{1500} = 2849.6 \text{ N} \cdot \text{m}$$

The shear area/bolt is $\frac{\pi}{4}d^2 = \frac{\pi}{4}(0.010)^2 = 7.854 \times 10^{-5}$ m². Therefore

$$n = \frac{T}{A_s S_s R} = \frac{2849.6}{7.854 \times 10^{-5} \times 70 \times 10^6 \times \frac{0.075}{2}} = 13.8$$

We would specify 14 bolts.

In the event that more than one row of bolts is used, it is necessary to be able to find the load taken by each row. When we considered the deflections in a shaft, we found them to be proportional to their distance from the center of the shaft. For bolt circles at differing radii from the center of the shaft, the deflections of the bolts and, consequently, the loads on the bolts will be proportional to their location relative to the center. We may write this relation in the following form for the situation shown in Fig. 5.8

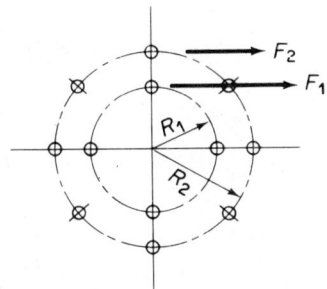

Fig 5.8 A coupling with two rows of bolts

$$\frac{F_1}{F_2} = \frac{R_1}{R_2} \qquad (5.16)$$

and

$$T = T_1 + T_2 = F_1 R_1 n_1 + F_2 R_2 n_2 \qquad (5.17)$$

Equations (5.16) and (5.17) can readily be modified to take into account three or more rows of bolts.

Illustrative problem 5.9

Twelve 12-mm-diameter bolts are arranged in two concentric bolt circles having diameters of 125 mm and 250 mm respectively. If there are four bolts on the inner bolt circle and eight bolts on the outer bolt circle, determine the torque they can transmit if the allowable shear stress is 70 MPa.

Solution:

Assume that the outer bolts are stressed to their allowable stress of 70 MPa. The shear force per bolt $= \frac{\pi}{4}(0.012)^2 \times 70 \times 10^6 = 7917$ N/bolt. On the inner row

$$F_1 = \frac{R_1}{R_2} F_2 = \frac{0.125/2}{0.250/2} \times 7917 = 3958 \text{ N}$$

the torque

$$T_1 = F_1 R_1 n_1 = 3958 \times \frac{0.125}{2} \times 4 = 990 \text{ N} \cdot \text{m}$$

$$T_2 = F_2 R_2 n_2 = 7917 \times \frac{0.250}{2} \times 8 = 7917 \text{ N} \cdot \text{m}$$

the total torque

$$T = T_1 + T_2 = 990 + 7917 = 8907 \text{ N} \cdot \text{m}$$

5.5 SPRINGS

Springs are used in many industrial applications to absorb energy (or shock), to control motions, to apply forces to elements of a mechanism, and to dampen and absorb vibrations. Due to the wide application for this device, it is in order in our study to investigate those springs that can be considered to be torsion members.

Let us consider a spring that can be termed a *linear spring* due to the fact that its deflection as a function of applied force is linear as shown in Fig. 5.9. This linear relation can be expressed in terms of the spring scale or spring constant (k)

$$k = \frac{F}{\delta} \quad (5.18)$$

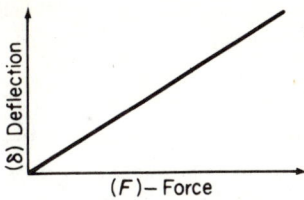

Fig 5.9 Force-deflection curve for a spring

The work done in deflecting the spring, which is also the energy stored in the deflected spring, is the area under the force deflection curve which, from Fig. 5.9, is

$$U = \tfrac{1}{2}F\delta \; (\text{N·m} \text{ or } \text{N·mm}) \tag{5.19}$$

or

$$U = \tfrac{1}{2}k\delta^2 \tag{5.19a}$$

For a member that is subject to torsion, the (angular) deflection θ is linearly related to the applied torque T, and using analogous reasoning, we can write for torsion

$$U = \frac{T\theta}{2} \tag{5.20}$$

In many spring applications, the torque applied to a member known as a *torsion bar* can be used to obtain the desired characteristic of load versus deflection as well as to save both weight and space. The torsion bar is usually fixed at one end and has a crank at the other end to which the force that places the bar in torsion is applied. The relations already developed in this chapter for the torsion of circular shafts can be applied to the torsion bar. The spring rate or spring constant of the torsion bar is expressed in terms of inch (or foot) pounds per angle of twist, giving

$$k = \frac{T}{\theta} \tag{5.21}$$

But θ is TL/JG or $TL/[(\pi d^4/32)G]$ for solid circular shafts. Therefore

$$k = \frac{T}{[TL/(\pi d^4/32G)]} = \frac{\pi d^4 G}{32L} \tag{5.22}$$

By using a compound system such as shown in Fig. 5.10(b), it is possible to "tailor" a torsion spring system to meet most applications.

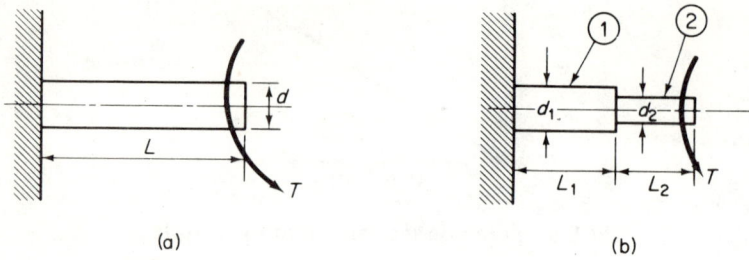

Fig 5.10 Torsion bars

Illustrative problem 5.10

A torsion bar is designed to be a spring. Determine the energy the spring can absorb and its angular deflection when subjected to a torque of 200 N·m. Assume the bar to be circular, 250 mm long, and 25 mm in diameter. Take $G = 84$ GPa.

Solution:

The angular deflection (in radians) is

$$\theta = \frac{TL}{JG} = \frac{200 \times 0.250}{\frac{\pi}{32}(0.025)^4 \times 84 \times 10^9} = 1.552 \times 10^{-2} \text{ radians}$$

Since

$$k = \frac{T}{\theta}, \quad k = \frac{200}{1.552 \times 10^{-2}} = 12\,887 \text{ N·m/radian}$$

The energy absorbed is

$$U = \frac{T\theta}{2} = \frac{200 \times 1.552 \times 10^{-2}}{2} = 1.552 \text{ N·m (J)}$$

Illustrative problem 5.11

Consider the torsion bar shown in Fig. 5.10(b) and take $L_1 = 125$ mm, $L_2 = 125$ mm, $d_1 = 50$ mm, $d_2 = 25$ mm, and $G = 84$ GPa. Determine the deflection and spring constant for this system when subjected to a torque of 400 N·m.

Solution:

Each section of this compound system is subjected to the same torque and angle of twist as is the entire torsion bar, and thus the result is the sum of the angular deflections of all parts. Let us therefore calculate the deflection in each part as

$$\theta_{1 \text{ (radians)}} = \frac{TL_1}{J_1 G} = \frac{400(0.125)}{[\pi(0.050)^4/32] \times 84 \times 10^9}$$
$$= 9.7 \times 10^{-4} \text{ radians}$$

$$\theta_2 = \frac{400 \times 0.125}{[\pi(0.025)^4/32] \times 84 \times 10^9} = 1.552 \times 10^{-2} \text{ radians}$$

$$\theta_{\text{total}} = \theta_1 + \theta_2 = 9.7 \times 10^{-4} + 1.552 \times 10^{-2}$$
$$= 1.649 \times 10^{-2} \text{ radians}$$

or

$$\theta_{\text{total}} = 0.0165 \text{ radians}$$

By definition

$$k = \frac{T}{\theta} = \frac{T}{\theta_1 + \theta_2} = \frac{400}{0.0165} = 24\,242 \text{ N·m/radian}$$

For this compound system, we can obtain a general expression for the system spring constant k in terms of k_1 and k_2 as follows

$$k = \frac{T}{\theta_1 + \theta_2} = \frac{1}{(\theta_1/T) + (\theta_2/T)} = \frac{1}{(1/k_1) + (1/k_2)}$$

One of the most common forms of spring is the *helical spring*. Figure 5.11 shows a compression spring with various types of ends, open or closed, plain or ground. Springs that are designed to operate as torsion (extension) members can have many different types of ends depending upon the specific applications for which they are intended. Some of the types of ends that are used on extension springs are shown in Fig. 5.12. Many times extension springs are coiled so as to place the coils in compression, and a load applied to a spring wound in this manner must

184   Strength of Materials for Engineering Technology

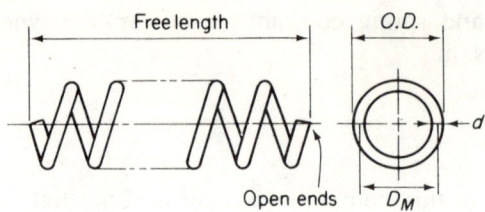

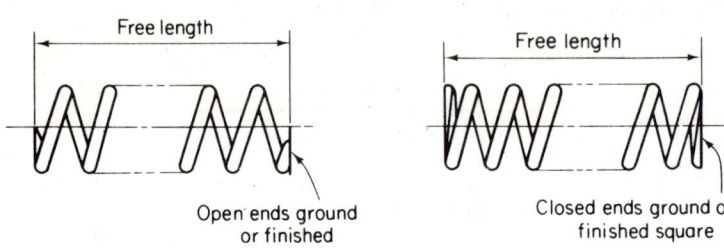

Fig 5.11   Compression springs with various ends

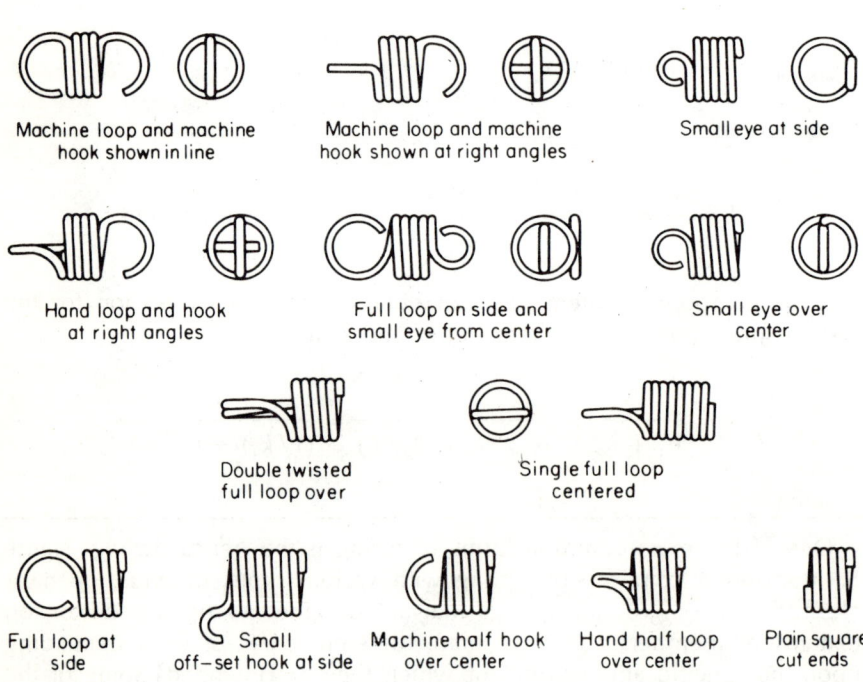

Fig 5.12   Types of ends used on extension springs

first overcome this pre-load before the coils of the spring open and spring action starts to occur.

Figure 5.13 shows a compression spring loaded axially. This axial force tends to rotate the wire as well as introduce a direct shear in the wire.

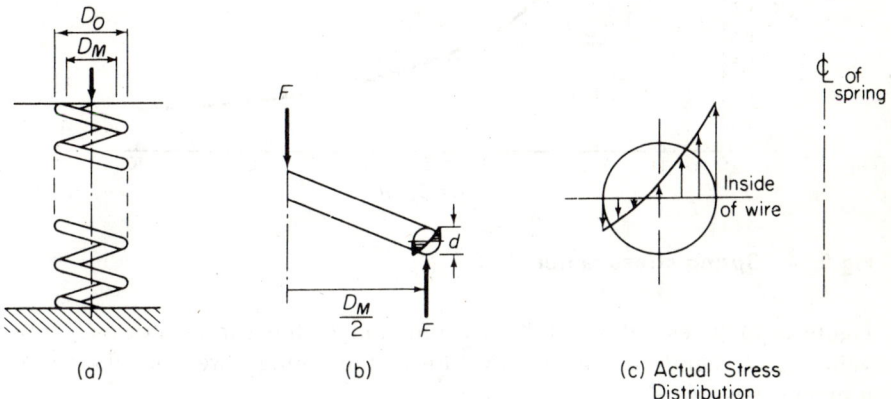

(a)   (b)   (c) Actual Stress Distribution

Fig 5.13   An axially loaded compression spring

In the following discussion, we will consider that the spring wire is circular, that torsion is the principal effect, and that the assumptions made for the torsion of straight circular shafts apply. We shall comment on the validity of these assumptions later on.

Referring to Fig. 5.13(b), the applied torque T is $FD_M/2$ and

$$S_{Smax} = \frac{TC}{J} = \frac{(FD_M/2)(d/2)}{\pi d^4/32} = \frac{8FD_M}{\pi d^3} \tag{5.23}$$

Equation (5.23) contains a built-in error since it neglects the transverse shear stress acting on the cross section of the wire, stresses due to the curvature of the wire in the helix, and stress arising from the component of the applied force along the wire. In order to include these effects, equation (5.23) can be rewritten as

$$S_{Smax} = \frac{K\,8FD_M}{\pi d^3} \tag{5.24}$$

where

$$K = \frac{4C-1}{4C-4} + \frac{0.615}{C} \quad \text{and} \quad C = \frac{D_M}{d} \tag{5.25}$$

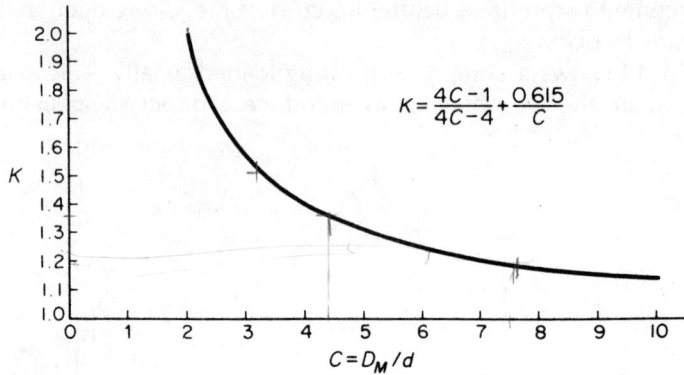

Fig 5.14 Spring stress factor

Figure 5.14 gives values of K as a function of the spring index C; low values of C signify a higher curvature of the spring wire and, therefore, a higher value of K.

Illustrative problem 5.12

A spring has a mean diameter of 75 mm and is made of a wire having a 12.5 mm diameter and an allowable stress in shear of 315 MPa. Determine the maximum load the spring can support.

Solution:

Applying equation (5.24)

$$S_{Smax} = \frac{K 8 F D_M}{\pi d^3}$$

$$F = \frac{S_{Smax}}{K} \left(\frac{\pi d^3}{8 D_M} \right)$$

$C = \dfrac{75}{12.5} = 6$ and from Fig. 5.14, $K = 1.25$. Therefore

$$F = \frac{315 \times 10^6 \times \pi (0.125)^3}{1.25 \, (8)(0.075)} = 2577 \text{ N}$$

Illustrative problem 5.13

A compression spring is to be designed to fit into a 75-mm-diameter cylinder and to carry a load of 1300 N. If the maximum design shear stress is 210 MPa, determine the size of the spring wire required.

Solution:

This problem cannot be solved readily since D_M and K both involve the diameter of the wire. A graphical solution is most suited for this type of problem. The procedure involves assuming several values of wire diameter d, evaluating D_M and K, and solving for d_1 from equation (5.24). When the assumed and calculated values are equal, the solution is satisfactory. For this problem let us make a first approximation by assuming $D_M = 75$ mm and $K = 1$. Thus, $d^3 = F8D_M/\pi S_s = 1.182 \times 10^{-6}$ and $d = 1.057 \times 10^{-2}$ m. Now assume $d = 1.0 \times 10^{-2}$ m, 1.05×10^{-2} m, 1.10×10^{-2} m, respectively. Proceeding as in the following table and then plotting the results in Fig. 5.15 shows we have $d = 10.88$ mm.

d(assumed) $\times 10^{-2}$ m	$D_M = D - d$ mm	$C = \dfrac{D_M}{d}$	K	$d\text{(calc)} = \sqrt[3]{\dfrac{FKD_M \times 8}{S_s \pi}}$ m
1.00	65.0	6.5	1.22	1.077×10^{-2}
1.05	64.5	6.14	1.25	1.083×10^{-2}
1.10	64.0	5.82	1.28	1.089×10^{-2}

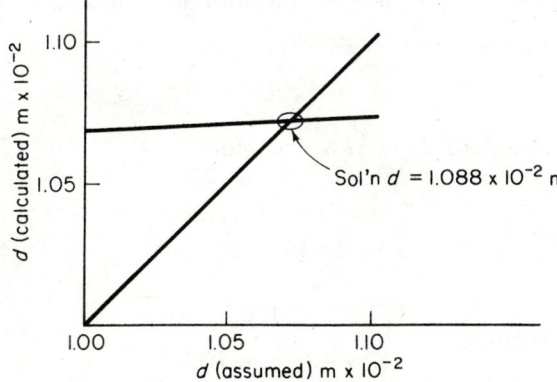

Fig 5.15 Illustrative problem 5.13

The solution is therefore a wire diameter of 10.88 mm (or the nearest standard value), and a mean coil diameter $D_M \cong 65.12$ mm or close to 65 mm.

The deflection of the spring will be sum of the torsional deflection of all coils of the spring. The length of each coil can be taken to be the length of its mean circumference or πD_M. If there are N active coils in the spring, the total length is $\pi D_M N$. Using equation (5.15) for the angular deflection

$$\theta = \frac{TL}{JG} = \frac{(FD_M/2)\pi D_M N}{(\pi d^4/32)G} = \frac{16 FD_M^2 N}{d^4 G} \qquad (5.26)$$

If θ is multiplied by the mean coil radius we obtain the coil deflection, δ

$$\delta = \frac{\theta D_M}{2} = \frac{8FD_M^3 N}{Gd^4} \qquad (5.27)$$

Earlier in this section we defined the spring constant k as being equal to the ratio of spring force divided by spring deflection. Therefore

$$k = \frac{F}{\delta} = \frac{Gd^4}{8D_M^3 N} \qquad (5.28)$$

Illustrative problem 5.14

Determine the number of active coils required for the spring designed in Illustrative Problem 5.13 if it must absorb 30 N·m and $G = 84$ GPa.

Solution:

The energy absorbed, $U = \frac{1}{2}F\delta$. Therefore

$$U = 30 = \frac{1}{2}(1300)\delta$$
$$\delta = 4.615 \times 10^{-2} \text{ m}$$

The spring constant $k = F/\delta = \dfrac{1300}{4.615 \times 10^{-2}} = 28\,169$ N/m deflection.

Using equation (5.28), we can write

$$N = \frac{Gd^4}{8D_M^3 k}$$

and

$$N = \frac{84 \times 10^9 \times (1.088 \times 10^{-2})^4}{8(0.065)^3 \times 28\,169} = 19.02, \text{ say 19 or 20 coils}$$

5.6 TORSION OF NONCIRCULAR SECTIONS

In the previous portions of this chapter, we studied the torsion of circular sections. The theory presented for circular shafts was based upon the following assumptions: a plane section before twisting remains plane after twisting; a diameter in the plane of twist prior to twisting remains a straight line after twisting. These assumptions are correct for either solid or concentric hollow sections but are incorrect for sections of any other shape. As the cross section deviates to a greater extent from circular, the discrepancy between the theory developed earlier and the actual shear stress in the member becomes greater. There is, however, one class of noncircular torsion members that can be treated by relatively simple considerations, namely the *thin-walled tube*.

Before taking up the problem of the torsion of noncircular thin-walled tubes, let us consider the free-body diagram of an element of volume in a simple member that is subjected to a shear stress S_s on the parallel faces (the *yz* faces), as shown in Fig. 5.16. For the body to be in equilibrium,

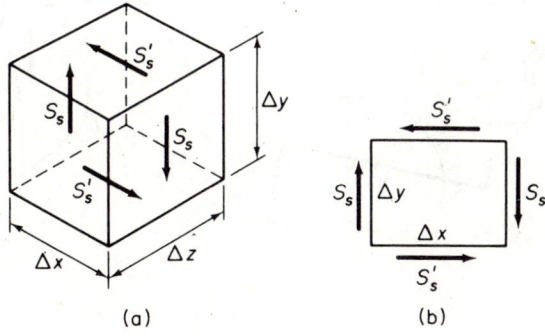

Fig 5.16 Stresses on an element

there must be an equal but opposite stress on the second *yz* face of the cube. Figure 5.16(b) shows these stresses on a plane taken perpendicular to the yz plane. Moments, taken about 0, of the forces due to these stresses indicate a net clockwise couple, which for the condition of equilibrium requires an equal counterclockwise couple to exist to insure a net zero moment. The stresses S'_s shown on both Fig. 5.16(a) and (b) are due to the forces required on the xz planes to create a net zero moment.

Since the area of the yz face is $(\Delta y)(\Delta z)$, the force on each such face is $S_s(\Delta y)(\Delta z)$. For the xz face, the area is $(\Delta x)(\Delta z)$, and the force on each of these faces is $S'_s(\Delta x)(\Delta z)$. Taking moments about 0 for these force systems yields

$$S_s(\Delta y)(\Delta z)(\Delta x) = S'_s(\Delta x)(\Delta z)(\Delta y) \tag{5.29}$$

from which we obtain

$$S_s = S'_s \tag{5.30}$$

From equation (5.30) we draw the important conclusion that a shearing stress cannot occur alone; it must always induce a numerically equal shearing stress on a perpendicular plane.

For the case of a circular shaft, we can demonstrate that the conclusion reached in equation (5.30) is also valid and general by referring to Fig. 5.17 and the free-body diagram shown in Fig. 5.17(b). S'_s acts on an

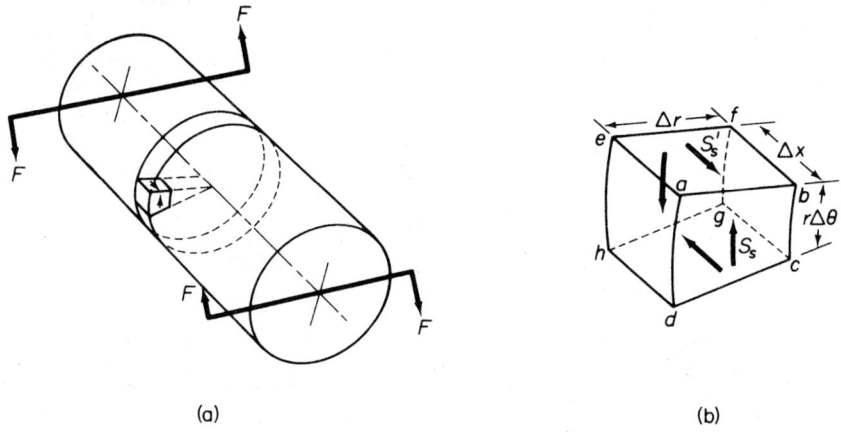

Fig 5.17 The induced stress in a circular shaft

area equal to $(\Delta r)(\Delta x)$, and S_s acts on an area $r(\Delta r)(\Delta \theta)$. Taking moments about h we have

$$S_s(r)(\Delta r)(\Delta \theta) = S'_s(\Delta r)(\Delta x) r \Delta \theta \tag{5.31}$$

which gives us

$$S_s = S'_s \tag{5.32}$$

We can therefore conclude that there is, in addition to the transverse shear stress, an induced longitudinal shear stress numerically equal to the transverse stress. At this point it is noted that the foregoing development has been covered for more than the immediate needs at hand. We shall make use of it later in our study of the shearing stresses in beams.

Consider the closed, thin-walled torsion member shown in Fig. 5.18. The element of volume being considered is shown enlarged in Fig. 5.18(b).

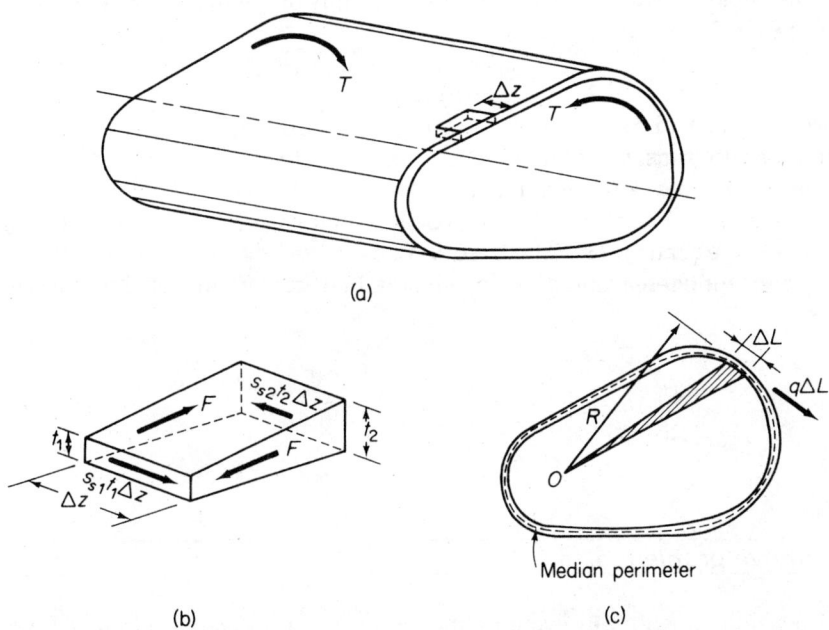

Fig 5.18 Torsion of a thin-walled member

The applied torque T causes shearing forces F to exist as shown, and as a consequence there must be equal forces induced on the perpendicular faces [also shown in Fig. 5.18(b)]. For equilibrium,

$$S_{s_1} t_1 \Delta z = S_{s_2} t_2 \Delta z \tag{5.33}$$

and therefore

$$S_{s_1} t_1 = S_{s_2} t_2 \tag{5.34}$$

If we denote the product of $S_s t$ as q, we have

$$q_1 = q_2 = q = \text{constant} \tag{5.35}$$

The term q is called the *shear flow*. The term shear flow may be compared to the steady flow of a fluid in a continuous pipe of varying diameter, hence the term shear flow.

The resisting torque per unit length of tube can be obtained from Fig. 5.18(c) as

$$\Delta T = q(\Delta L) R \tag{5.36}$$

But the shaded area on Fig. 5.18(c) is simply a triangle whose area A is $(\Delta L)R/2$. Thus we have

$$\Delta T = q 2 \Delta A \tag{5.37}$$

Summing all terms, we obtain

$$T = q(2A) \tag{5.38}$$

where A is the area enclosed by the median center line of the section. Thus

$$q = S_s t = \frac{T}{2A} \quad \text{or} \quad S_s = \frac{T}{2At} \tag{5.39}$$

Illustrative problem 5.15

Using equation (5.39), evaluate the stress in a thin-walled circular tube having a thickness t and a mean diameter D. Compare this result with the results obtained by applying equation (5.5), if the O.D. is D_o and the I.D. is D_i.

Solution:

The area enclosed by the mean perimeter is $\pi D^2/4$. Thus, from equation (5.39)

$$S_s = \frac{T}{2At} = \frac{T}{(\pi D^2/4)(t)} = \frac{4T}{\pi D^2 t}$$

Using equation (5.5)

$$S_s = \frac{TC}{J} = \frac{(T)(D/2)}{(\pi/32)(D_o^4 - D_i^4)}$$

but

$$D_o^4 - D_i^4 = (D_o^2 + D_i^2)(D_o^2 - D_i^2) = (D_o^2 + D_i^2)(D_o + D_i)(D_o - D_i)$$

where

$$D_o^2 + D_i^2 \cong 2D^2, \quad D_o + D_i \cong 2D, \quad \text{and} \quad D_o - D_i = t.$$

Thus

$$S_s = \frac{T(D/2)}{(\pi/32)(2D^2)(2D)(t)} = \frac{4T}{\pi d^2 t}$$

the same result as was obtained above.

The determination of torsional stress and angular deformation in non-circular sections is a complex calculation at best. For several members having simple cross sections, the analysis (approximate for most areas) has been performed and the results are given in Table 5.1. In each case the maximum stress occurs at the outer surface nearest the axis of the member; but in the case of rectangular and triangular sections, the stress is zero at the corners.

Illustrative problem 5.16

Compare the maximum stress in a solid circular shaft and a square shaft when both are subjected to a torque T if: (a) the cross-sectional areas are equal, and (b) the diameter of the circle equals the side of the square.

Solution:

(a) Denoting the side of the square as b and the diameter of the circle as D we have, for equal area

$$b^2 = \frac{\pi}{4}D^2$$

Table 5.1 Approximate Formulas for Torsional Shearing Stress and Angle of Twist, Obtained from Mathematical Analysis*

Cross-section	Relation between Shearing Unit-stress and Twisting Moment	Relation between Angle of Twist and Twisting Moment
Ellipse ($2b \times 2h$)	$S_s = \dfrac{2T}{ab}$ $= \dfrac{2T}{\pi h b^2}$	$\phi = 4\pi^2 \dfrac{J}{a^4} \cdot \dfrac{T}{G}$ $= \dfrac{h^2 + b^2}{\pi h^3 b^3} \cdot \dfrac{T}{G}$
Equilateral triangle	$S_s = \dfrac{20T}{b^2}$	$\phi = \dfrac{80}{b^4 \sqrt{3}} \cdot \dfrac{T}{E} = \dfrac{46.2}{b^4} \cdot \dfrac{T}{G}$
Rectangle ($h > b$)	$S_s = \dfrac{8\beta T}{\mu h b^2}$	$\phi = \dfrac{16}{\mu h b^3} \cdot \dfrac{T}{G}$
	Saint Venant gives the following approximate formulas for β and μ which yield values that agree within 4 per cent of the exact values: $\beta = \dfrac{3}{8}\left(1 + 0.6\dfrac{b}{h}\right)\mu \qquad \mu = \dfrac{16}{3} - 3.36\dfrac{b}{h}\left(1 - \dfrac{1}{12}\dfrac{b^4}{h^4}\right)$ Using these values we find	
	$S_s = \left(\dfrac{15h + 9b}{5h^2 b^2}\right) T$	$\phi = \dfrac{16}{\left[\dfrac{16}{3} - 3.36\dfrac{b}{h}\left(1 - \dfrac{1}{12}\dfrac{b^4}{h^4}\right)\right] h b^3} \cdot \dfrac{T}{G}$ $= \dfrac{4\pi^2 J}{a^4} \cdot \dfrac{T}{G}$ approximate. (Error is small if $h > 2b$)
Square	For a square the above formulas reduce to: $S_s = \dfrac{24}{5} \dfrac{T}{b^3}$	$\phi = \dfrac{7.11}{b^4} \cdot \dfrac{T}{G}$
Any compact section *not* containing re-entrant angles	$S_s = \dfrac{4\pi^2 J}{a^4 y} \cdot T$ y = distance of most remote edge from center of bar	$\phi = k \dfrac{J}{a^4} \cdot \dfrac{T}{G}$ $k = 4\pi^2$ for ellipse $k = 40$ to 42 for rectangles

T = Twisting moment ϕ = Angle of twist per unit of length a = Area of cross-section. J = Polar moment of inertia G = Shearing modulus of elasticity

*Reproduced with permission from Advanced Mechanics of Materials by F. B. Seely, John Wiley and Sons, Inc., New York, 1932.

Table 5.1 *(continued)*

Cross-section	Relation between Shearing Unit-stress and Twisting Moment	Relation between Angle of Twist and Twisting Moment
(hollow circle, radius r, thickness t)	$S_s = \dfrac{T}{2\pi r^2 t}$	$\phi = \dfrac{T}{2\pi r^3 t G}$
(slit hollow circle, radius r, thickness t)	$S_s = \dfrac{3r}{t} \cdot \dfrac{T}{2\pi r^2 t}$ $= \dfrac{3T}{2\pi r t^2}$	$\phi = \dfrac{3r^2}{t^2} \cdot \dfrac{T}{2\pi r^3 t G}$ $= \dfrac{3T}{2\pi r t^3 E_s}$
(elliptical tube, $2h \times 2b$, outer $2h_0 \times 2b_0$)	$S_s = \dfrac{T}{2\pi k h_0 b_0^2 (1+k)}$ $k = \dfrac{h - h_0}{h_0} = \dfrac{b - b_0}{b_0}$	$\phi = \dfrac{h_0^2 + b_0^2}{4\pi k h_0^3 b_0^3} \cdot \dfrac{T}{G}$
(rectangular tube, $h \times b$, thickness t, t_1)	$S_s = \dfrac{T}{2t(h-t)(b-t_1)}$	$\phi = \dfrac{h t + b t_1 - t^2 - t_1^2}{2 t t_1 (h-t)^2 (b-t_1)^2} \cdot \dfrac{T}{G}$

T = Twisting moment. ϕ = Angle of twist per unit of length. a = Area of cross-section J = Polar moment of inertia G = Shearing modulus of elasticity

and

$$b = D\sqrt{\dfrac{\pi}{4}} = 0.886D$$

From Table 5.1, the stress in the square is given as

$$S'_s = \dfrac{24}{5}\dfrac{T}{b^3} = \dfrac{24}{5}\dfrac{T}{(\pi/4)D^2(0.866D)} = 7.06\dfrac{T}{D^3}$$

For the circle

$$S_s = \frac{16T}{\pi D^3} = 5.09 \frac{T}{D^3}$$

Thus, the ratio of the stress in the square to the stress in a circle having the same area as the square is

$$\frac{S'_s}{S_s} = \frac{7.06T/D^3}{5.09T/D^3} = 1.39$$

(b) If the square has a side equal to D, the stress in the square is

$$S'_s = \frac{24}{5}\frac{T}{D^3} = 4.8\frac{T}{D^3}$$

For the circle

$$S_s = \frac{16T}{\pi D^3} = 5.09\frac{T}{D^3}$$

and the ratio of S'_s is

$$\frac{S'_s}{S_s} = \frac{4.8T/D^3}{5.09T/D^3} = 0.943$$

From the foregoing we can conclude that the shear stress in a square shaft can be conservatively approximated by using the circle having a diameter equal to the side of the square. The error is on the order of 5%, which is within the limitations of the assumptions made in the mathematical analysis of the noncircular sections.

5.7 CLOSURE

In most applications, the ability of materials to resist an applied torque can yield a machine element that is useful. Examples of such elements have been studied in this chapter so that we could understand their principles of operation and be able to design them. It must be apparent to the student that it has not been possible to cover all facets of each topic comprehensively. We have limited ourselves to those basic considerations appropriate to the level of our study. The student desiring more information on these topics will find that he will be led to texts on advanced mechanics of

PROBLEMS

5.1 A weight of 400 N is lifted at the rate of 30 m/sec. What horsepower is being used?

5.2 A motor runs at N R.P.M. and transmits H.P. horsepower. If the motor is lifting a weight, show that the weight can be expressed as $W = 60 \times 746 \times H.P./\pi DN$, where D is the shaft diameter in metres.

5.3 A motor is required to transmit a torque of 100 N·m at 1000 R.P.M. What is its rating in H.P.?

5.4 What will be the rated speed in R.P.M. of a motor that is rated at 25 H.P. that is required to transmit a torque of 125 N·m?

5.5 Determine the torque transmitted in each portion of the shaft shown in Fig. P5.5. The shaft rotates at 150 R.P.M.

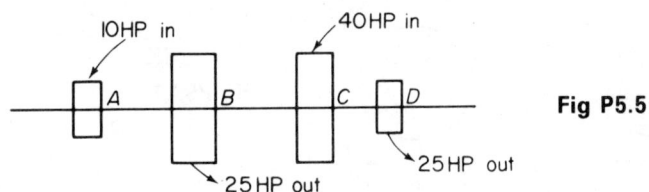

Fig P5.5

5.6 A solid circular shaft is used to transmit 75 H.P. at 250 R.P.M. If the allowable shear stress is 56 MPa, determine the diameter of the shaft.

5.7 A solid circular shaft is 50 mm in diameter. What horsepower can it transmit at 200 R.P.M. if the allowable shear stress is 70 MPa?

5.8 A solid circular shaft is to transmit 100 H.P. at 110 R.P.M. If the allowable shearing stress is 70 MPa, determine the shaft diameter. G is 84 GPa.

5.9 Determine the angular twist of the shaft in Problem 5.8 if the shaft is 1.5 m long.

5.10 A hollow circular shaft is stressed in shear to 65.5 MPa due to torsion. If the outside diameter of the shaft is 112.5 mm and the inside diameter is 50 mm, determine the power transmitted when the shaft is operated at 125 R.P.M.

5.11 A hollow steel shaft has a 100-mm O.D. and a 50-mm I.D. It is designed to transmit 500 H.P. at 500 R.P.M. Determine the maximum shear stress in the shaft. G is 84×10^9 Pa.

5.12 A hollow circular shaft transmits 1000 N·m. If the outside diameter is 75 mm and the inside diameter is 40 mm, what is the shear stress in the shaft at the outside if it operates at 500 R.P.M.?

5.13 Determine the angular twist per metre length of the shaft in Problem 5.11.

5.14 Determine the angle of twist per metre of shaft in Problem 5.12 if $G = 84$ GPa.

5.15 It is desired to limit the angular deflection of a shaft to $\frac{3°}{4}$ per metre of shaft. If $G = 77$ GPa and the applied torque is 750 N·m, determine the diameter of the shaft.

5.16 A torque of 1000 N·m is applied to a circular shaft having a diameter of 0.05 m. What is the deflection per metre of shaft if $G = 80$ GPa?

5.17 Show that the stress in a hollow shaft of outside diameter d and inside diameter $\frac{1}{2}d$ is approximately 6% greater than a solid shaft of d, but the hollow shaft will weigh approximately 25% less than the solid shaft.

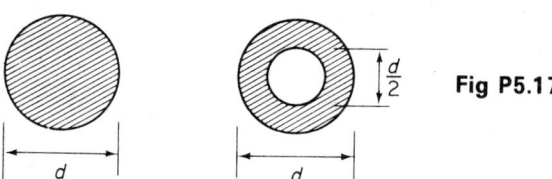

Fig P5.17

5.18 A torsion testing machine is used to test an unknown material. If a specimen 19 mm in diameter and 0.3 m in length is used, and it is experimentally found that a torque of 180 N·m produces an angular deflection of 5°, determine G.

5.19 A torque of 500 N·m produces an angular deflection of 0.03 radians/m in a shaft that has a diameter of 50 mm. What is G of the material?

5.20 A torque of 100 N·m is applied to the shaft shown in Fig. P5.20. Determine the total angle of twist of the free end. Take G to be 70 GPa for both sections.

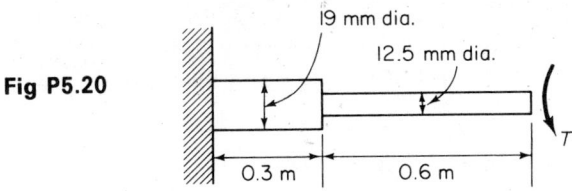

Fig P5.20

5.21 The shafting shown in P5.21 has a power input and take-off as indicated. Determine the relative angle of twist between pulley A and B.

(*HINT:* The angle of twist between A and B is the arithmetic sum of the angles between A and C, and C and B with due respect to the direction of twist.)

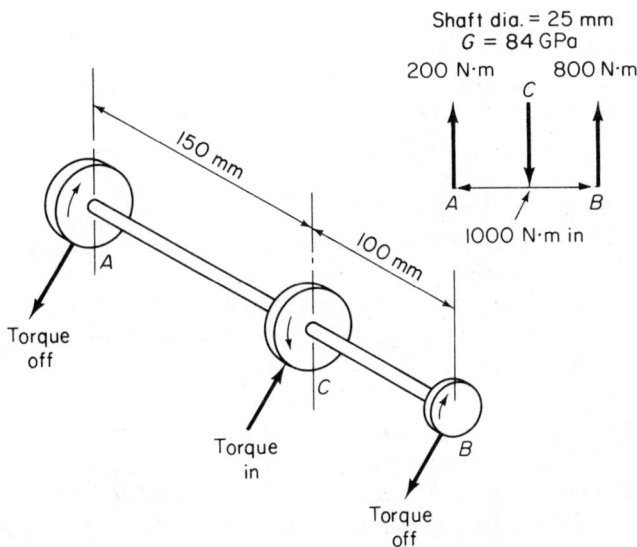

Fig P5.21

5.22 Solve Problem 5.21 if the torque take-off at B is in a direction opposite to that shown.

5.23 A large marine propeller shaft, 15 m long, is to be made of steel ($G = 84$ GPa). If the outside diameter of the shaft is 0.5 m and the inside diameter is 0.3 m, determine its angle of twist when subjected to a torque of 100 000 N·m.

5.24 Show that both the angular deflection and maximum shear stress can be specified for a given shaft if

$$\frac{S}{\theta} = \frac{GC}{L}$$

5.25 The shaft shown in Fig. P5.25 is subjected to a torque of 100 N·m pounds clockwise at A. Determine the torque absorbed by each portion of the shaft and their angular deflection. $G = 77$ GPa.

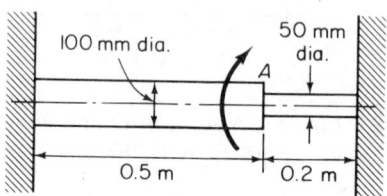

Fig P5.25

5.26 A shaft coupling having a single circle of six 12.5-mm bolts is used to couple two shafts operating at 500 R.P.M. If the bolt circle is 100 mm in diameter, what horsepower can this coupling transmit? The maximum bolt shear stress is 105 MPa.

5.27 A shaft coupling has a single circle of eight, 10-mm bolts. If the bolt circle diameter is 200 mm, what horsepower can be transmitted? Assume that $G = 140$ MPa and that the connecting shafts operate at 75 R.P.M.

5.28 Determine the bolt circle diameter for a coupling to transmit 250 H.P. at 100 R.P.M. if eight 12.5-mm diameter bolts having a design S_s of 84 MPa are used.

5.29 A coupling is required to transmit 100 H.P. at 100 R.P.M. What bolt circle diameter is required if ten, 10-mm-diameter bolts are used having a design shear stress of 50 MPa?

5.30 A coupling uses two rows of bolts on bolt circles of 75-mm diameter and 200-mm diameter, respectively. If six 12.5-mm diameter bolts are used in each row, determine the torque capacity of the coupling. Take the maximum shear stress in the bolts to be 105 MPa.

5.31 A shaft coupling has two rows of bolts on a bolt circle diameter of 50 mm and 100 mm, respectively. If four, 15-mm-diameter bolts are used in each row, determine the torque capacity of the coupling. Use a maximum allowable shear stress of 84 MPa.

5.32 A shaft coupling has twelve 12.5-mm-diameter bolts arranged on a bolt circle with a diameter of 0.2 m. What torque can this coupling transmit if the allowable shear stress is 42 MPa in the bolts?

5.33 A shaft coupling is made with eight 16-mm-diameter bolts located on a bolt circle having a 150-mm-diameter. If the allowable shear stress in the bolts is 56 MPa, determine the torque that can be transmitted. If the shaft rotates at 125 R.P.M., calculate the horsepower that can be transmitted.

5.34 A torsion bar used as a spring consists of a 19-mm-diameter rod, 0.3 m long. If $G = 84$ GPa, determine its spring constant.

5.35 Determine the spring constant for the torsion bar shown in Fig. P5.35 if G is 84 GPa.

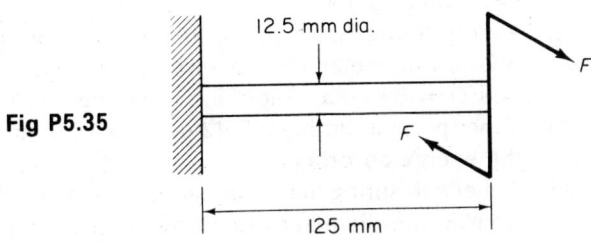

Fig P5.35

5.36 Determine the energy absorbed by a torsion bar spring, 12.5 mm in diameter and 0.3 m long, when subjected to a torque of 50 N·m. Assume $G = 77$ GPa.

5.37 Determine the spring constant of the arrangement shown in Fig. P5.37 if G is 80 GPa.

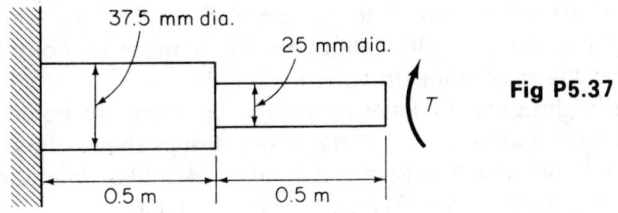

Fig P5.37

5.38 Determine the spring constant for the arrangement shown in Fig. P5.38 if G is 77 GPa.

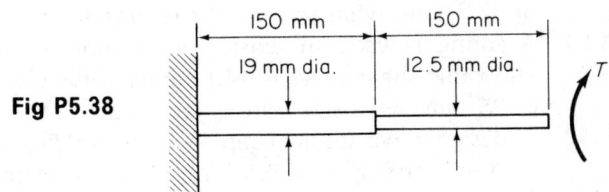

Fig P5.38

5.39 If the arrangement shown in Fig. P5.38 is subjected to a torque of 100 N·m, determine the angular deflection of the free end.

5.40 A solid circular shaft is 1.25 m long and must transmit 50 H.P. at 200 R.P.M. If the permissible shearing stress is 56 MPa and the permissible angular deflection is 1° for the shaft, determine the diameter of the shaft. G is 84 GPa.

5.41 Determine the diameter of the shaft required in Problem 5.38 if G is equal to 150 GPa.

5.42 A coil spring has a mean diameter of 50 mm and is made of a 3.2-mm-diameter circular wire. If the spring has 12 active coils and $G = 84$ GPa, determine its spring constant.

5.43 If the spring in Problem 5.42 deflects 12.5 mm under load, determine the energy absorbed.

5.44 A helical spring has a mean diameter of 100 mm and is made from 5-mm-diameter circular wire. The spring has ten active coils and G is 150 GPa. Determine its spring constant.

5.45 A compression spring is made of 6.5-mm-diameter wire and is coiled over a rod that is 75 mm in diameter. If it is subjected to a load of 450 N, determine the maximum stress in the spring.

5.46 A helical spring is made of 12.5-mm-diameter rod and has a mean coil diameter of 100 mm. If the spring constant is 87.6 kN/m, determine the maximum stress when a 4.5 kN axial extension load is applied to it. $G = 77$ GPa.

5.47 A helical spring has a mean diameter of 125 mm and is made of 16-mm-diameter wire. If the spring deflects 50 mm when a 9 kN weight is hung on it, determine the number of coils in the spring and the maximum stress. G is 80 GPa.

5.48 A spring is designed to have its mean coil diameter equal to 10 times its wire diameter. If the maximum design shear stress is 245 MPa when the spring supports a load of 4.5 kN, determine the mean spring diameter and wire size. G is 77 GPa.

5.49 A spring is designed to have an outer diameter of 63.5 mm and to carry a load of 650 N. Determine the necessary wire size if the allowable shear stress is 175 MPa.

5.50 A tension spring is designed to support a load of 225 N. Using a design allowable shear stress of 175 MPa and a mean coil diameter of 37.5 mm, what size wire is required?

5.51 A spring is used in tension to support a load of 100 N. If the allowable shear stress is 140 MPa and the coil has a mean diameter of 25 mm, what size wire is required?

5.52 A designer wishes to keep the ratio of D_M/d in a spring at 6 to 1. If the spring is a tension spring and supports a load of 330 N with the maximum shear stress not to exceed 175 MPa, determine the wire size and mean diameter of the coil.

5.53 Two springs are arranged as shown in Fig. P5.53. What is the equivalent spring constant of this system?

Chapter 5, Torsion   203

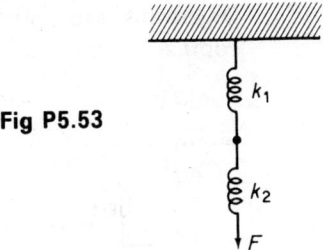

Fig P5.53

5.54 Two springs are arranged as shown in Fig. P5.54. What is the equivalent spring constant of this system?

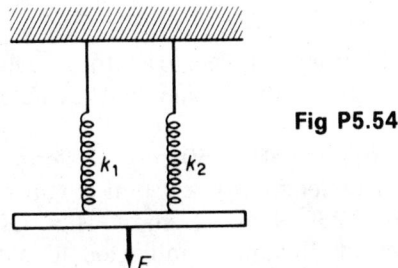

Fig P5.54

5.55 Which configuration (Problem 5.53 or 5.54) does the system shown in Fig. P5.55 correspond to?

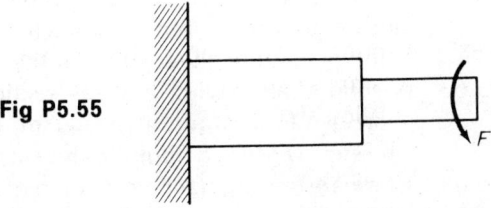

Fig P5.55

5.56 Does the system in Fig. P5.56 correspond to Fig. P5.53 or P5.54?

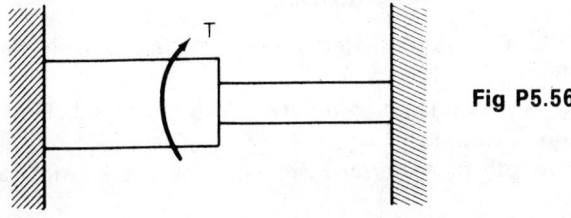

Fig P5.56

5.57 Determine the spring constant of the configurations shown and compare them with those in Problems 5.53 and 5.54.

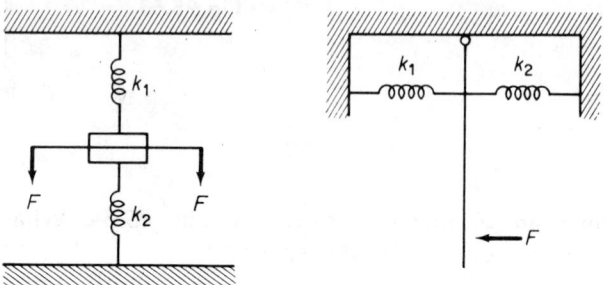

Fig P5.57

5.58 A square shaft having sides of 50 mm is subject to a torque. Determine the maximum torque it can carry if the maximum allowable shear stress is 84 MPa.

5.59 A solid square shaft has sides equal to 50 mm. Determine the maximum stress and angular deflection for a 1 mm length of shaft that is subjected to a torque of 250 N·m. Assume G is 80 GPa.

5.60 A square shaft having sides of 75 mm is subjected to a torque of 3000 N·m. Determine the maximum shear stress in the shaft.

5.61 A thin, hollow, rectangular section whose outside dimensions are h and b, and whose thickness is uniformly t is subject to torsion. Determine the shear stress in the material. Compare your result with the data in Table 5.1.

5.62 A solid shaft having a cross section of an equilateral triangle with sides equal to 50 mm is subjected to a torque of 200 N·m. What is the maximum shear stress in the shaft?

5.63 A solid shaft having a cross section of an equilateral triangle is subjected to a torque of 100 N·m. If the maximum allowable shear stress is 70 MPa, determine the side of the triangle.

5.64 Compare the angular deformation of a circular shaft having a diameter d with the angular deformation of a square shaft having sides equal to d.

REFERENCES

Arges, K. P. and A. E. Palmer. *Mechanics of Materials*. New York: McGraw-Hill Book Co., 1963.

Black, P. H. and O. E. Adams, Jr. *Machine Design*. 3rd ed. New York: McGraw-Hill Book Co., 1968.

Breneman, J. W. *Strength of Materials*. 3rd ed. New York: McGraw-Hill Book Co., 1965.

Faires, V. M. *Design of Machine Elements.* 4th ed. New York: Macmillan Co., 1965.

Higdon, A., E. E. Ohlsen, and W. B. Stiles. *Mechanics of Materials.* 3rd ed. New York: John Wiley & Sons, 1976.

Jensen, A. and H. H. Chenoweth. *Applied Strength of Materials.* 3rd ed. New York: McGraw-Hill Book Co., 1971.

Levinson, Irving J. *Mechanics of Materials.* 2nd ed. Englewood Cliffs, Cliffs, NJ: Prentice-Hall, Inc., 1970.

Olsen, G. A. *Elements of Mechanics of Materials.* 3rd ed. Englewood Cliffs, NJ: Prentice-Hall, Inc., 1974.

Seely, F. B. and N. E. Ensign. *Analytical Mechanics for Engineers.* New York: John Wiley & Sons, Inc.

Seely, F. B. and J. O. Smith. *Advanced Mechanics of Materials.* 2nd ed. New York: John Wiley & Sons, Inc., 1952.

Singer, F. L. *Strength of Materials.* 2nd ed. New York: Harper and Row, 1962.

6

Shear and bending moments in beams

6.1 INTRODUCTION

In the previous chapters of this book, we were primarily concerned with the behavior of structural components subjected to axial forces or torques, not transverse forces. When a member, such as a rivet, was subjected to a transverse force, it was assumed that the member was short and that bending could be neglected. This chapter and the following two chapters concern themselves with a different type of structural member, the *beam*. A beam is a structural member that resists transverse forces normal to its longitudinal axis. These transverse forces cause the beam to bend in the plane of the applied loads, and internal stresses are set up in the material as it resists these loads. Chapters 7 and 8 are concerned with the evaluation of the stresses and deflections in beams and are based upon the work of this chapter. None of the material in this chapter should be novel to the student who has taken a course in statics, and thus it will be presented in a manner that will make our subsequent studies easier. Also, certain conventions are adopted in this chapter that will be retained throughout the rest of this text. Thus it must be emphasized that learning these conventions at this time is an endeavor that will be rewarding not only for itself but also because it is necessary for the mastery of latter portions of this book.

6.2 BEAM SUPPORTS AND BEAM LOADINGS

Before proceeding to study the shear and bending moments in beams, let us first classify beams as to the types of support and the types of loads that are placed on them.

6.2a Supports

The classification of beams as to type of support they have can be done in terms of conventional designations as:

1. Simple beams

A *simple beam* is one that is freely supported (rests on supports) at each end. It is possible to have a roller support at one end and also, at most, one hinged end. Figure 6.1 shows several types of simple beams subject to a single concentrated load. In Fig. 6.1(a) and (b), the support reactions are assumed to be vertical, while in Fig. 6.1(c) the left, hinged support resists the horizontal component of the oblique load and has a horizontal component. In every case we can apply the conditions for equilibrium: $\Sigma F_x = 0$, $\Sigma F_y = 0$, $\Sigma M = 0$ and fully determine the support reactions. For this reason a simple beam is said to be *statically determinate*.

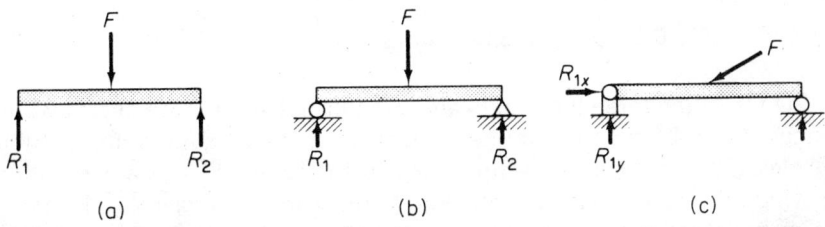

Fig 6.1 Simple beams

2. Cantilever beams

A *cantilever beam* is a beam that is unsupported at one end and is built-in at the other end so as to be able to support moments and rotation. Figure 6.2(a) shows a cantilever beam with a concentrated load at its free end. The conditions that exist at the fixed end are complex, but we can, in principle at least, replace the force system at the fixed end by either a moment and a force as shown in Fig. 6.2(b) or by the support

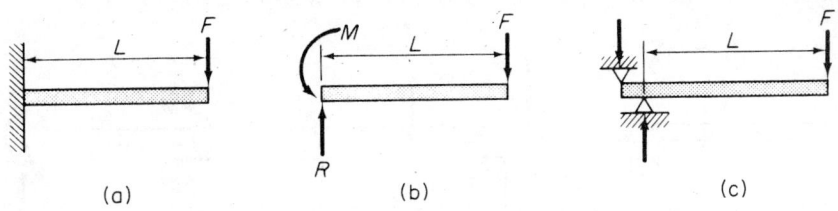

Fig 6.2 Cantilever beams

system shown in Fig. 6.2(c). Figure 6.2(b) represents a free-body diagram of the beam and we can readily see that the three conditions for equilibrium suffice to fully determine the support reactions—this beam is statically determinate.

3. Overhanging beams

An *overhanging beam* is a simple beam with one or both of its ends extending beyond the supports, as shown in Fig. 6.3. All support reactions are assumed to be vertical and can be evaluated from the conditions for static equilibrium. This type of overhanging beam is also statically determinate.

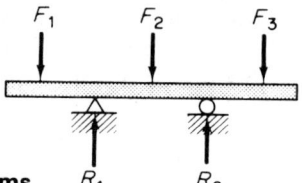

Fig 6.3 Overhanging beams

The foregoing three types of beams are illustrative of those beams whose reactions can be found from the conditions of static equilibrium, and thus we designate them as *statically determinate beams*. There are situations in which the conditions for static equilibrium are insufficient to determine all of the reactions at the supports of a beam. The following three types of beams are illustrative of conditions in which there are either more supports or conditions at the supports than can be evaluated from the three conditions for equilibrium. These beams are called *statically indeterminate beams* and can only be solved by adding additional conditions evaluated from the elastic deformations (deflections) under load. We shall discuss this further in Chapter 8. Continuing with our classification of beam types:

4. Built-in beams

A *built-in beam* has both ends built into masonry with each end fully restrained. It can be considered (schematically) to be a double-ended cantilever beam with the end conditions shown in Fig. 6.4(b) or (c).

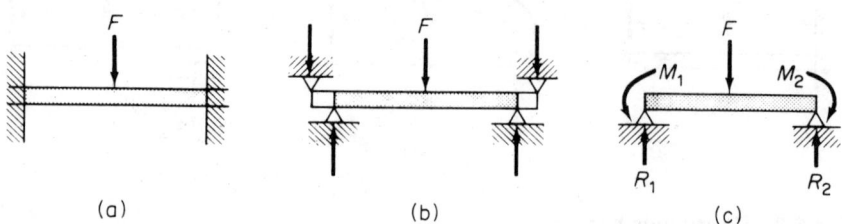

Fig 6.4 Built-in beams

5. Propped beams

The *propped beam* is a cantilever beam with a support under the free end, as shown in Fig. 6.5(a). The equivalent beams, from a force and moment viewpoint, are shown in Fig. 6.5(b) and (c).

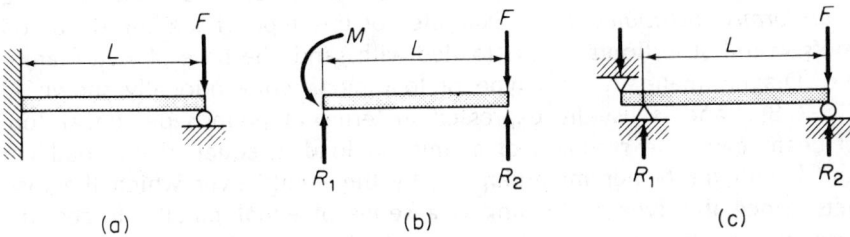

Fig 6.5 Propped beams

6. Continuous beams

The *continuous beam,* a beam having multiple supports, is our last example of statically indeterminate beams. In this case there are more supports than are necessary (from a consideration of statics only) and conditions insufficient to explicitly evaluate all of the reactions.

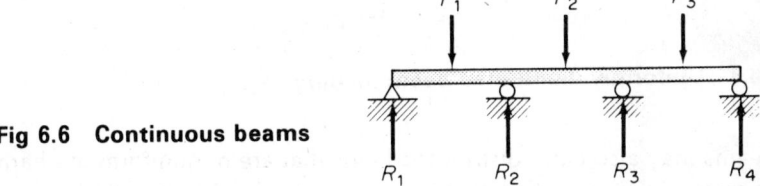

Fig 6.6 Continuous beams

6.2b Loads

In the previous discussion we have used the concept of a concentrated load without defining it. A weight hung from a wire is an example of

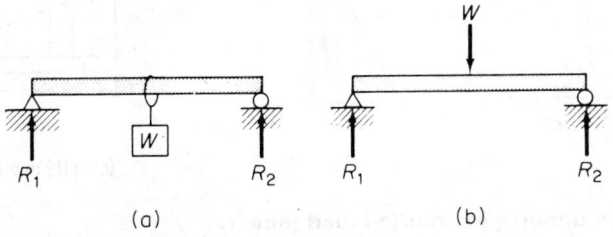

Fig 6.7 A concentrated load

a concentrated load; that is, a *concentrated load* is a load that is being supported by an area sufficiently small so that the load can be considered as being carried by a single point on the beam (as shown in Fig. 6.7). In addition to the concentrated load, we shall deal with other types of beam loads.

A load that is uniformly distributed over a portion of a beam is called a *uniformly distributed load*. Examples of this type of loading occur on roofs, warehouse floors, trucks loaded with sand, the base of water tanks, etc. Diagrammatically, this type of loading is conventionally shown in Fig. 6.8(a) and is usually expressed in terms of pounds per linear foot of beam (w). The resultant of a uniform load is equal to the load per unit length (or N per m) multiplied by the length over which the load acts. Since this type of loading is a series of equal parallel forces, the resultant can be taken to act at the midpoint of the loaded length when determining external moments. Figure 6.8 illustrates these concepts.

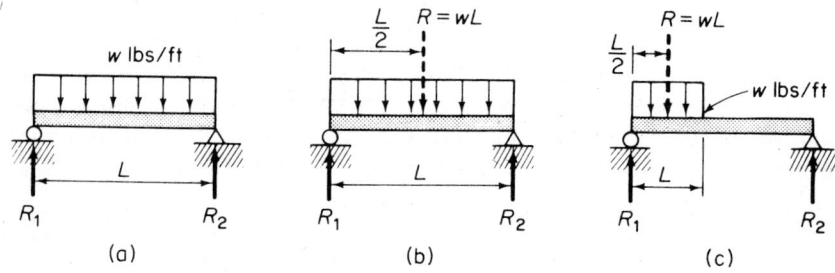

Fig 6.8 Uniformly distributed beam loading

Beams may also carry distributed loads that are nonuniform in character. One example of a well defined, *nonuniform load* is the load carried by a vertical wall (cantilever) used as a dam. In this case the loading varies linearly from zero at the top of the wall to a maximum value at the bottom of the wall. This loading is shown diagrammatically in Fig. 6.9.

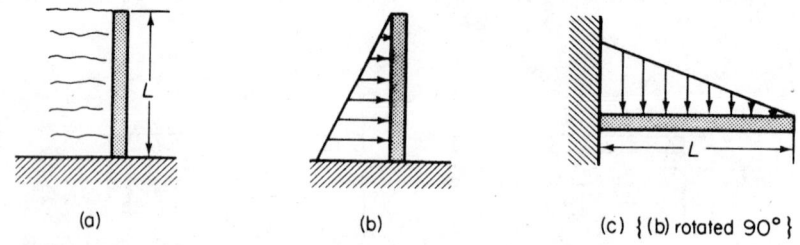

Fig 6.9 Non-uniform distributed load (linear)

Chapter 6, Shear and Bending Moments in Beams   211

The final type of loading that we shall consider is the *moment*. The term moment as used here is synonymous with the term *couple* used in Chapter 5 and is presumably understood from your prior study of statics. An example of this type of loading is shown in Fig. 6.10 where the applied forces constitute a couple whose magnitude is $F \times d$.

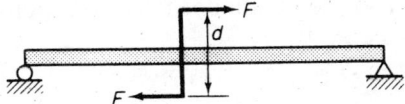

Fig 6.10   Beam subject to pure bending

6.3   SHEAR AND BENDING MOMENT

When a beam is subjected to any of the loadings previously discussed, either singly or in any arbitrary combination, the beam must resist these loads and remain in equilibrium. In order for the beam to remain in equilibrium, an internal force system must exist within the beam to resist the applied forces and moments. In Chapter 7 we shall concern ourselves with the internal stresses in beams arising from these resisting forces; at present our concern is to define the external shear forces and moments applied to beams.

For the present, let us consider the beam shown in Fig. 6.11 to be rigid and inelastic. When subjected to the concentrated load shown in Fig. 6.11(a), a section of the beam will tend to shear out of the beam as shown in Fig. 6.11(b). If we draw a free-body diagram of the section that is being sheared (we are neglecting the effect of bending at this point),

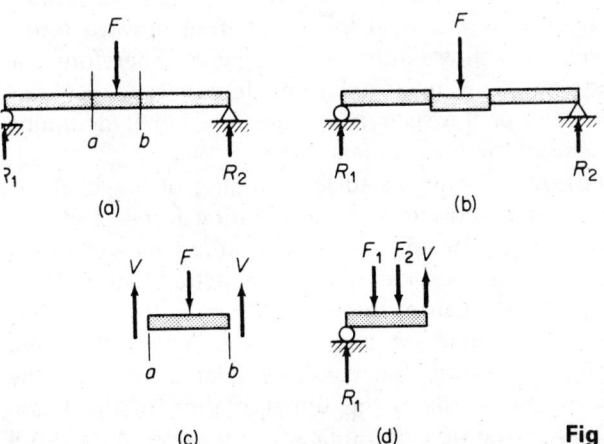

Fig 6.11   Shear of a beam

we obtain the diagram in Fig. 6.11(c), where the resultant of the internal shear forces at each surface being sheared is designated as V. Let us now consider the free-body diagram (still neglecting bending) from the left support to section a of our hypothetical beam [Fig. 6.11(d)]. For equilibrium, $V = R_1$ since $\Sigma F_y = 0$. If in place of the force F, the forces F_1 and F_2 [dotted in Fig. 6.11(d)] were placed on the beam, our argument would be the same, and the conditions for equilibrium yield

$$V = R_1 - F_1 - F_2 \tag{6.1}$$

In other words, the shearing force on any section is equal to the summation of the vertical forces on one side of the section (we have denoted this force to be V). It is important to emphasize that the shearing force V is the resultant external transverse shear force acting on the beam at a given section of the beam. This shearing force is equal in magnitude, but opposite in direction, to the resultant of the internal shearing forces in the beam. By convention and for consistency, we shall adopt a standard designation for positive and negative shear. For this purpose we shall consider the portion of the beam to the *left* of the section in question and state that shear is positive when its effect is to push the portion of the beam to the left upward with respect to the portion on the right. Figure 6.12

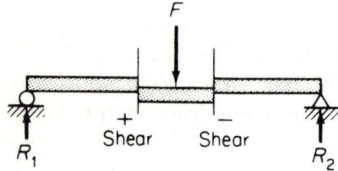

Fig 6.12 Shear convention

illustrates this convention. An alternate (but equivalent) sign convention for vertical shear can be simply obtained by considering upward forces to be positive and downward acting forces to be negative. Therefore, the sign of vertical shear (external as distinguished from internal) is the algebraic sum of the forces to the left of the section in question. Either definition is satisfactory and both should be fully understood.

In the foregoing discussion, we neglected the bending of the beam in the interest of simplicity. Let us now ignore the shearing forces and turn our attention to bending of the beam. In order to study the action of the beam in bending, we must consider its elastic action under load. The force F shown in Fig. 6.13(a) causes the beam to bend. The free-body diagram of Fig. 6.13(b) shows that the beam must provide a moment M at the cut end in order to satisfy the condition that $\Sigma M = 0$. The elastic deformation of the beam places the upper portion of the beam in compression and the lower portion of the beam in tension. As a result

Chapter 6, Shear and Bending Moments in Beams   213

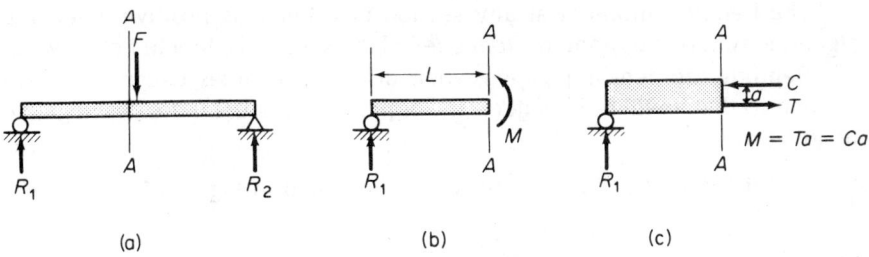

Fig 6.13 Bending of a beam

of this action we have the resultant internal force vectors C and T shown in Fig. 6.13(c). These forces constitute a couple M, which resists the rotation of the beam. In order for equilibrium to exist, this internal couple must equal the external moment acting on the beam at this section. The external moment at section A is

$$M = R_1 L \qquad (6.2)$$

which must equal the internal couple Ta or Ca. At this point we shall define the term *bending moment* to be the algebraic sum of the moments acting on either side of the section in question. Consequently, the internal moment will have the same magnitude but will be of the opposite sense to the bending moment.

Consistency of notation requires us to now define positive and negative bending moment. Several equally correct definitions of positive and negative bending moment are as follows:

1. Bending moment is positive when the curvature of the member causes compression in the fibers of the beam on the loaded side and tension on the side opposite the loads.

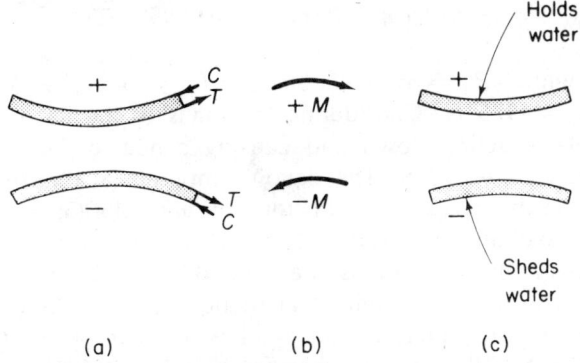

Fig 6.14 Bending moment convention

2. The bending moment at any section of a beam is positive when the algebraic sum of the moments *to the left* of the section is directed clockwise.
3. Bending moment is positive when downward forces cause the beam to bend so that it will hold water—the opposite curvature indicates negative bending moment.

The three definitions given above are illustrated in Fig. 6.14.

Illustrative problem 6.1

Determine the shear and bending moment at sections a-a, b-b, and c-c for the beam shown in Fig. 6.15.

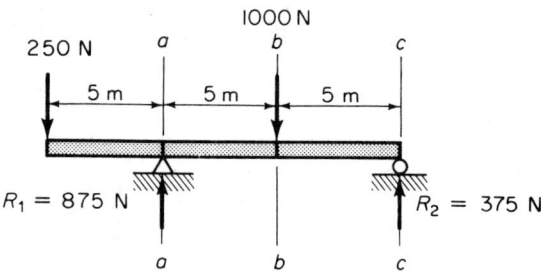

Fig 6.15 Illustrative problem 6.1

Solution:

The first step in the solution is to determine the reactions R_1 and R_2. Taking moments about R_2, we have for $\Sigma M = 0$

$$-250(15) + R_1(10) - 1000(5) = 0$$

and consequently, $R_1 = 875$ N. Since $\Sigma F_y = 0$, $R_1 + R_2 = 250 + 1000$ and $R_2 = 375$ N. Considering the loads to the left of section a-a we have 250 N acting down and causing a negative curvature. Thus, the shear must be -250 N. The bending moment at this section is also negative due to the curvature and is found to be $-250(5) = -1250$ N·m. To the left of section b-b, we find 250 N acting down and 875 N acting up. The shear at this section is therefore $875 - 250 = 625$ N acting up (+ 625). The bending moment at section b-b is $-250(10) + (875)5 = +1875$ N·m. The final section c-c has a shear of $-250 + 875 - 1000$ or -375 N. The bending moment is thus $-250(15) + 875(10) - 1000(5) = 0$. To summarize:

Section	Shear (N)	Bending Moment (N·m)
a-a	−250	−1250
b-b	+625	+1875
c-c	−375	0

6.4 SHEAR AND BENDING MOMENT IN CANTILEVER BEAMS

In order to continuously portray the shear and bending moment at any section of a beam, it is convenient to plot curves of these quantities along the beam. A *shear diagram* is a curve (plotted at every section of the beam) whose ordinate is the shear acting on the beam at any given section in both magnitude and sense. A *bending moment diagram* is a curve (plotted at every section of the beam) whose ordinate is the bending moment acting on the beam at any given section in both magnitude and sense. These two definitions are quite simple, and the interpretations of the resulting shear and bending moment diagrams by engineers are the same universally.

It will be found to be most advantageous to begin our study of shear and bending moment diagrams by examining the cantilever beam under various types of loadings and subsequently to generalize our study to any type of beam subject to arbitrary loadings. The four types of loadings that we had defined earlier in this chapter and that must be considered on cantilever beams are the concentrated load, constant moment, uniformly distributed load, and uniformly varying (triangular) load.

The cantilever beam illustrated in Fig. 6.16 is shown subjected to a concentrated load at its free end. The shear at any section x has a constant value, equal to $-F$. At the free end the shear goes discontinuously from

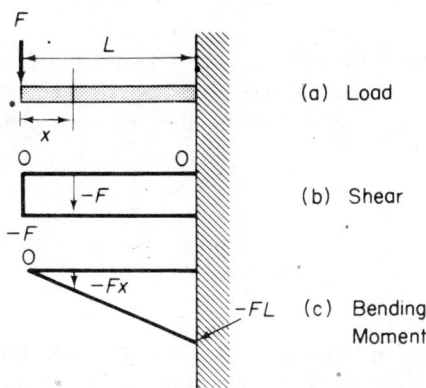

Fig 6.16 **Cantilever with concentrated load**

0 to $-F$, stays constant along the beam, and at the built-in end goes discontinuously from $-F$ to 0. This is depicted in Fig. 6.16(b). From these facts we can generalize the following:

1. At a concentrated load there is an abrupt change in the shear diagram that is portrayed as a vertical discontinuity at the section of the beam where the force is applied.
2. The shear diagram starts at zero on the left and ends at zero at the right. We shall find this to be always true, regardless of the loading and the manner in which the beam is supported.
3. If there are no loads on a section of a beam, the shear remains constant on this section of the beam.

The moment at any section of the beam is $-Fx$, increasing (in absolute magnitude) to a value of $-FL$. The moment diagram is therefore a straight line with constant slope, starting at zero on the left and increasing (in absolute magnitude) to a maximum value of $-FL$ at the right.

The bending moment curve is shown in Fig. 6.16(c). Summarizing the bending moment curve:

1. A concentrated load yields a bending moment with a constant slope that increases linearly from zero at the point of application of the load.
2. The moment of a given concentrated force F is simply Fx, where x is the distance from the point of application of the load to the section in question. It is imperative that the proper attention be given to the sign convention that we have adopted; namely, a positively directed force (upward) yields a positive bending moment, and a negatively directed force (downward) yields a negative bending moment in the cantilever beam.

Illustrative problem 6.2

A cantilever beam is loaded by a downward acting concentrated load of 1000 N at its free end. If the beam is 10 m long, draw the shear and moment diagrams, and determine the maximum shear and maximum bending moment.

Solution:

Figure 6.16 is repeated below with the numerical data for this problem noted on it. As previously discussed, the shear diagram is as shown in

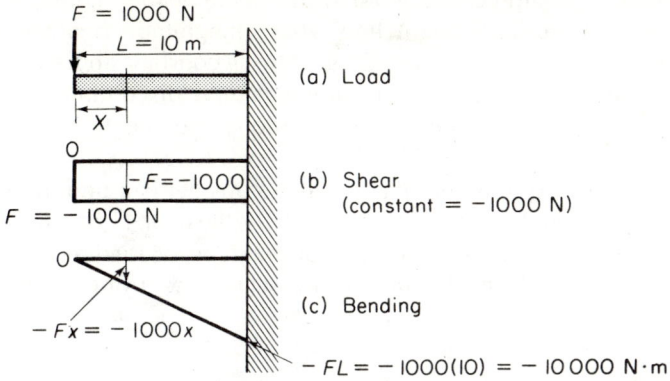

Fig 6.16 Repeated for illustrative problem 6.2

Fig. 6.16(b) with the value of the shear constant over the length of the beam and equal to -1000 N. The free end and the fixed end shears undergo a discontinuous change from zero to -1000 N and from -1000 N to zero, respectively. The moment diagram varies linearly from zero at the free end of the beam to a value of $-1000(10) = -10\,000$ N·m at the fixed end of the beam. At any section x, the bending moment is $-1000(x)$ N·m.

The second type of loading that we shall consider on a cantilever beam is that of a pure moment at its free end. This loading is shown in Fig. 6.17(a) as a moment M or a couple Fd at the free end. At any section

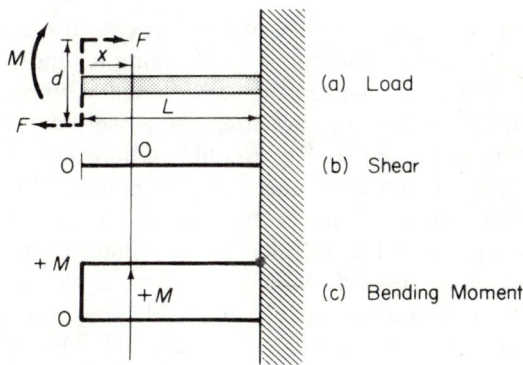

Fig 6.17 Cantilever beam with constant moment

of the beam x there is no net force acting transverse to the beam. Therefore, the shear curve is a horizontal straight line whose magnitude is always zero. The bending moment everywhere on the beam is constant and equal to M, the applied moment. Thus, the bending moment curve discontinuously rises to M at the free end and stays constant along the length of the beam.

The third type of cantilever loading that we shall consider is a uniformly distributed load that is denoted as being equal to w lb/ft or N/m. Roof loads and floor loads are common examples of this type of loading. Fig. 6.18(a) shows the loading on the beam. At any section, x, from the left

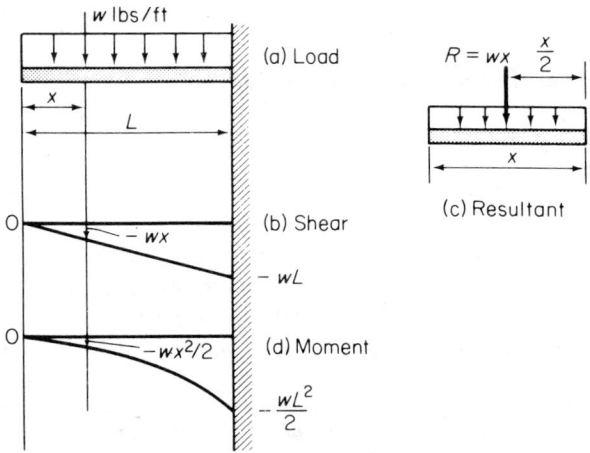

Fig 6.18(a)   Cantilever with uniformly distributed load

end of the beam we find that the total downward force equals the sum of the individual constant loads of w lb/ft acting for the distance x. Thus the shear at section x is $-wx$ and it varies linearly from zero at the left end to a maximum value (in the absolute sense) of $-wL$ at the right built-in end. This is shown in Fig. 6.18(b), where the shear is denoted as negative in accordance with our sign convention. The moment curve can be established by considering the section of the beam to the left of section x as shown in Fig. 6.18(c). The resultant of all of the uniform forces is wx acting downward, with its point of application at $x/2$. Thus, the moment of the uniformly distributed load is $-wx(x/2)$ or $-wx^2/2$, and it varies from zero at the left end to a maximum value of $-wL^2/2$ at the right end of the beam. The equation ax^n is the equation of a parabola where the exponent n denotes the degree or order of the parabola; i.e., ax^3 is a cubic parabola, etc. Note that a uniform load always yields a linear shear variation and a parabolic moment variation.

Illustrative problem 6.3

A cantilever beam is loaded with a uniform load of 500 N/m over 5 metres of its length as shown in Fig. 6.18 (modified for Illustrative Problem 6.3). Plot the shear and moment diagrams, and determine the maximum shear and maximum bending moment.

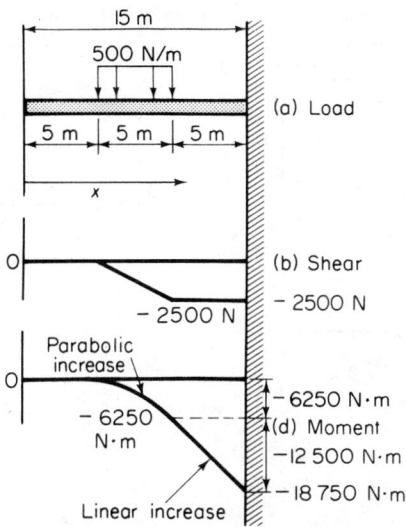

Fig 6.18(b)   Modified for illustrative problem 6.3

Solution:

As is shown in the figure, the shear diagram is zero up to the point of the beginning of the load. From this point to the end of the load, the shear diagram is a linearly varying curve from zero to $-500(5) = -2500$ N·m. From the end of the uniform load to the end of the beam there are no other loads, and the shear remains constant at its maximum numerical value of -2500 N. The moment diagram is zero up to the beginning of the load. For the length of the load, the bending moment varies parabolically to a maximum numerical value of $-500(5)^2/2 = -6250$ N·m. From the end of the uniform load to the end of the beam, the moment increases linearly to its maximum numerical value of $-18\,750$ N·m. [From $x = 10$ m to $x = 15$ m the *increase* in moment is $(x - 10)(-2500)$ N·m.]

As the fourth and last case of cantilever beam loading, let us consider a load that varies linearly from zero at the left (free) end of the beam to a maximum value of w lb/ft or N/m at the built-in end of the beam. This type of loading is also called a triangular load, and it is conventionally portrayed as in Fig. 6.19(a). The shear at any section x can be obtained by reference to Fig. 6.19(b). The loading diagram is a triangle, and the

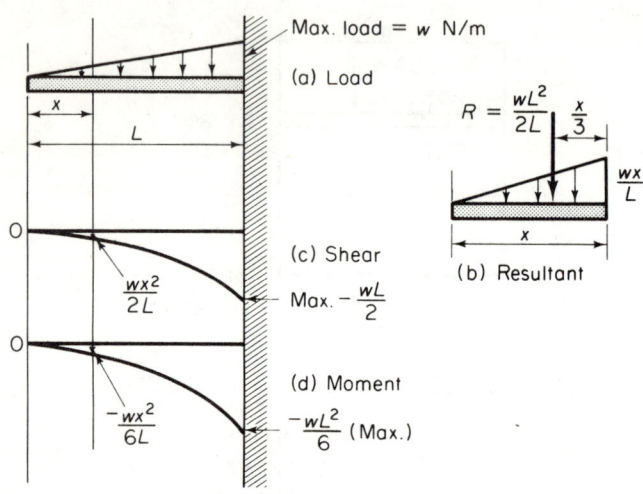

Fig 6.19 Cantilever beam with triangular loading

total downward acting load (shear) is the area of the load triangle having a base x and an altitude wx. The shear is therefore $(-wx/L)(x/2) = wx^2/2L$, varying from zero at the free end to a maximum of $-wL/2$ at the built-in end. The complete shear diagram is shown in Fig. 6.19(c). We can also construct the moment diagram by referring once again to Fig. 6.19(b). The total loading acting to the left of section x is $-wx^2/2L$ and, since the loading can be taken as acting at the location of the center of gravity of the triangular area, $(\frac{x}{3})$ the moment is $(-wx^2/2L)(x/3) = -wx^3/6L$, varying from zero at the left end of the beam to a maximum value of $-wL^2/6$ at the built-in end. The triangular loading therefore gives us a parabolic shear variation and a parabolic bending moment variation. One word of caution must be advanced at this time: For the triangular loading, w is the maximum load per foot or metre, and it occurs at the built-in end.

Chapter 6, Shear and Bending Moments in Beams   221

Table 6.1 summarizes our findings concerning cantilever beams subjected to the four types of loading that we have studied. In addition, we have had to use the properties of several curves in the previous discussion.

Table 6.1   Cantilever Beams

Loading	Load and Shear Curve	Load and Moment Curve
Constant moment	Load: M ⟵ ────── ▨ ; Shear: zero	Load: M ⟵ ────── ▨ ; Moment: + , +M
Concentrated load	Load: F↓ ────── ▨ ; Shear: −F	Load: F↓ ────── ▨, L ; Moment: −FL
Uniform load	Load: w/unit length, L ; Shear: −wL	Load: w/unit length, L ; Moment: $-\dfrac{wL^2}{2}$
Triangular load	Load: Max. w/m, L ; Shear: $-\dfrac{wL}{2}$	Load: Max. w/m, L ; Moment: $-\dfrac{wL^2}{6}$ cubic parabola

Since we shall have occasion to use the properties of the areas bounded by these curves, their properties have been tabulated in Table 6.2. Table 4.1 and Table 6.2 will provide us with all of the properties of areas that we shall require.

Table 6.2 Properties of Areas

Shape and Center of Gravity	Area
Rectangle — $\frac{b}{2} = \bar{x}$	bh
Triangle — $\frac{b}{3} = \bar{x}$	$\frac{bh}{2}$
Second order parabola — $y = ax^2$, Vertex, $\frac{b}{4} = \bar{x}$	$\frac{bh}{3}$
Third order parabola (Cubic Parabola) — $y = ax^3$, Vertex, $\frac{b}{5} = \bar{x}$	$\frac{bh}{4}$
N^{th} order parabola — $y = ax^m$, Vertex, $\frac{b}{m+2} = \bar{x}$	$\frac{bh}{m+1}$
N^{th} order parabola — Vertex, $\frac{b}{2}\left(\frac{m+1}{m+2}\right) = \bar{x}$	$bh\left(\frac{m}{m+1}\right)$

Illustrative problem 6.4

A cantilever beam is subjected to the loading shown in Fig. 6.20. Develop the shear and bending moment curves, and determine the maximum values of the shear and bending moment.

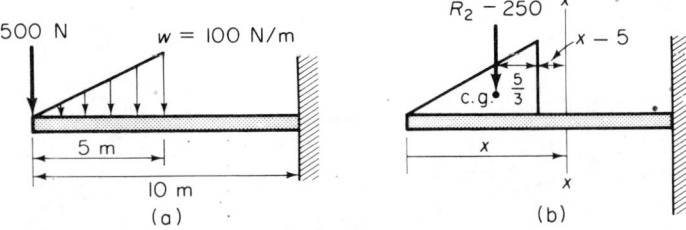

Fig 6.20 Illustrative problem 6.4

Solution:

One method of solving shear and bending moment problems is to consider the effect of each item of loading separately and then to add them together. Consider the 500-N load; its effect at the end of the beam is to produce a shear and bending moment curve such as was obtained in Fig. 6.16. The maximum moment due to this load is $-500(10)$ or -5000 N·m at the built-in end. The shear will be constant and equal to -500 N due to the concentrated load. The triangular load will give us shear and bending moment curves such as the ones obtained in Fig. 6.19, with the maximum value of shear equal to $-100(5)/2 = -250$ N and a moment at the 5-m section from the left equal to $-wL^2/6$ or $-100(5^2)/6 = -2500/6$ N·m. In the portion of the beam where there is no load, the shear will remain constant, but the bending moment will not. The center of gravity (centroid) of the triangular loading is $\frac{5}{3}$ m from the section that is 5 m from the left end of the beam and the resultant load is -250 N. As shown in Fig. 6.20(b), the moment arm will be located at a distance of $x - 5 + \frac{5}{3}$ m when a section to the right of the 5-m location is considered. Therefore the moment will be $-250(x - 5 + \frac{5}{3})$ and it will vary linearly to a maximum (absolute) value of $-5000/3$ N·m. Figure 6.21 has been plotted to show the individual effects as well as the composite shear and bending moment curves.

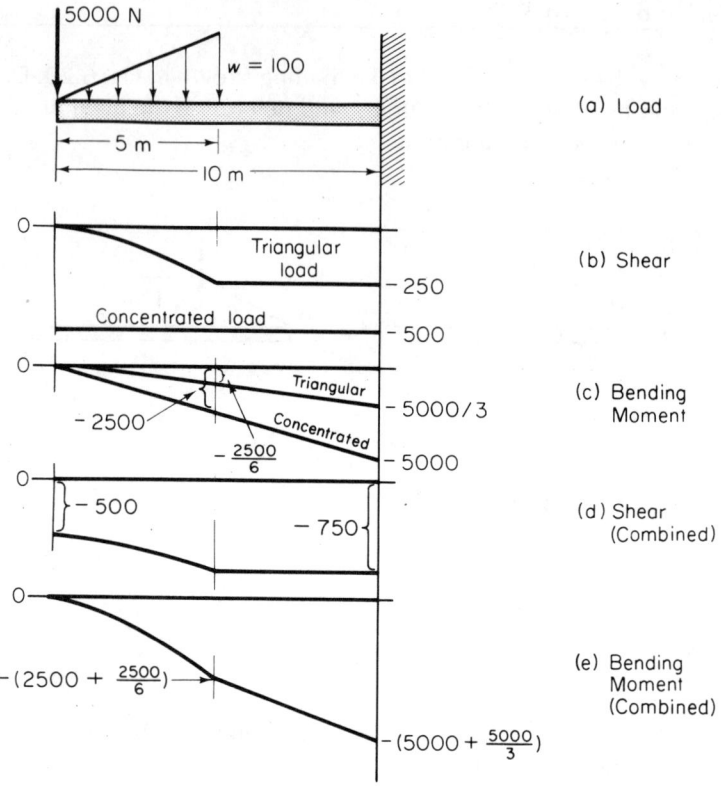

Fig 6.21 Illustrative problem 6.4

6.5 GENERAL RELATION BETWEEN SHEAR AND BENDING MOMENT: APPLICATIONS

The cantilever beam studied in the preceding section represents one of the simplest beams for which shear and bending moment curves can be constructed. More complicated loadings on beams with several different types of supports would require a time-consuming effort if we used the method of step-by-step calculation of these curves. It is possible, however, to derive certain general mathematical relations that we shall find most useful in the construction of these diagrams. In addition, these general expressions will give us an insight into the location of those sections of the beam where the maximum bending moment occurs. For this purpose, let us take a beam subject to an arbitrary loading and consider a free-body

Chapter 6, Shear and Bending Moments in Beams

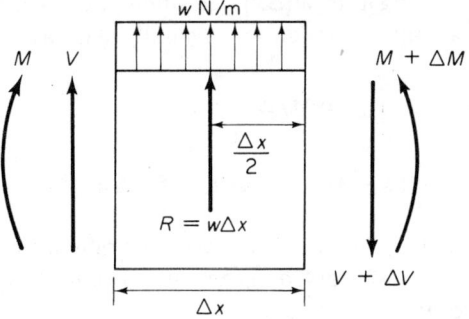

Fig 6.22 General shear and bending

diagram of a short section of the beam of length (Δx). If the section is taken to be sufficiently short, the loading can be effectively considered to be uniform. As shown in Fig. 6.22, there is a shear V and a moment M on the left portion of the element. In the interval Δx, there is a uniformly distributed vertical load causing the shear on the right to increase to $V + \Delta V$ and the moment to increase to $M + \Delta M$. Since the element is in equilibrium, $\Sigma F_y = 0$, which yields

$$V - w\Delta x - (V + \Delta V) = 0 \qquad (6.3)$$

and therefore

$$\frac{\Delta V}{\Delta x} = w \qquad (6.4)$$

Since $\Delta V/\Delta x$ is the slope of the shear diagram, we can generalize and state that *the slope of the shear curve is equal to the intensity (magnitude) of the loading at that point.* If we now rewrite equation (6.4), we obtain

$$V_2 - V_1 = \Sigma \Delta V = \Sigma w \Delta x \qquad (6.5)$$

Equation (6.5) can be interpreted to give us the fact that *the change in shear between any two sections of a beam is equal to the area of the load diagram between these sections.*

The conditions for equilibrium also require that $\Sigma M = 0$; which, for Fig. 6.22, yields equation (6.6), when taken from the right edge of the section.

$$M + V\Delta x + (w\Delta x(\Delta x)/2) - (M + \Delta M) = 0 \qquad (6.6)$$

If we neglect the resulting term $w(\Delta x)^2/2$, since $(\Delta x)^2$ is very small compared to (Δx) (which was initially taken to be small), we have

$$\Delta M / \Delta x = V \qquad (6.7)$$

and

$$M_2 - M_1 = \Sigma \Delta M = \Sigma V \Delta x \qquad (6.8)$$

Equations (6.7) and (6.8) yield the following significant conclusions:
(1) *The slope of the moment diagram at a point, $\Delta M/\Delta x$, must equal the shear at the point.*
(2) *The change in moment between any two sections of a beam is equal to the area of the shear diagram between these two sections.*
(3) *The maximum and minimum moments will occur at the section where the shear is zero.*

We have developed all of the information that will be required to draw any shear or bending moment curve. The following summary is a recapitulation of these points, which should be studied and understood before proceeding further in this chapter.

1. Shear is positive when the resultant of all of the loads to the left of the section acts upward.
2. Bending moment is positive when the algebraic sum of the moments to the left of the section is directed clockwise.
3. The shear curve for a beam that supports only concentrated loads is a series of horizontal and vertical straight lines.
4. The change in shear between two sections of a beam is equal to the sum of the loads between the sections.
5. Where the shear diagram is horizontal, the moment diagram will be a straight line. The only exception that will occur is the case in which there is a couple on a beam, which will cause an abrupt change in the moment diagram that is not indicated by the shear diagram.
6. Discontinuities occur in shear curves at sections where concentrated loads or reactions act.
7. An abrupt change in shear (as at a concentrated load) indicates a discontinuity of the slope of the moment diagram.
8. Under a uniformly distributed load, the shear curve is a sloping straight line.
9. Where the shear diagram is a sloping straight line, the moment diagram will be a segment of a parabola.
10. For concentrated loads, the moment curve is a series of connected straight lines. The slope of the curve changes at each concentrated load or reaction.
11. The ordinate of the moment curve at any section of the beam equals

the area under the shear curve between that section and a section at which the moment is zero.
12. The slope of the moment curve at any section of the beam is equal to the shear at that section.
13. The slope of the moment curve is greatest where the shear is greatest.
14. The maximum moment always occurs at a section of the beam at which the shear is zero.
15. The change in moment between any two sections of a beam is equal to the area of the shear diagram between these two sections.
16. The slope of the shear curve is equal to the intensity of the loading at that point.

To illustrate the preceding discussion, we shall consider a number of problems. As an aid in doing these and other problems, the following procedure will be utilized; it is suggested that the student employ it in doing problems.

1. Calculate the reactions. Where symmetry exists, use it to minimize the arithmetical computations.
2. Determine the shear at sections at which the load changes.
3. Based upon the type of loading, the principles enumerated in the preceding paragraphs, and Tables 6.1 and 6.2, sketch the shear diagram. Accurately determine the sections of zero shear.
4. Compute the bending moment at sections at which the load changes and at sections having zero shear.
5. Using the principles enumerated in the preceding paragraphs and Tables 6.1 and 6.2, sketch the bending moment curve.

Illustrative problem 6.5

Draw the shear and bending moment curves for a beam, simply supported and carrying a uniform load of w lb/ft or N/m.

Solution:

The total load on the beam is wL. By symmetry, the load will be equally divided and the support reactions will each equal $wL/2$. For a uniform load, the slope of the shear curve will be a straight line (item 8) connecting each end reaction. Zero shear (and by item 14, maximum moment) occurs at the center of the beam. We evaluate the maximum moment at the center by taking moments to the left of the center, which gives us $(wL/2)(L/2) - w(L/2)(L/4) = wL^2/8$. By item 9, the resultant bending moment curve will be a parabola having its vertical axis in the middle of the beam.

228   Strength of Materials for Engineering Technology

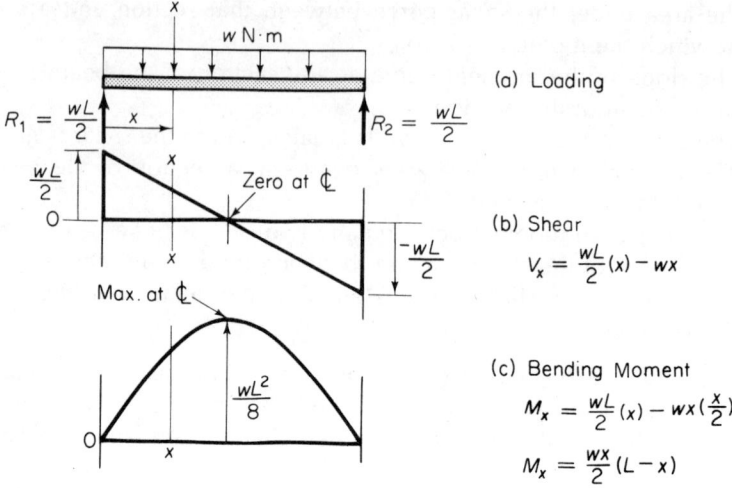

Fig 6.23 Illustrative problem 6.5

The shear and bending moment curves for the simply supported beam with a uniform load are shown in Fig. 6.23. The general expressions for the shear and bending moment at any section x are also given in Fig. 6.23.

Illustrative problem 6.6

Construct the shear and bending moment curves for the beam of Illustrative Problem 6.1.

Solution:

In the solution of Illustrative Problem 6.1, we determined that $R_1 = 875$ lb and $R_2 = 375$ N. Using these values and noting that there are only concentrated loads on this beam, we can draw a shear curve consisting of a series of horizontal and vertical straight lines. The shear will pass through zero at the left support and at the section where the 1000-N load is applied. We must investigate both of these sections to determine the location of the maximum moment. The moment at the left support is

$-250(5) = -1250$ N·m. The moment goes from zero to this value linearly (item 10). At the support there is a shear change (item 6), which causes a change in the slope of the moment curve (item 7). We can calculate

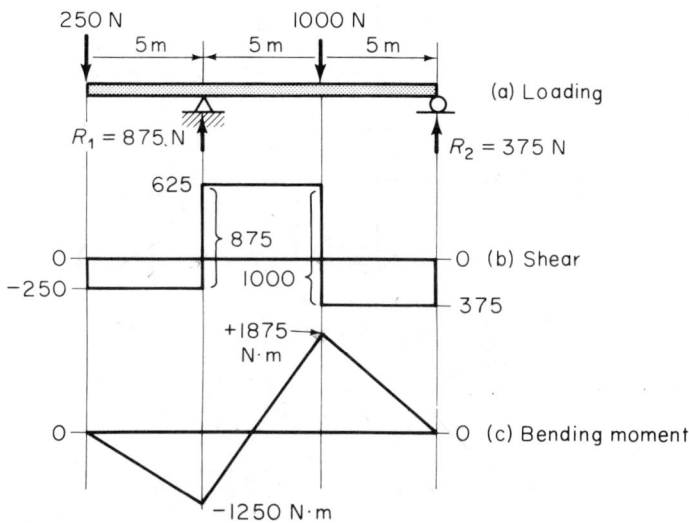

Fig 6.24 Illustrative problem 6.6

the moment at the 1000-N load by taking moments to the left of the load. Thus $-250(10) + 875(5) = 1875$ N·m. We can also obtain the moment at this section by summing the area under the shear curve between the left end (zero moment) and the section (item 11). Thus, $-250(5) + (875 - 250)(5) = 1875$ N·m. The moment curve is therefore a straight line between the value of -1250 N·m at the left support and 1875 N·m under the 1000-N load. The value of $+1875$ N·m is therefore the maximum moment. Since the shear curve between the 1000-N load and the right support is a horizontal line having a negative value, the moment curve will decrease linearly from $+1875$ to 0 N·m at the right support. The complete shear and bending moment curves are shown in Fig. 6.24.

Illustrative problem 6.7

A simply supported beam is loaded with a linear load as shown in Fig. 6.25. Determine the shear and bending moment curves.

230   Strength of Materials for Engineering Technology

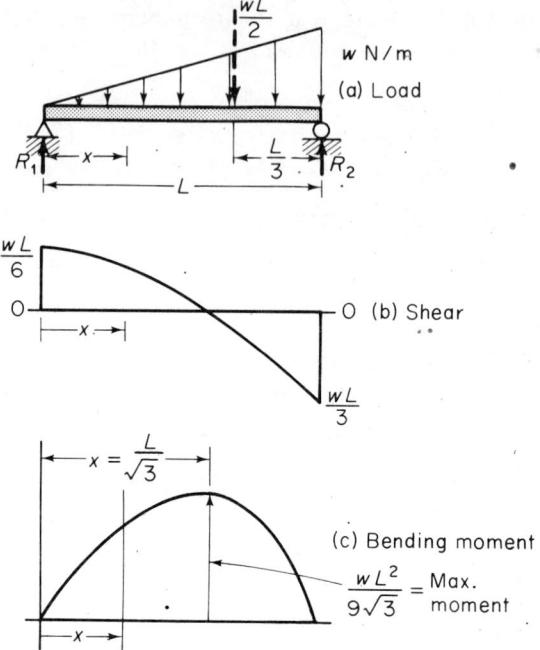

Fig 6.25   Illustrative problem 6.7

Solution:

Our first step in the solution of this problem is to determine the forces at the supports. Since the total load is the area under the load curve, $wL/2$

$$R_1 + R_2 = wL/2 \tag{6.9}$$

In order for the beam to be in equilibrium, $\Sigma M = 0$. The center of gravity of the triangular load from the right-hand side of the beam is found from Table 6.2 to be located at $L/3$. Taking moments about the right-hand support we obtain

$$R_1(L) - (wL/2)(L/3) = 0 \tag{6.10}$$

from which $R_1 = wL/6$. Substituting this value in equation (6.9), $R_2 = wL/3$. The shear at any section x from the left end can be found by subtracting from the concentrated end load $wL/6$, the effect of the triangular load. The effect due to the triangular load is obtained from our discussion of cantilever beams (see Fig. 6.19) as $wx^2/2L$. The shear is therefore

$$V_x = (wL/6) - (wx^2/2L) = \frac{w}{6L}(L^2 - 3x^2) \tag{6.11}$$

Chapter 6, Shear and Bending Moments in Beams   231

We can obtain the location of the maximum moment by determining the section at which the shear is zero. Solving equation (6.11) for V_x equal to zero yields

$$x = L/\sqrt{3}$$

We may find the moment at any section x in an analogous manner, i.e., by subtracting the moment of the triangular load from the moment due to the concentrated load at the support. Figure 6.19 gives us the effect of the triangular load as $wx^3/6L$. The moment at x is therefore

$$M_x = (wL/6)(x) - (wx^3/6L) \qquad (6.12)$$

The maximum moment is obtained by placing $x = L/\sqrt{3}$ in equation (6.12), which yields a maximum moment of $wL^2/9\sqrt{3}$.

Illustrative problem 6.8

Construct the shear and bending moment curves for the beam shown in Fig. 6.26.

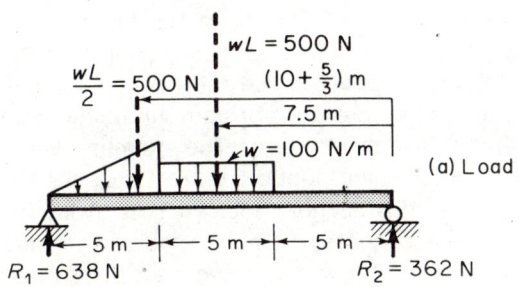

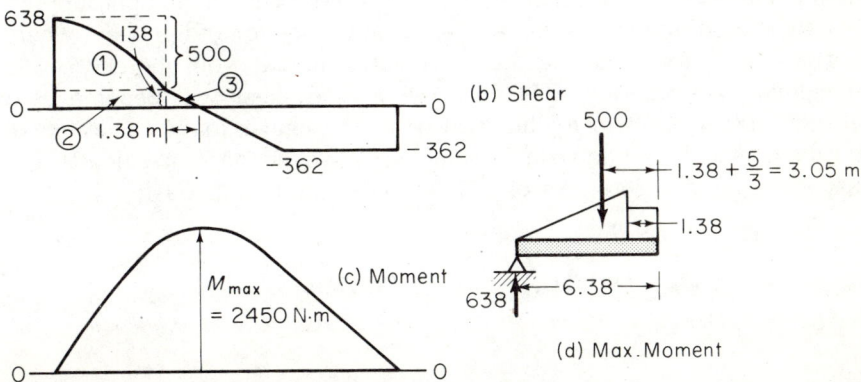

Fig 6.26 Illustrative problem 6.8

Solution:

As before, we shall initially compute the value of each of the support reactions. The area under the load curve for the triangular load is $wL/2 = 200(5)/2 = 500$ N, and the area under the uniform load is $100(5) = 500$ N. Thus

$$R_1 + R_2 = 1000 \text{ N}$$

Taking moments about the right support and noting that the centroid of the triangular load is $(10 + \frac{5}{3})$ m from the right support, and the centroid of the uniform load is 7.5 m from the right support, yields

$$15R_1 = 500(10 + \tfrac{5}{3}) + 500(7.5)$$

Solving, we get R_1 equal to 638 N and R_2 equal to 362 N.

Starting at the left support, we have a positive value of shear of 638 N. The triangular load causes a parabolic decrease in shear of 500 N, which gives us 138 N at the section located 5 m from the left support. The shear decreases linearly at the rate of 100 N/m to a value of -362 N at the section 10 m from the left support. Since there is no load in the beam over its last 5 m, the shear remains constant until the right support is reached at which point it goes to zero. The moment curve increases as a cubic parabola from the left support until it reaches the section where the load changes. At this section, located 5 m from the left support, the slope of the curve changes since it is now a second order parabola. The moment curve reaches a maximum value and starts to decrease. In the portion of the beam where the shear is constant, the moment undergoes a linear change until it reaches zero at the right support. The location of the maximum moment is at the section of the beam where the shear is zero. Since the shear decreases linearly from the end of the triangular load where the load is 138 N, the shear will become zero at $138/100 = 1.38$ m to the right of the triangular load. The value of the moment at this section can be obtained by two independent calculations. Referring to Fig. 6.26(d), we obtain the maximum moment as

$$6.38(638) - 500(3.05) - [1.38(100)(1.38)/2] = 2450 \text{ N} \cdot \text{m}$$

An alternate, independent calculation of the maximum bending moment is obtained by calculating the area under the shear curve between the left support (where the moment is zero) and the section under consideration. Using Table 6.2 and the sections 1, 2, and 3 indicated in Fig. 6.26(b):

$$
\begin{aligned}
&\text{For 1} \quad \text{area} = \tfrac{2}{3}(bh) = \tfrac{2}{3}(500)5 = 1667 \\
&\text{For 2} \quad \text{area} = 138(5) \hphantom{= \tfrac{2}{3}(500)5} = 690 \\
&\text{For 3} \quad \text{area} = 138(1.38)/2 \hphantom{= \tfrac{2}{3}} = 95 \\
&\hphantom{For 3 area = 138(1.38)/2 =00} \overline{2452 \text{ N} \cdot \text{m}}
\end{aligned}
$$

Both values obviously are in excellent agreement.

It will be recalled that Illustrative Problem 6.4 was solved by considering each load independently and then summing these individual effects to arrive at the final result. This procedure has been used (in principle) several times in other illustrative problems and is known as the *superposition method*. This method can sometimes simplify a specific problem and, if properly executed, will always lead to the complete shear and bending moment diagrams for any beam. Illustrative Problem 6.9* illustrates the application of this method applied to a beam subjected to both concentrated and uniform loads.

Illustrative problem 6.9*

Draw the shear and bending moment diagrams for the beam shown in Fig. 6.27 using the superposition method.

---

*Used with permission from *Mechanics of Materials* by H. D. Conway, Prentice-Hall, Inc., Englewood Cliffs, N.J., 1950. Revised to SI units.

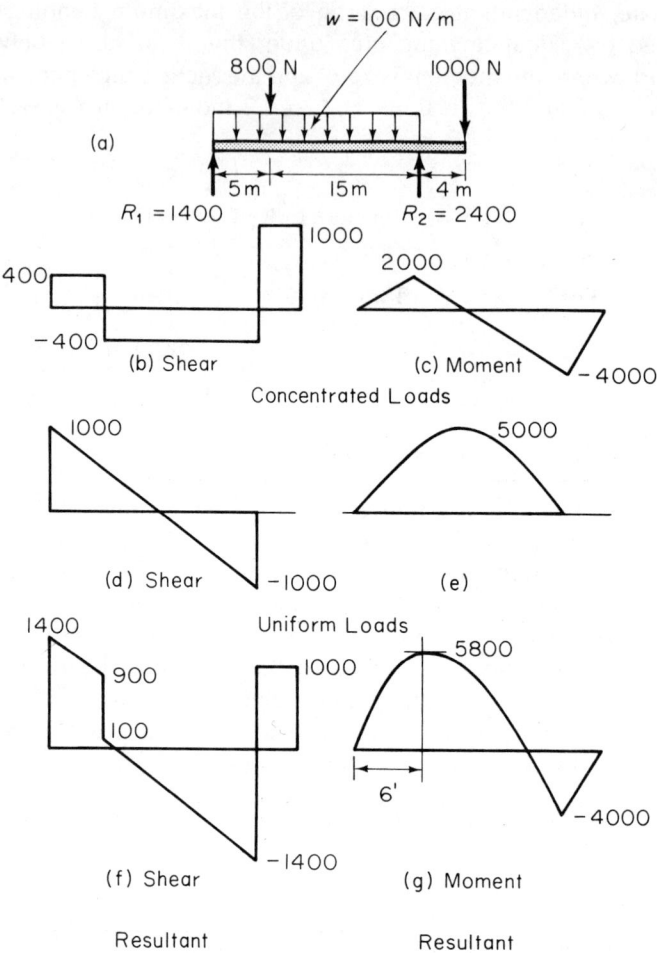

Fig 6.27 Illustrative problem 6.9

Solution:

The reactions at the supports are calculated as in previous problems. The next step in the solution is to construct the shear and bending moment curves for the concentrated loads, which gives us Fig. 6.27(b) and (c), respectively. The shear and bending moment curves for the distributed loads are plotted in Fig. 6.27(d) and (e), respectively. The ordinates in Fig. 6.27(b) and (d) are then added, and the resultant shear diagram shown in Fig. 6.27(f) is drawn. Similarly, the ordinates in Fig. 6.27(c) and (e) are added, and the resultant moment diagram shown in Fig. 6.27(g) is obtained.

In Fig. 6.27(f) the shear is zero at a cross section of the beam located 6 m from the left end. It follows that the bending moment is a maximum at this section. The value of the bending moment at this section is

$$1400(6) - 800(1) - [100(6^2)/2] = 5800 \text{ N} \cdot \text{m}$$

6.6 MOVING LOADS

A unique situation occurs when we consider a vehicle such as a train, a truck, or other moving load as it passes over a beam. The problem of finding the maximum moment is complicated by the fact that as a load moves along a beam, the moment caused by this load changes with its position. The same situation is true of all of the loads that move across the span as a group. Additionally, it is necessary to evaluate the maximum shear, which varies with the position of the load group on the span. Let us examine the system of loads shown in Fig. 6.28. The loads F_1 and

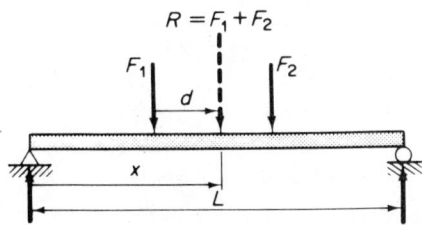

Fig 6.28 Moving load on a beam

F_2 move across the span as a fixed unit with the distance between them kept constant. The bending moment is maximum under one of the loads since the shear must go through zero under one of these concentrated loads. On Fig. 6.28 the resultant of these loads is indicated as being a variable distance x from the left end. The magnitude of R is $F_1 + F_2$, and its location is determined by taking moments about F_1; i.e.,

$$F_2 a = Rd = (F_1 + F_2)d \quad (6.13)$$

and

$$d = \frac{F_2 a}{F_1 + F_2} \quad (6.14)$$

If we now consider that the only load on the beam is R, we can obtain the reactions readily. The left reaction is

$$R_1 = \left(\frac{L-x}{L}\right)R \qquad (6.15)$$

The moment under the load F_1 is M_1, which is found to be

$$M_1 = R_1(x-d) = \left(\frac{L-x}{L}\right)(x-d)R \qquad (6.16)$$

If we expand the terms on the right-hand side of equation (6.16)

$$M_1 = -\frac{R}{L}[x^2 - (L+d)x + Ld] \qquad (6.17)$$

Since we wish to determine when M_1 is maximum, it is necessary to determine when the quadratic expression in the brackets is maximum. It will be recalled that the solution of a quadratic equation is

$$x = \frac{-b}{2a} \pm \frac{\sqrt{b^2 - 4ac}}{2a} \qquad (6.18)$$

when the equation is of the form

$$ax^2 + bx + c = 0 \qquad (6.19)$$

The only time that x has a single real value is when $x = -b/2a$. This value of x yields the maximum value of M_1 when substituted into equation (6.17). Therefore for maximum moment x is

$$x = \frac{L}{2} + \frac{d}{2} \qquad (6.20)$$

We may generalize this result to state that the maximum bending moment under a particular load occurs when the bisector of the distance between that load and the resultant of all of the loads then on the span coincides with the center of the span. The maximum shear force occurs at one of the supports and is equal to the maximum support force. This occurs at the left support when the leftmost load is over it or at the right support when the rightmost load is over the right support. Of these two possibilities, the greater reaction will occur at that support nearest the resultant of the group of loads. If one or more of the loads falls off the span, it is necessary to investigate this condition separately.

Illustrative problem 6.10

A truck having a wheelbase of 15 m carries a load of 12000 N distributed between the front and back wheels as shown in Fig. 6.29. Determine the maximum moment and the maximum shear as the truck moves over the span.

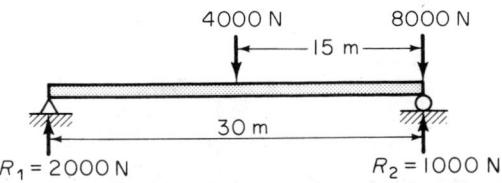

Fig 6.29 Illustrative problem 6.10

Solution:

The resultant of the two axle loads is 12 000 N, which is located at $d = (\frac{8000}{12000})(15) = 10$ m from the left load. The position of the wheels necessary for the maximum moment to occur under the left load is as shown in Fig. 6.30. Taking moments about R_1

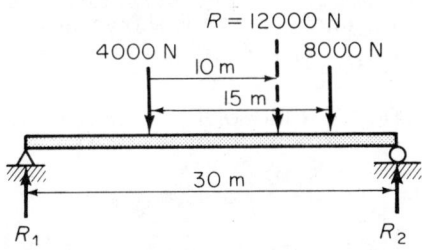

Fig 6.30 Illustrative problem 6.10

$$20(12000) - 30 R_2 = 0$$

which gives us R_2 equal to 8000 N and R_1 equal to 4000 N. The moment about the 4000-N load is therefore $4000(10) = 40\ 000$ N·m, the maximum value for this load.

Now consider the 8000-N load. For maximum moment due to this load, the truck must have moved to the position shown in Fig. 6.31. Again taking moments about R_1

$$12\,000(12.5) - 30\,R_2 = 0$$

which gives us $R_2 = 5000$ N and $R_1 = 7000$ N. The moment about the 8000-N load is $17.5(7000) - 4000(15) = 62500$ N·m. The maximum

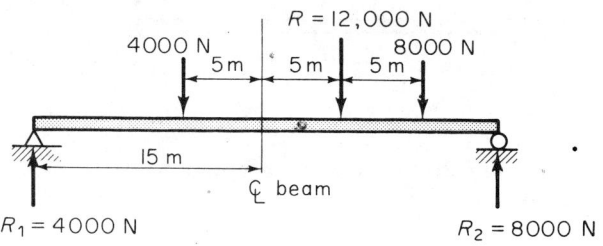

Fig 6.31 Illustrative problem 6.10

moment due to the 8000-N load is therefore the maximum moment that this moving load group imposes on the beam.

Since the resultant of the load group is nearest the 8000-N load, the maximum shear will occur at the right support when the 8000-N load is over the support. This configuration is shown in Fig. 6.32. Once again taking moments about R_1

$$4000(15) + 8000(30) - 30(R_2) = 0$$

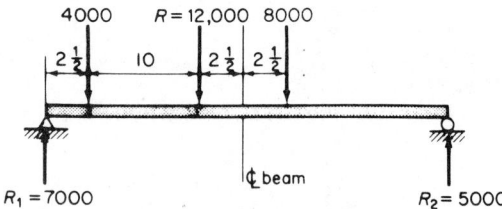

Fig 6.32 Illustrative problem 6.10

and R_2 is found to equal 10 000 N. This is the maximum shear on the beam due to the truck moving over the beam.

6.7 CLOSURE

Having defined the terms shear and bending moment, we have been able to determine these quantities at any section of a beam using the principles of statics. In addition, general relations have been obtained relating load, shear, and bending moment. The end result of these studies has been the development of shear and bending moment diagrams. These diagrams permit one to readily identify the critical sections of a beam and to evaluate the shear and bending moment at these sections.

In the introduction to this chapter it was noted that the information developed herein would be used in later sections of this book. This point cannot be stressed strongly enough—*this material must be thoroughly understood before proceeding.* Merely doing problems by rote will not give the student the mastery required for further studies of the stresses and deflections in beams.

PROBLEMS

6.1 A concentrated load is placed on a simple beam as is shown in Fig. P6.1. Determine the reactions R_1 and R_2. Also, determine the shear and bending moment at section x-x if this section is to the left of the force as shown in the figure.

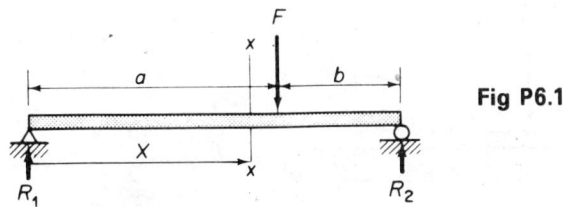

Fig P6.1

6.2 Determine the shear and bending moment at section x-x if it is to the right of the load as shown in Fig. P6.2.

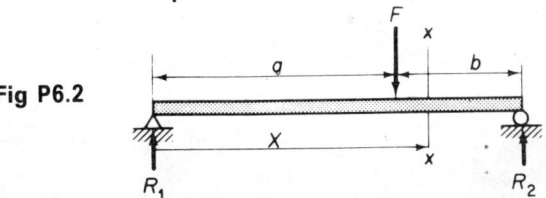

Fig P6.2

6.3 A cantilever beam carries the load shown. Determine the shear and bending moment at the built-in end.

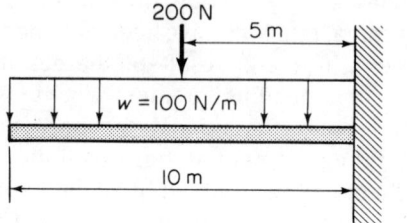

Fig P6.3

6.4 A cantilever beam carries the load shown. Determine the shear and bending moment at the built-in end.

Fig P6.4

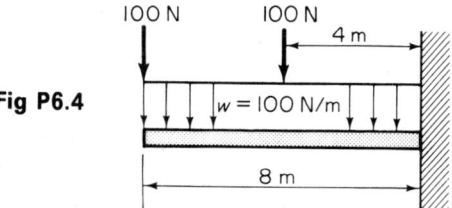

6.5 A cantilever beam carries the load shown. Calculate the shear and bending moment at the built-in end.

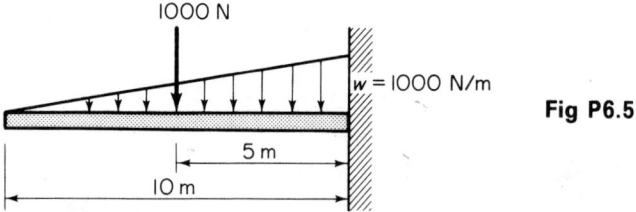

Fig P6.5

6.6 For the cantilever beam shown, calculate the shear and bending moment at the built-in end.

Fig P6.6

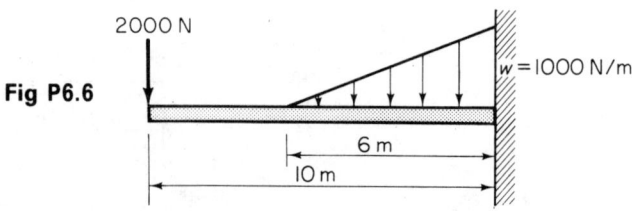

6.7 Determine the shear and bending moment at the built-in end of the cantilever beam shown in Fig. P6.7.

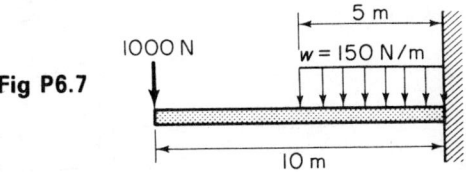

Fig P6.7

6.8 Determine the shear and bending moment at the built-in end of the cantilever beam shown in Fig. P6.8.

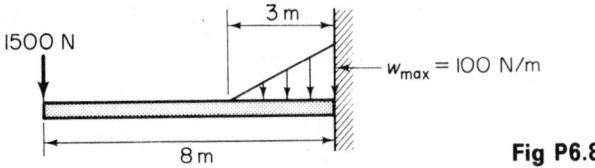

Fig P6.8

6.9 A cantilever beam carries a trapezoidal load. Determine the shear and bending moment at the built-in end.

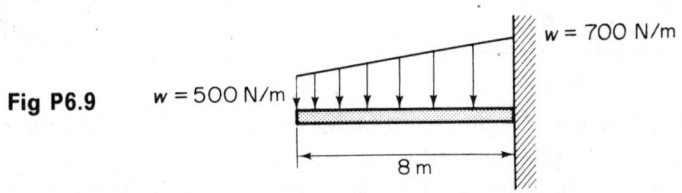

Fig P6.9

6.10 A simple beam carries a trapezoidal loading as shown in Fig. P6.10. Calculate the support reactions R_1 and R_2. Determine the location of the section of zero shear.

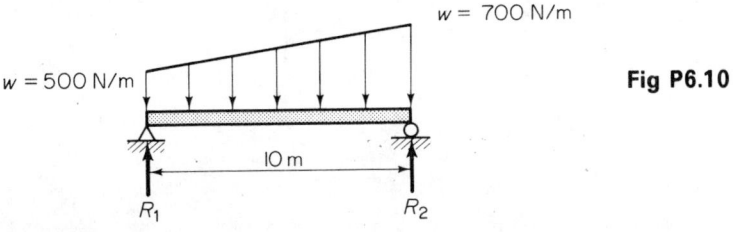

Fig P6.10

6.11 Determine the shear and bending moment at the built-in end of the cantilever beam shown in Fig. P6.11.

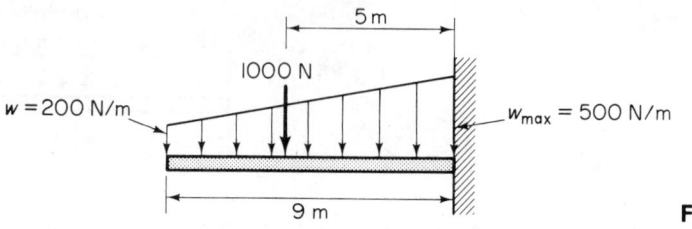

Fig P6.11

6.12 A cantilever beam has a concentrated load at its free end and a distributed load over a portion of its length. Determine the shear and bending moment at the built-in end.

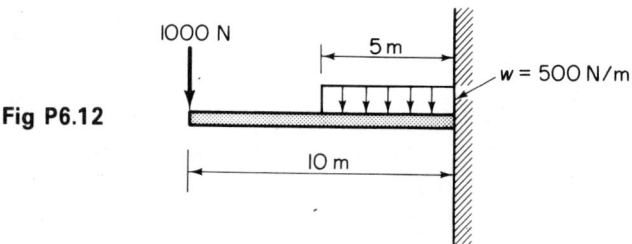

Fig P6.12

6.13 Draw the shear and bending moment curves for the cantilever beam of Problem 6.12.
6.14 Determine R_1 and R_2, and plot the shear and bending moment curves for the beam shown in Fig. P6.14.

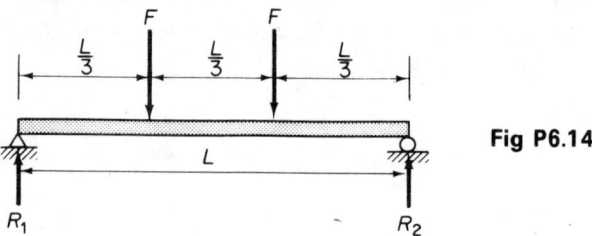

Fig P6.14

6.15 Determine the reactions and draw the shear and moment diagrams for the beam shown in Fig. P6.15.

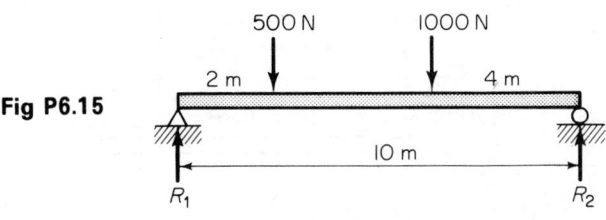

Fig P6.15

6.16 Determine the reactions and draw the shear and moment diagrams for the beam shown in Fig. P6.16.

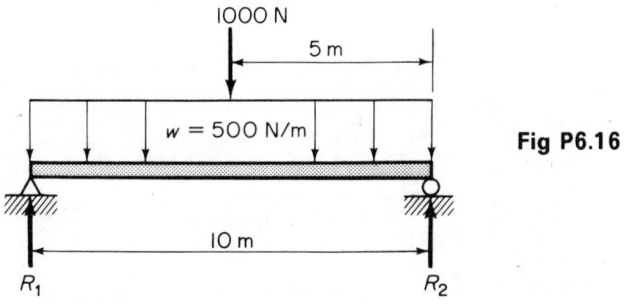

Fig P6.16

6.17 Determine the reactions and draw the shear and moment diagrams for the beam shown in Fig. P6.17.

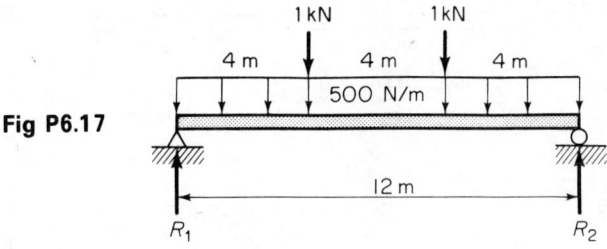

Fig P6.17

6.18 Determine the reactions and draw the shear and moment diagrams for the beam shown in Fig. P6.18.

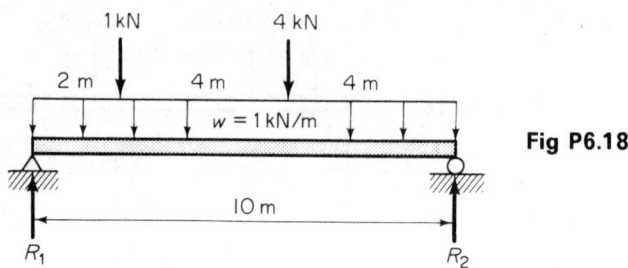

Fig P6.18

6.19 Determine the support reactions, the maximum moment, and the location of the maximum moment for the simple beam shown in Fig. P6.19.

Fig P6.19

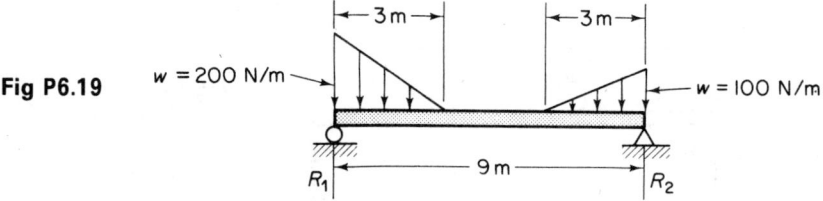

6.20 Determine the reactions, and draw the shear and moment diagrams for the beam shown in Fig. P6.20.

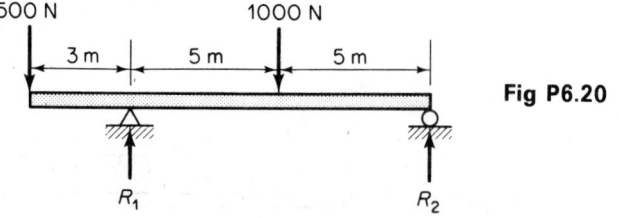

Fig P6.20

6.21 Determine the support reactions, the maximum moment, and the location of the maximum moment for the simple beam shown in Fig. P6.21.

Chapter 6, Shear and Bending Moments in Beams   245

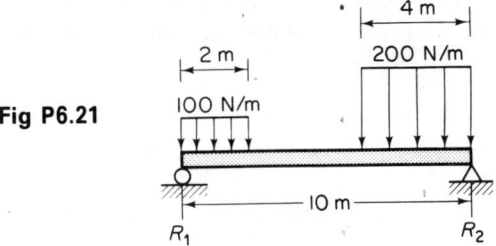

Fig P6.21

6.22 Determine the support reactions, the maximum moment, and the location of the maximum moment for the simple beam shown in Fig. 16.22.

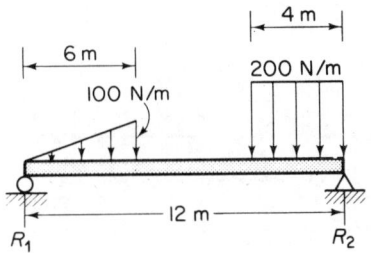

Fig P6.22

6.23 Determine the support reactions, the maximum moment, and the location of the maximum moment for the simple beam shown in Fig. P6.23.

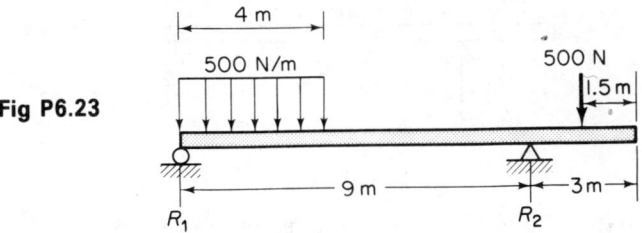

Fig P6.23

6.24 Determine the support reactions, the maximum moment, and the location of the maximum moment for the simple beam shown in Fig. P6.24.

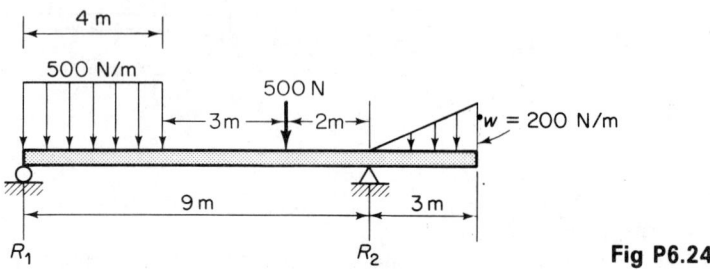

Fig P6.24

6.25 Determine the reactions, and draw the shear and bending moment curves for the beam shown in Fig. P6.25.

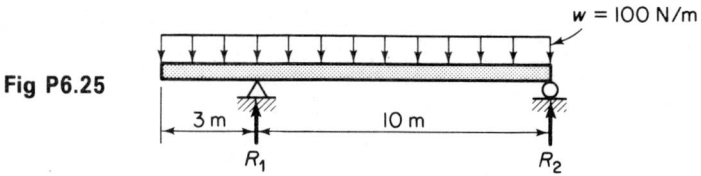

Fig P6.25

6.26 Determine the reactions and draw the shear and bending moment curves for the beam shown in the Fig. P6.26.

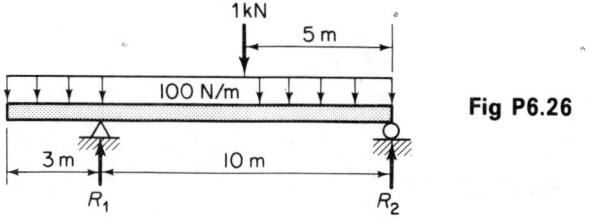

Fig P6.26

6.27 Determine the reactions and draw the shear and bending moment curves for the beam shown in Fig. P6.27.

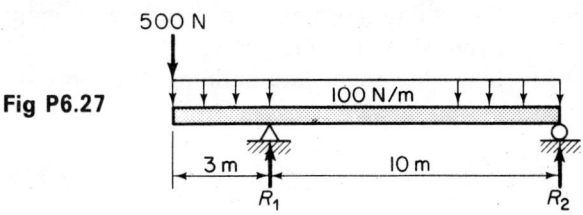

Fig P6.27

6.28 Determine the reactions, and draw the shear and bending moment curves for the beam shown in Fig. P6.28.

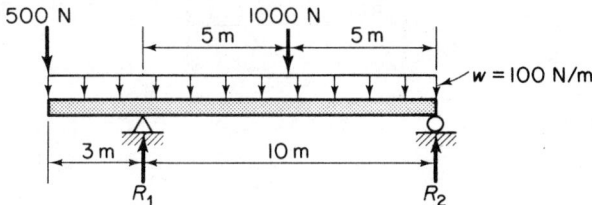

Fig P6.28

6.29 Determine the reactions, and draw the shear and moment diagrams for the beam shown in Fig. P6.29 if the couple is placed at the center of the beam.

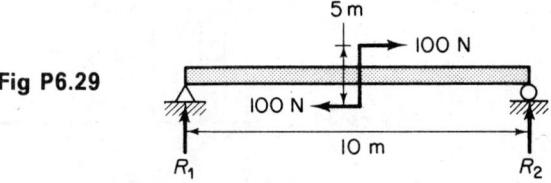

Fig P6.29

6.30 Draw the shear and bending moment curves for the beam shown in Fig. P6.30.

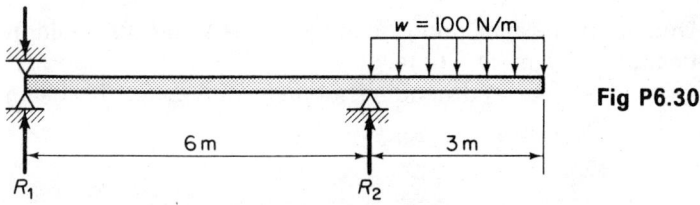

Fig P6.30

6.31 Determine the maximum bending moment and its location for the beam shown in Fig. P6.30.

6.32 Draw the shear and bending moment curves for the beam shown in Fig. P6.32.

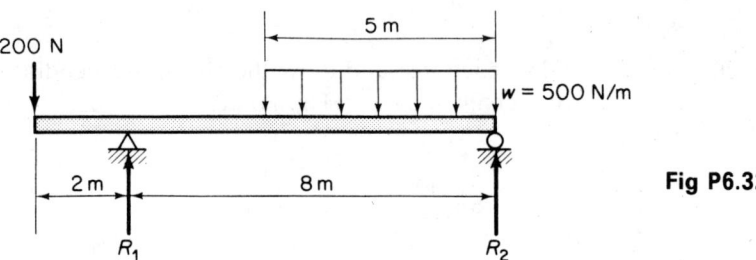

Fig P6.32

6.33 Determine the maximum bending moment and its location for the beam shown in Fig. P6.32.

6.34 Draw the shear and bending moment curves for the beam shown in Fig. P6.34.

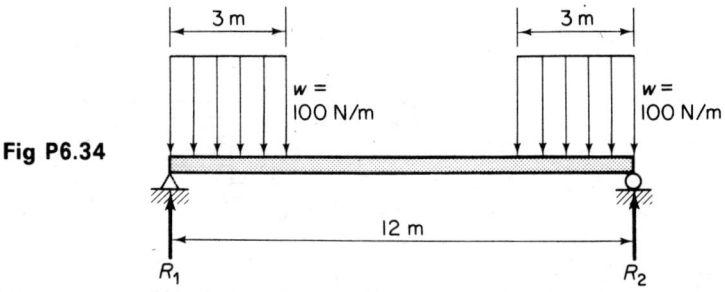

Fig P6.34

6.35 Determine the maximum bending moment and its location for the beam shown in Fig. P6.34.

6.36 Draw the shear and bending moment curves for the beam shown in Fig. P6.36.

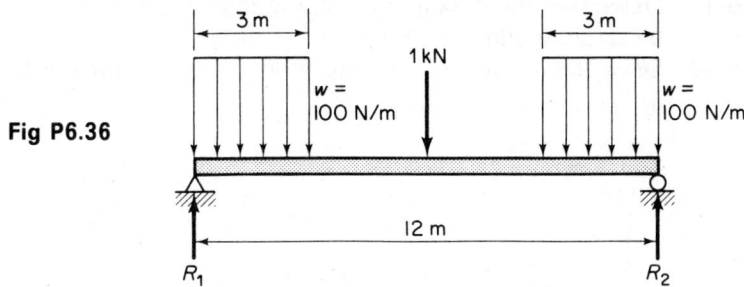

Fig P6.36

6.37 Determine the maximum bending moment and its location for the beam shown in Fig. P6.36.

6.38 Draw the shear and bending moment curves for the beam shown in Fig. P6.38.

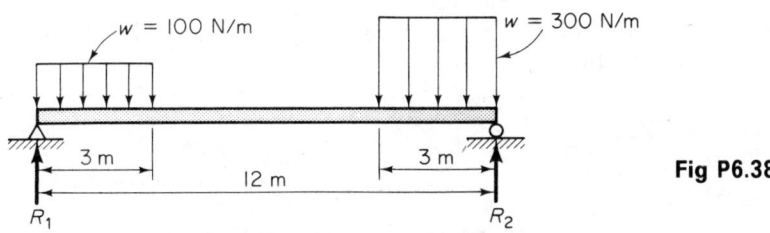

Fig P6.38

6.39 Determine the maximum bending moment and its location for the beam shown in Fig. P6.38.

6.40 Draw the shear and bending moment curves for the beam shown in Fig. P6.40.

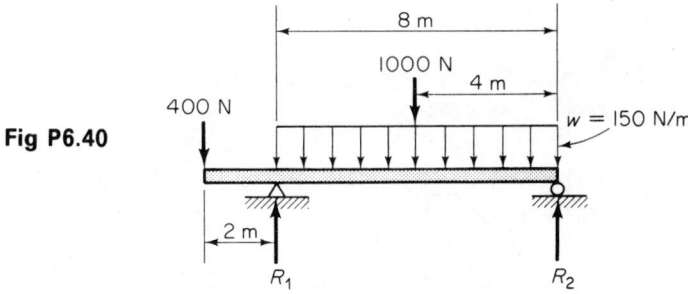

Fig P6.40

6.41 Determine the maximum bending moment and its location for the beam shown in Fig. P6.40.

6.42 Draw the shear and bending moment curves for the beam shown in Fig. P6.42.

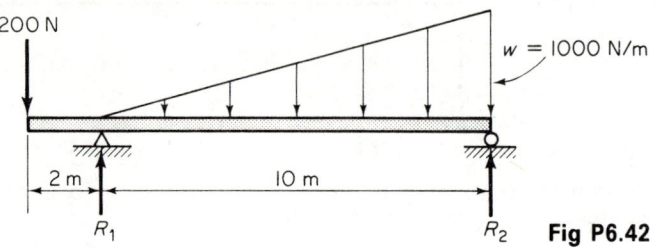

Fig P6.42

6.43 Determine the maximum bending moment and its location for the beam shown in Fig. P6.42.

6.44 Determine the maximum bending moment in the beam shown in Fig. P6.44.

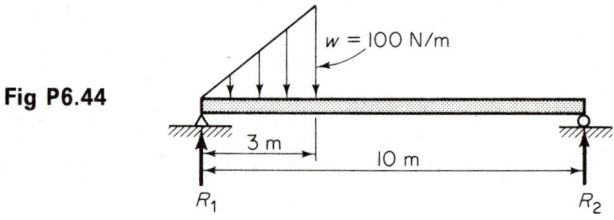

Fig P6.44

6.45 Determine the maximum shear and bending moment in the beam shown in Fig. P6.45.

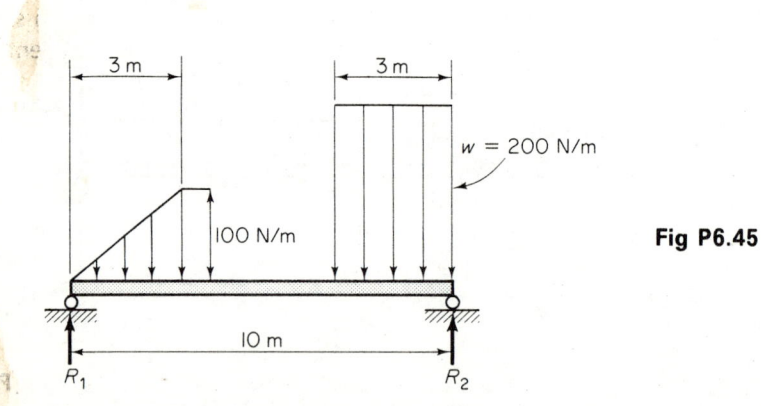

Fig P6.45

6.46 Solve Problem 6.45 if an additional load of 1000 N is placed at the center of the beam shown in Fig. P6.45.

6.47 Draw the shear and bending moment diagrams for the beam shown in Fig. P6.47, and calculate the maximum bending moment and maximum shear.

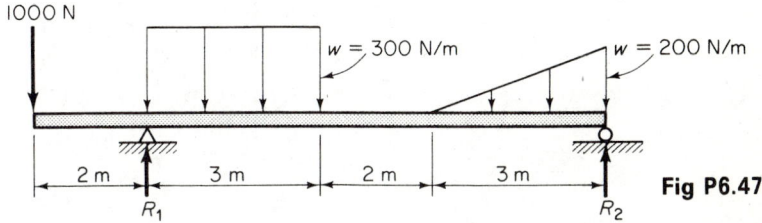

Fig P6.47

6.48 Draw the shear and bending moment diagrams for the beam shown in Fig. P6.48, and calculate the maximum bending moment and maximum shear.

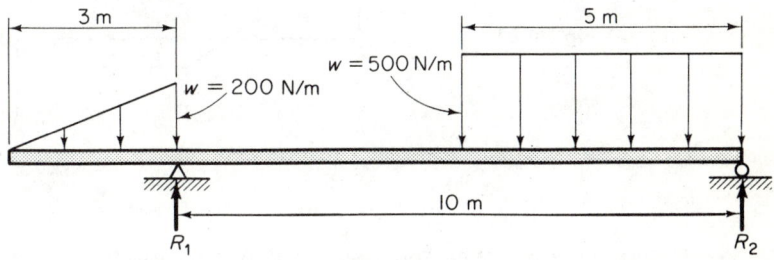

Fig P6.48

6.49 Draw the shear and bending moment diagrams for the beam shown in Fig. P6.49 and calculate the maximum bending moment and maximum shear.

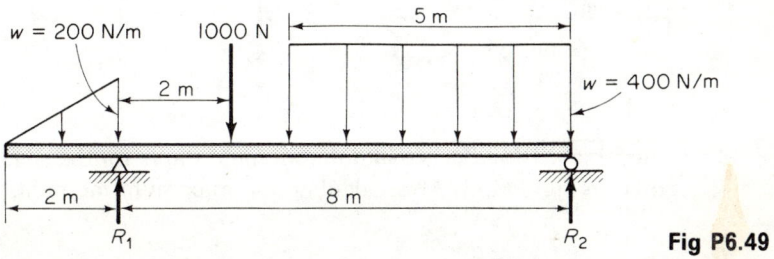

Fig P6.49

6.50 The shear diagram is shown for a cantilever beam. Determine the load diagram and the bending moment diagram for this beam.

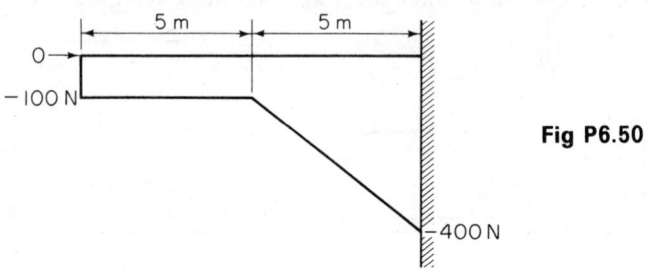

Fig P6.50

6.51 Determine the load and bending moment curves for the beam having the shear curve shown in Fig. P6.51.

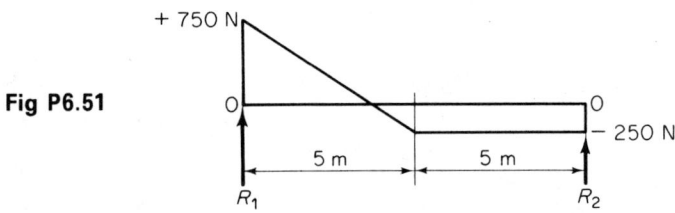

Fig P6.51

6.52 Determine the shear and loading diagrams for the beam having the bending moment diagram shown in Fig. P6.52.

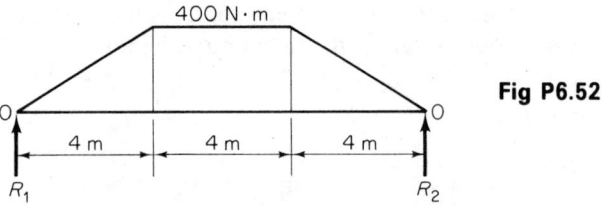

Fig P6.52

6.53 Determine the load curve and the moment curve for the shear curve shown in Fig. P6.53. Also calculate the maximum moment.

Chapter 6, Shear and Bending Moments in Beams

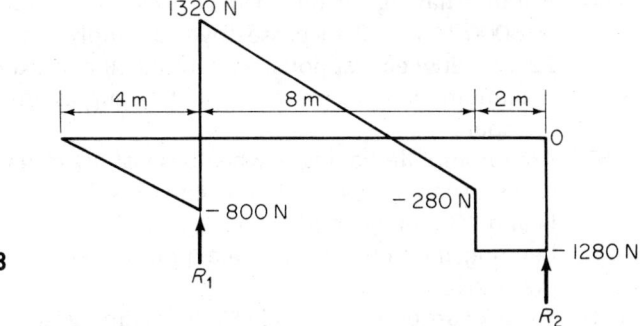

Fig P6.53

6.54 Determine the load diagram, the moment diagram, and the maximum bending moment for the beam having the shear curve shown in Fig. P6.54.

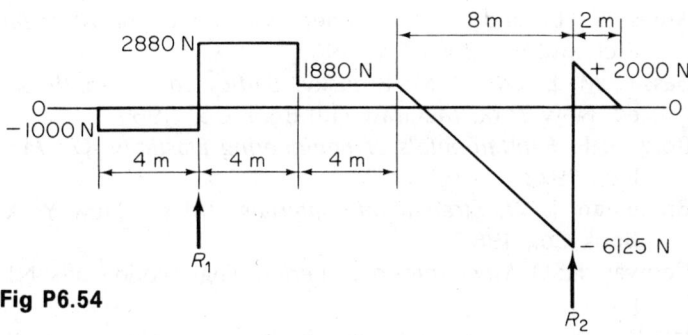

Fig P6.54

6.55 Determine the support reactions, the maximum moment, and the location of the maximum moment for the simple beam shown in Fig. P6.55.

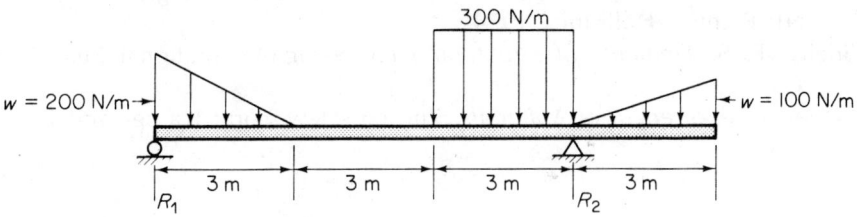

Fig P6.55

6.56 A trailer having a front load of 4000 N separated from a rear load of 8000 N by 12 m passes over a simply supported span that has 22 m between supports. Determine the maximum shear and the maximum bending moment that occur as the trailer passes over the span.

6.57 An automobile having a wheelbase of 10 m moves across a bridge whose span is 30 m. If the car plus its occupants weighs 5000 N and 40% of the load is on the rear wheels, calculate the maximum bending moment and the maximum shear as the car travels over the bridge.

6.58 A truck carries a load of 15 000 N including its own weight. Assuming that the load is distributed to the front and rear wheels in the ratio of 40% and 60%, respectively, determine the maximum bending moment and the maximum shear as the truck crosses a bridge having a span of 50 m. The wheelbase of the truck is 18 m.

REFERENCES

Arges, K. P. and A. E. Palmer. *Mechanics of Materials.* New York: McGraw-Hill Book Co., 1963.

Bassin, M. E. and S. M. Brodsky. *Statics and Strength of Materials.* 2nd ed. New York: McGraw-Hill Book Co., 1969.

Borg, S. F. *Fundamentals of Engineering Elasticity.* D. Van Nostrand Co., Inc., 1973.

Breneman, J. W. *Strength of Materials.* 3rd ed. New York: McGraw-Hill Book Co., 1965.

Conway, H. D. *Mechanics of Materials.* Englewood Cliffs, NJ: Prentice-Hall, Inc.

Higdon, A., E. E. Ohlsen, and W. B. Stiles. *Mechanics of Materials.* 3rd ed. New York: John Wiley & Sons, 1976.

Jensen, A. and H. H. Chenoweth. *Applied Strength of Materials.* 3rd ed. New York: McGraw-Hill Book Co., 1971.

Levinson, Irving J. *Mechanics of Materials.* 2nd ed. Englewood Cliffs, NJ: Prentice-Hall, Inc., 1970.

Olsen, G. A. *Elements of Mechanics of Materials.* 3rd ed. Englewood Cliffs, NJ: Prentice-Hall, Inc., 1974.

Sheiry, E. S. *Elements of Structural Engineering.* International Textbook Co.

Singer, F. L. *Strength of Materials.* 2nd ed. New York: Harper and Row, 1962.

7

Stresses in beams

7.1 INTRODUCTION

Using the principles of statics, we have seen that it is possible to determine the shear and moment acting on any section of a beam in terms of the applied loads. Our next concern is to determine the beam's ability to resist these loads. In order to evaluate the resistive strength of the beam, it will be necessary to address ourselves to the following problems:

1. stresses due to bending only
2. stresses due to shear loads
3. stresses in beams of several materials

The stresses of concern arise internally in a beam as it deforms and accommodates itself to the loading. The exact distribution of the stresses at any section of a beam is complex and its calculation in detail is more properly relegated to studies in the theory of elasticity. It is possible, however, to develop mathematical expressions that are suitable for engineering calculations of the bending and shear stresses in beams and that are based upon simplifying, yet quite accurate, assumptions. The results of these calculations have been tested against experiments and found to be in good accord with the test data.

7.2 PURE BENDING—THE FLEXURE FORMULA

Consider a portion of a beam that is subjected to pure bending only by couples (designated as M) at each end as shown in Fig. 7.1(a). Since

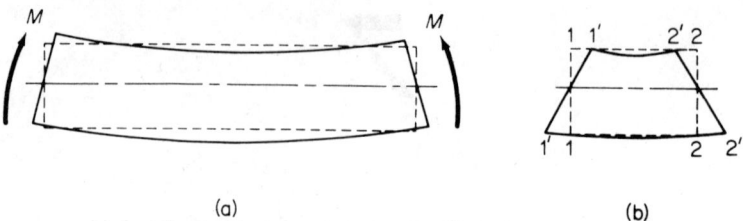

Fig 7.1 Beam subject to bending

the beam is in equilibrium, the moments at each end will be numerically equal, but of opposite sense. Due to these couples, the beam is bent from its original straight position (dashed) to the (exaggerated) curved shape indicated. Due to this bending action, the length of the upper parts of the beam decrease, while the bottom parts of the beam undergo a lengthening. This action has the effect of placing the upper portion of the beam in compression and the lower portion of the beam in tension. In the following discussion we will make certain assumptions concerning the section of beam subjected to pure bending. These are:

1. The beam is straight before bending and has a uniform cross section throughout its length.
2. The material of the beam is homogeneous and its stress-strain curve is linear; i.e., stress is proportional to strain. It obeys Hooke's Law.
3. The beam material has the same modulus of elasticity in compression as it does in tension.
4. A transverse plane remains a plane after bending. This latter assumption can best be illustrated by reference to Fig. 7.1(b). Planes 1-1 and 2-2 have rotated to new positions 1'-1' and 2'-2', respectively, but are still planes.

If we consider the beam to be a series of thin rods or fibers, we note that the action shown in Fig. 7.1(b) leads to the following conclusions:

1. The upper fibers are shortened and are in compression.
2. The lower fibers are lengthened and are in tension.
3. At some depth in the beam there is a fiber that is neither lengthened nor shortened. Since this fiber does not undergo any strain, it is free of bending stresses. This fiber is called the *neutral fiber*, and the plane of all such fibers is called the *neutral plane*.

Figure 7.2 shows the stress distribution in a beam of rectangular cross section subjected to the indicated end couples. Since plane sections remain plane after bending, a fiber located a distance y above the neutral plane will be compressed an amount equal to the extension of a fiber located

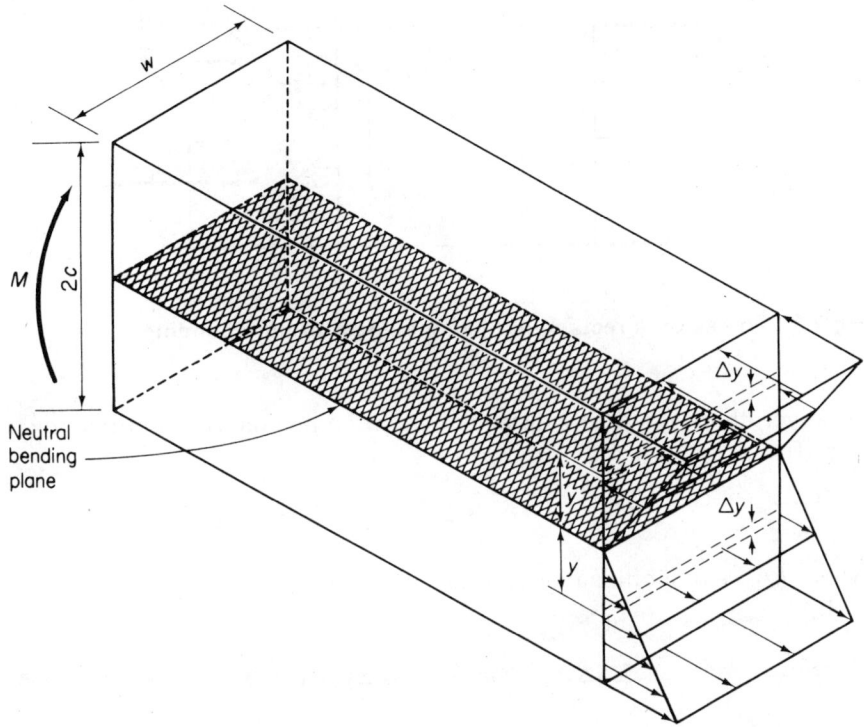

Fig 7.2 Stresses in a rectangular beam subject to pure bending

a distance y below the neutral plane. Since it is assumed that the material obeys Hooke's Law, we can state that the stress on a fiber at a distance y above the neutral plane is equal to the stress on a fiber at a distance y below the neutral plane. We can also conclude that the stress varies linearly from the neutral plane to the outermost fiber in either compression or tension. The portion of the beam shown in Fig. 7.2 must be in equilibrium; it is therefore necessary that:

1. The sum of the longitudinal forces acting on the end planes be zero.
2. The resisting moment of the beam equals the applied moment M.

The condition of zero longitudinal force can be evaluated by reference to Fig. 7.3. The compressive force F_c must equal the tensile force F_t for equilibrium.

If we denote the element of area at any distance from the neutral axis y as ΔA and the stress on it as S, the requirement for equilibrium of forces yields

$$\Sigma F = 0 = \Sigma S \Delta A \tag{7.1}$$

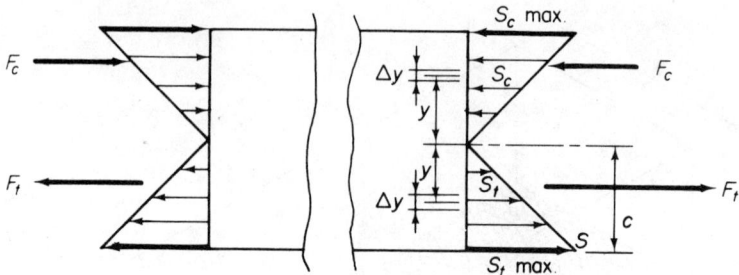

Fig 7.3 Forces on a rectangular beam subject to pure bending

However, the stress at any distance y from the neutral axis is proportional to y. Thus,

$$S = ky \tag{7.2}$$

where k is a proportionality constant. Placing equation (7.2) in equation (7.1),

$$\Sigma S \Delta A = \Sigma k y \Delta A = k \Sigma y \Delta A = 0 \tag{7.3}$$

where k has been taken out of the summation since it is a constant. However, we note that this summation term is the first moment of area as we defined it in Chapter 4. Denoting $\bar{y}$ as the distance from the neutral axis to the centroid (center of gravity) of the area

$$\bar{y} A = \Sigma y \Delta A \tag{7.4}$$

and

$$\bar{y} = \frac{\Sigma y \Delta A}{A} \tag{7.5}$$

From equation (7.3), $\Sigma y \Delta A$ must equal zero since k is not zero. Thus, the numerator of equation (7.5) is zero; but the denominator, A, is not zero which leads to

$$\bar{y} = 0 \tag{7.6}$$

Thus, the neutral axis for bending coincides with the centroid of the section.

The second condition for equilibrium requires that the resisting moment of the beam be equal to the applied moment. Again let us refer to Figs.

Chapter 7, Stresses in Beams   259

7.2 and 7.3. The force at any distance y above the neutral axis is $S\Delta A$. The moment at this same distance is the moment arm multiplied by this force. The requirement that the sum of all of these moments equal the applied moment M gives us

$$M = \Sigma y S \Delta A \qquad (7.7)$$

We have already noted that the stress is proportional to the distance of the fiber from the neutral axis, i.e., $S = ky$. Inserting this in equation (7.7)

$$M = \Sigma(ky)y\Delta A = k\Sigma y^2 \Delta A \qquad (7.8)$$

If we denote the stress at the outer fiber as S_{max} and its distance from the neutral axis as C (as shown in Fig. 7.3), we have

$$S_{max} = kC \qquad (7.9)$$

and

$$k = \frac{S_{max}}{C} \qquad (7.10)$$

Placing equation (7.10) in equation (7.8)

$$M = \frac{S_{max}}{C} \Sigma y^2 \Delta A \qquad (7.11)$$

However, we defined the second moment of area (moment of inertia, I) in Chapter 4 to be equal to $\Sigma y^2 \Delta A$. Therefore

$$M = \frac{S_{max}}{C}(I) \qquad (7.12)$$

or in terms of the maximum stress

$$S_{max} = \frac{MC}{I} \qquad (7.13)$$

Equation (7.13) is known as the *Flexure Formula* and it relates the maximum stress in the beam to the applied moment and the properties of the cross-sectional area of the beam. In equation (7.13), S is the normal stress in pounds per square inch, M is the applied moment in inch pounds,

C is the distance to the outermost fiber from the neutral axis in inches, and I is the moment of inertia of the cross-sectional area with respect to the neutral axis in inches⁴, all in English units. In SI units, S is in Pa, C is in metres, and I is in m⁴.

Illustrative problem 7.1

A steel beam having an allowable maximum stress of 140 MPa in tension and compression is to have a rectangular cross section whose height is twice its width. Determine the beam's dimensions if it is to resist a bending moment of 25 000 N·m.

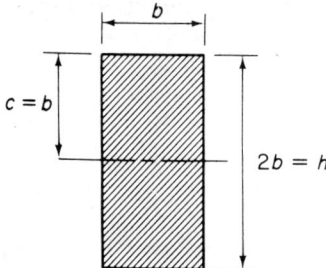

Fig 7.4   Illustrative problem 7.1

Solution:

Designating the width of the beam as b, the height becomes $2b$. From Chapter 4, $I = \dfrac{bh^3}{12}$ and $C = \dfrac{h}{2}$. However $h = 2b$ in this problem. Therefore

$$S = \frac{MC}{I} = 140 \times 10^6 = \frac{(25\,000)(b)}{b(2b)^3/12} = \frac{37\,500}{b^3}$$

and

$$b^3 = \frac{37\,500}{140 \times 10^6}$$

$b = 0.0645$ m $= 64.5$ mm

$h = 2b = 2(0.0645) = 0.129$ m $= 129$ mm

When a beam has equal strength in both tension and compression, as was the case in Illustrative Problem 7.1, the most economical section is one that is symmetrical about the neutral axis since this would cause the beam to be stressed equally in both tension and compression. Thus for materials such as steel, which has very much the same properties in tension and compression, the beam sections are usually symmetrical to the neutral axis. However, for a material such as cast iron, which is strong in compression and weak in tension, the section should not be symmetrical about the neutral axis, but should be proportioned to yield C distances in the same ratio as the ratios of the compressive and tensile strengths.

In the Flexure Formula [equation (7.13)], the geometric properties of the cross-sectional area of the beam enter the equation as the distance from the neutral axis to the outermost fiber (C) and the moment of inertia (I) about the neutral axis. The ration of I/C is known as the *section modulus* of the beam and is most commonly denoted as Z. Thus

$$Z = \frac{I}{C} \qquad (7.14)$$

and

$$S = \frac{M}{Z} \qquad (7.15)$$

The A.I.S.C. uses the symbol S for section modulus. Because of the possible confusion with the same symbol for stress, Z is used for section modulus in this text and in many other texts on strength of materials. Values of Z, the section modulus, are tabulated for standard beam shapes and will be found in Appendix B.

The data for structural shapes in Appendix B is from the Seventh Edition of the *AISC Steel Construction Manual,* 1970. In this edition, newer grades of high strength steels having yield strengths of up to 100,000 psi (689.5 MPa) have been approved for construction use, as contrasted to the 50,000 psi (344.75 MPa) maximum limit found in the sixth edition *Manual.* At the same time, a new system of designations of structural shapes has been adopted by the steel industry. Since the old designations still persist, Table 7.1 is an abridged comparison between the old and new designations.

Strength of Materials for Engineering Technology

TABLE 7.1 DESIGNATION OF STRUCTURAL SHAPES

Group	Old Designation	New Designation	Type of Shape
Wide Flange Shapes	*24 WF76	W24 × 76	W Shape
Miscellaneous Light Beams	14 B26	W14 × 26	W Shape
American Std. Beams	24 I 100	S24 × 100	S Shape
Miscellaneous Light Beams	8 M 18.5	M8 × 18.5	M Shape
Junior Beams	10 Jr. 9.0	M10 × 9	M Shape
Miscellaneous Light Columns	8 × 8 M 34.3	M8 × 34.3	M Shape
American Std. Channel	12 ⊏ 20.7	C12 × 20.7	
Equal Leg Angle	6 × 6 × $\frac{3}{4}$	L6 × 6 × $\frac{3}{4}$	
Unequal Leg Angle	6 × 4 × $\frac{5}{8}$	L6 × 4 × $\frac{5}{8}$	

*Note: A W24 × 76 beam has a nominal overall depth (d) of 24 in. and a nominal weight of 76 lb/ft. Similarly a S24 × 100 beam has a nominal depth of 24 in. and a nominal weight of 100 lb/ft. Linear dimensions in inches should be multiplied by 0.0254; in.^{2} by (0.0254)^2; in.^{3} by (0.0254)^3; in.^{4} by (0.0254)^4; and lb by 4.448 to convert to consistent SI units.

Illustrative problem 7.2

Select a W10 beam suitable to support a bending moment of 50 000 N·m, if the allowable stress for steel in both tension and compression is 140 MPa.

Solution:

First evaluate Z

$$Z = \frac{M}{S} = \frac{50\ 000}{140 \times 10^6} = 3.57 \times 10^{-4}\ m^3 = \frac{3.57 \times 10^{-4}(1000)^3}{(25.4)^3}$$
$$= 21.8\ in^3$$

From Appendix B, a W10 × 25 beam having a Z = 26.5 in.^{3} is satisfactory.

A note of caution must be voiced at this point: Throughout this chapter we shall ignore the weight of the beam in our calculations. This is a common practice since the weight of the beam usually represents a small fraction of the maximum moment on the beam. For example, assume that the beam in Illustrative Problem 7.2 is 7 m long. The maximum moment due to the beam's own weight = $wL^2/8$ = [25 × 3.28 × 4.448 × 7^2/8] = 2234 N·m or $4\frac{1}{2}$% of the moment of 50 000 N·m that the beam was selected for. This is less then the difference between the minimum section modulus and the one selected, i.e., [(26.5 − 21.8)/21.8] × 100 = 21%. There will be cases in which this procedure might lead to difficulty, and the student is cautioned to remember that we have neglected the weight of the beam only in the interest of simplicity.

If a beam is symmetrical about the neutral axis (as noted earlier this is usually the case for steel), the calculations are not difficult. Where a beam is not symmetrical about the neutral axis (as say, for cast iron), it is desirable (if possible) to have the maximum stress in tension and compression occur simultaneously. This can be accomplished in some cases, but in others it is necessary to investigate both the maximum compressive stress and the maximum tensile stress.

Illustrative problem 7.3

A "U" beam is used to carry a moment that causes the upper part of the beam to be in compression. Assume that the material has an allowable compressive stress of 70 MPa and an allowable tensile stress of 35 MPa. Determine w for the maximum tensile and compressive stresses to occur simultaneously.

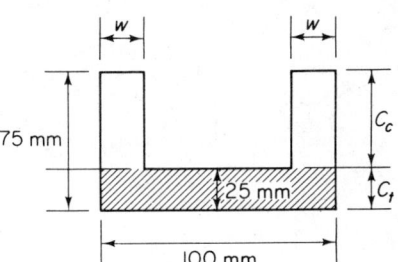

Fig 7.5   Illustrative problem 7.3

Solution:

For optimum design

$$\frac{C_c}{C_t} = \frac{S_c}{S_t} = \frac{70 \times 10^6}{35 \times 10^6} = \frac{2}{1}$$

This locates the neutral axis since $C_c + C_t = 75$. Thus, we have $C_t = 25$ and $C_c = 50$. Since the neutral axis coincides with the centroid, the moment of the shaded portion of Fig. 7.5 about the neutral axis must equal the moment of the unshaded portion about the neutral axis. This gives us

$$w \times 50 \times 25 + w \times 50 \times 25 = 100\,(25) \left(\frac{25}{2}\right)$$

$$\text{and} \quad w = \frac{25}{2} = 12.5 \text{ mm}$$

It should be noted that had the beam been turned with the legs of the facing downward ∩, it would not be possible to have the maximum compressive and maximum tensile stresses occur simultaneously.

7.2a Flooring

The flooring in a building and its support is an illustration of the loading of beams in series. This is shown quite clearly in Fig. 7.6 where the

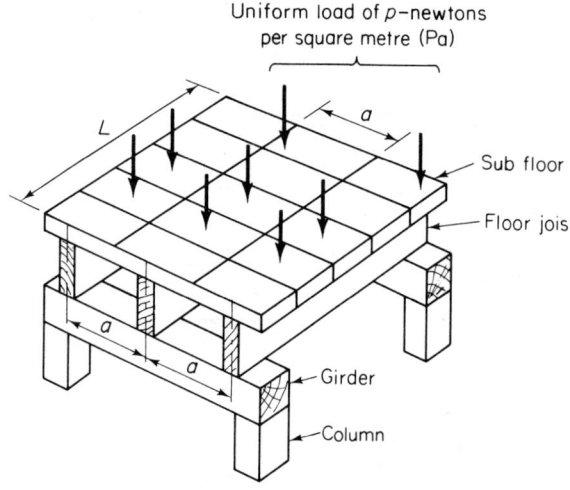

Fig 7.6 Floor framing. Reproduced with permission from *Strength of Materials* by F. L. Singer, Harper and Bros. Publishers.

sub-flooring is uniformly loaded at p-newtons per square metre (Pa), which load is in turn transmitted to the floor joists as a uniform load, which is then transmitted to the girder as uniformly spaced concentrated loads. The columns, as shown, ultimately take all of the loading. A numerical illustration will serve to show the various loadings.

Illustrative problem 7.4

A room is 3 m × 6 m with the 3 m dimension being the "L" direction shown in Fig. 7.6. Show the loadings on the sub-floor, joists, and girders if the floor loading is 2400 Pa. Assume the sub-floor to be 25-mm × 150-mm boards, the joists to be 75 mm × 200 mm and the girder to be 100 mm × 250 mm. (Use full sizes even though lumber comes in dressed sizes, i.e., a 3 in. × 8 in. is really $2\frac{5}{8}$ in. × $7\frac{1}{2}$ in. At present these are no SI equivalents for these sizes and conversions must be made until SI standards are adopted.)

Solution:

Each board in the sub-floor is subject to a load equal to the uniform floor loading multiplied by the board's area. For this problem we have the load per metre of sub-floor as 2400 × 0.15 × 1 = 360 N/m. The beam is loaded and supported as shown in Fig. 7.7(a). The load on the

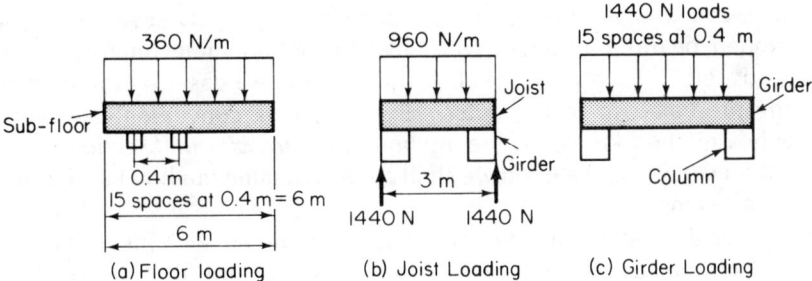

Fig 7.7 Illustrative problem 7.4

joists is equal to a uniform load of 2400 Pa multiplied by the area of the floor and divided by the number of joists. This is the same as applying the uniform load to a repeating section whose area is $a \times L$, as shown in Figure 7.6. Thus the loading on the joist is 0.4 × 1 × 2400 = 960 N/m since a = 0.4 m from Fig. 7.7(a). This leads to an end reaction of 1440 N that must be carried by the girder. This load is placed on the girder at intervals of 0.4 m as indicated in Fig. 7.7(c) to give a total load of 21 600 N per girder. Since there are two girders, they support a total load of 43 200 N. As a check, this should equal the floor loading multiplied by the floor area, i.e., 2400 × 3 × 6 = 43 200 N, which agrees with our value of 43 200 N. At this point, each unit, the sub-floor, the joists, and the girders can be individually checked for bending stresses using the principles developed thus far in this chapter and in Chapter 6. As an example, the maximum moment in the joists will be $\frac{960\,(3)^2}{8}$ = 1080 N·m (see Chapter 6 or Case 9, Table 8.2). The maximum bending stress is given by

$$S = \frac{MC}{I} = \frac{1080 \times 0.1}{[0.075(0.2)^3]/12} = 2.16 \text{ MPa},$$

which is satisfactory. For the girders, see Problem 7 at the end of this chapter.

7.3 BEAMS OF SEVERAL MATERIALS

The use of beams of two or more materials is a common industrial practice, and it is assumed that the student has probably seen such cases as wooden beams reinforced with steel plates, concrete reinforced with steel rods, etc. It is entirely possible to analyze these cases by the methods already presented in this chapter. However, it is convenient and easier to apply a method known as the *method of the transformed section*, which we will now develop. Later on we shall apply this same method to reinforced concrete beams.

Consider the rectangular wooden beam reinforced with steel plates on top and bottom as shown in Fig. 7.8(a). If we assume that the steel and

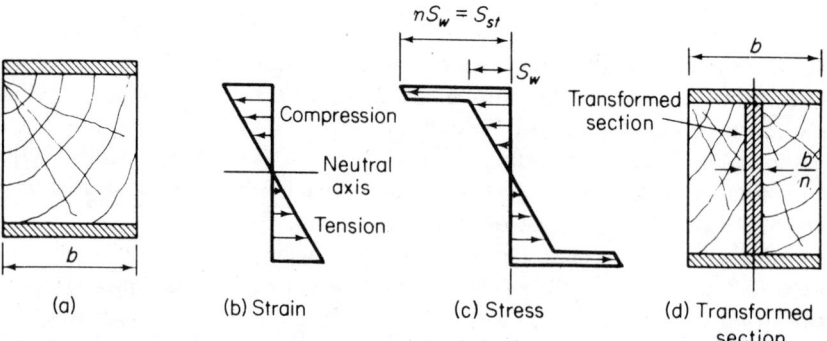

Fig 7.8 **Stresses and strains in composite beams**

wood are firmly fastened to each other and are physically constrained to move as a single unit, then each must undergo the same deflections at their interfaces. In other words, the strain of the steel and wood at their junctions is the same. Assume both materials are not stressed above their proportional limits and Hooke's Law is applicable to both

$$S_w = \varepsilon E_w \quad \text{and} \quad S_{st} = \varepsilon E_{st} \tag{7.16}$$

where S_w is the stress in the wood, E_w and E_{st} are the moduli of elasticity of the wood and steel, respectively, and ε is the common strain at the interface. Thus

$$\frac{S_{st}}{S_w} = \frac{E_{st}}{E_w} = n \tag{7.17}$$

and even though the strains are equal, the stress in the steel is n times greater than the stress in the wood at the same distance from the neutral axis of the beam. Thus, even though the strain is linear throughout the beam as shown in Fig. 7.8(b), the stress curve is as shown in Fig. 7.8(c).

If the wooden portion of the beam were replaced with a steel section such that the width of the steel replacement was equal to the width of the wood divided by n, we would have the situation shown in Fig. 7.8(d). This transformed beam would have the same characteristics of stress and strain that the actual beam had. In other words, we may treat the transformed beam exactly as we have treated beams of a single material. For the case of a wooden beam reinforced only at the top, the beam would be transformed as noted, and the Flexure Formula would be applied to the transformed section. It is necessary in this case to find the neutral axis of the transformed section and the moment of inertia of the transformed section about this neutral axis.

After the transformed section has been solved using the Flexure Formula, it is necessary to calculate the stresses in the actual beam using equation (7.17). The following four problems illustrate these procedures.

Illustrative problem 7.5

A 100-mm × 200-mm wooden beam is reinforced by 12-mm steel plates on top and bottom and is subjected to a moment of 50 000 N·m. If $E_w = 7$ GPa, determine the maximum stress in each material. $E_{st} = 210$ GPa.

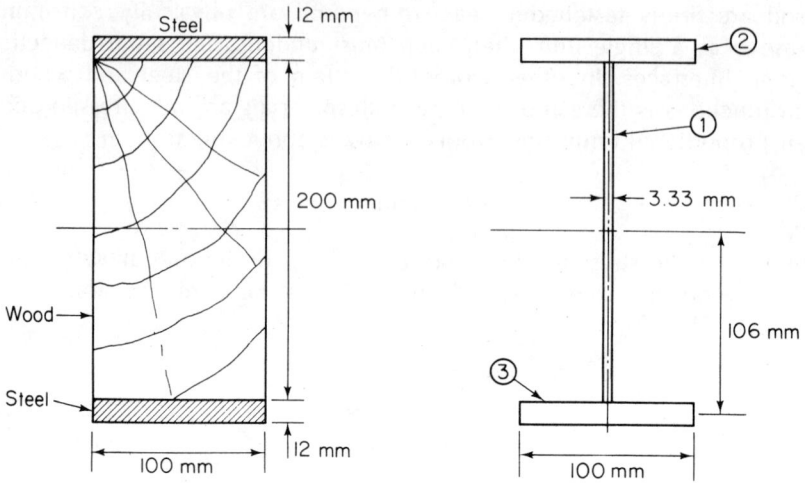

Fig 7.9 Illustrative problem 7.5

Solution:

For ease of solution, let us transform the wood to steel. Therefore the width of the transformed section is $100 \times 7 \times 10^9 / 210 \times 10^9 = 3.33$ mm. The moment of inertia of the transformed beam is

For ① $\qquad \dfrac{bh^3}{12} = \dfrac{0.0033 \times (0.200)^3}{12} = 2.2 \times 10^{-6} \text{ m}^4$

For ② $\qquad \dfrac{bh^3}{12} + Ad^2 = \dfrac{(0.100)(0.012)^3}{12} + (0.100)(0.012)(0.106)^2$

$\qquad\qquad\qquad\qquad = 13.5 \times 10^{-6} \text{ m}^4$

For ③ $\qquad\qquad\qquad\qquad = 13.5 \times 10^{-6} \text{ m}^4$

The total $I = 29.2 \times 10^{-6}$ m^4. At the outer fiber of the steel,

$$S = \frac{MC}{I} = \frac{50\,000 \times 0.112}{29.2 \times 10^{-6}} = 1.92 \times 10^8 \text{ Pa}$$

At the wood-steel interface, the stress in the steel is proportional to the distance from the neutral axis to the interface. Therefore $S_{\text{interface}} = 1.92 \times 10^8 \times \dfrac{0.100}{0.112} = 1.71 \times 10^8$ Pa. This figure of 1.71×10^8 Pa is the stress in the steel—not the stress in the wood. The stress in the wood is $1/n^{th}$ the stress in the steel. Thus the maximum stress in the wood is $(7 \times 10^9 / 210 \times 10^9)\, 1.71 \times 10^8 = 5.7$ MPa.

Illustrative problem 7.6

Solve Illustrative Problem 7.5 by transforming the steel to wood.

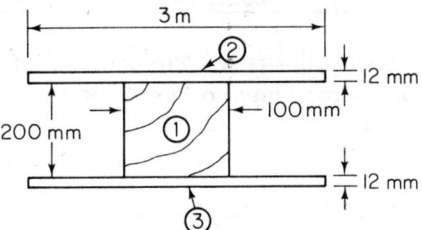

Fig 7.10 Illustrative problem 7.6

Solution:

The width of the transformed steel will be n times the width of the actual steel = 100 $(210 \times 10^9 / 7 \times 10^9)$ = 3000 mm = 3 m. The moment of inertia of the transformed beam is

For ① $\dfrac{bh^3}{12} = \dfrac{0.100\,(0.200)^3}{12} = 6.67 \times 10^{-5}$ m^4

For ② $\dfrac{bh^3}{12} + Ad^2 = \dfrac{3(0.012)^3}{12} + 3 \times 0.012 \times (0.106)^2$

$\qquad\qquad\qquad = 40.49 \times 10^{-5}$ m^4

For ③ $\qquad\qquad\qquad = 40.49 \times 10^{-5}$ m^4

$I = 87.65 \times 10^{-5}$ m^4. The maximum stress at the outside of the transformed beam is

$$S = \frac{MC}{I} = \frac{(50\,000)(0.112)}{87.65 \times 10^{-5}} = 6.39 \times 10^6 \text{ Pa}$$

However, the real beam is not wood at this section—it is steel. The actual stress is therefore $6.39 \times 10^6\,(210 \times 10^9 / 7 \times 10^9) = 1.92 \times 10^8$ Pa in the steel, which agrees with our solution to Illustrative Problem 7.5.

Continuing, the maximum stress in the actual wood occurs at the wood-steel interface and must equal the stress at the outermost fiber of the transformed beam multiplied by the ratio of the interface distance from the neutral axis to the outermost distance from the neutral axis. Thus, $S_{wood\,max} = 6.39 \times 10^6 \times \dfrac{0.100}{0.112} = 5.71$ MPa, which also agrees with the results of Illustrative Problem 7.5.

Illustrative problem 7.7

Solve Illustrative Problem 7.5 if only the bottom steel plate is used for reinforcement of the wood.

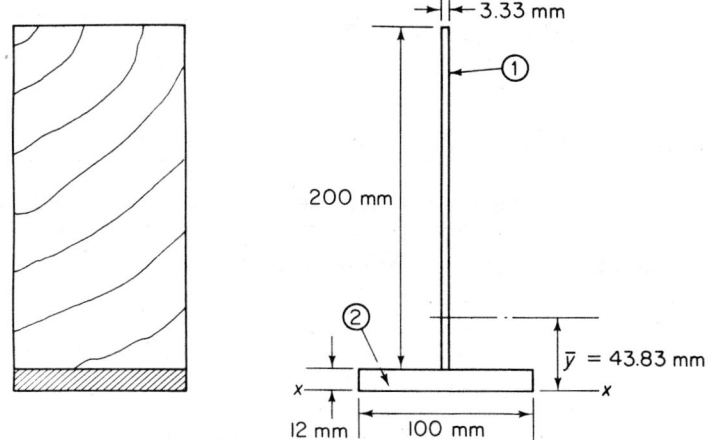

Fig 7.11 Illustrative problem 7.7

Solution:

As before, we shall transform the wood to steel and solve the transformed beam. The web of the beam becomes $100 \times 7 \times 10^9 / 210 \times 10^9 = 3.33$ mm.

We shall now have to locate the neutral axis of the transformed beam. Using X-X as the reference axis and dimensions in mm,

$$\bar{y} = \frac{(200 \times 3.33 \times 112 + 100 \times 12 \times 6)}{200 \times 3.33 + 100 \times 12} = 43.83 \text{ mm}$$

The moment of inertia of the transformed section about the neutral axis is

For ①
$$\frac{bh^3}{12} + Ad^2 = \frac{(0.0033)(0.200)^3}{12} + (0.0033 \times 0.200)(0.100 - 0.0438)^2$$
$$= 4.28 \times 10^{-6} \text{ m}^4$$

For ②
$$\frac{bh^3}{12} + Ad^2 = \frac{(0.100)(0.012)^3}{12} + 0.100(0.012)(0.0438 - 0.006)^2$$
$$= \frac{1.73 \times 10^{-6} \text{ m}^4}{6.01 \times 10^{-6} \text{ m}^4}$$

Since the beam is not symmetrical, it is necessary to investigate both the compressive and tensile stresses. For the top of the transformed beam

$$S = \frac{MC}{I} = \frac{50\,000 \times (0.212 - 0.0438)}{6.01 \times 10^{-6}} = 1.40 \times 10^9 \text{ Pa} = 1.40 \text{ GPa}$$

in the transformed beam. For the real beam

$$S_{top} = 1.40 \times 10^9 \times \frac{7 \times 10^9}{210 \times 10^7} = 46.6 \text{ MPa}$$

At the bottom of the beam

$$S = \frac{MC}{I} = \frac{(50\,000)(0.0438)}{6.01 \times 10^{-6}} = 364.4 \text{ MPa}$$

At the wood-steel interface

$$S = 364.4 \times 10^6 \times \frac{(43.83 - 12)}{43.83} = 264.6 \text{ MPa}$$

which, for the wood, is

$$264.6 \times 10^6 \times \frac{7 \times 10^9}{210 \times 10^9} = 8.82 \text{ MPa}$$

The maximum stresses in the actual beam are therefore 46.6 MPa in the wood and 364.4 MPa in the steel. These are so high that we may have exceeded the yield points of both materials, which would invalidate the solution. If this were so, it would be necessary to determine the maximum moment that the beam could support without exceeding the yield point of either material.

Illustrative problem 7.8

If the yield point of wood is taken as 24.5 MPa and that of the steel is taken to be 210 MPa, determine the maximum moment that the beam in Illustrative Problem 7.7 can support.

Solution:

All of the items pertaining to the properties of the transformed section are as evaluated in Illustrative Problem 7.7. For the outermost top fiber, using the values given

$$S_{\text{top wood}} = 24.5 \text{ MPa}; \quad S_{\text{top steel}} = 24.5 \times 10^6 \times \frac{210 \times 10^9}{7 \times 10^9} = 735 \text{ MPa}$$

The moment producing this stress in the transformed beam is therefore

$$M = \frac{SI}{C} = \frac{735 \times 10^6 \times 6.01 \times 10^{-6}}{(0.212 - 0.0438)} = 26\,262 \text{ N·m}$$

At the bottom

$$M = \frac{SI}{C} = \frac{210 \times 10^6 \times 6.01 \times 10^{-6}}{0.0438} = 28\,815 \text{ N·m}$$

At the wood-steel interface, the stress in the wood (based upon the steel being stressed to 210 MPa) is

$$210 \times 10^6 \times \frac{(43.83 - 12)}{43.83} \times \frac{7 \times 10^9}{210 \times 10^9} = 5.08 \text{ MPa}$$

The limiting moment is therefore 26 262 N·m, and this limit is due to the wood reaching its yield point before the steel reaches its limiting stress.

7.4 REINFORCED CONCRETE BEAMS

Concrete is widely used as a material of construction since it possesses such desirable characteristics as ease of handling, fireproof qualities, etc. There is one characteristic of concrete that is highly undesirable—it possesses almost no tensile strength. Because of this, concrete alone is not suitable as a beam material and it must be reinforced on the tensile side with a structural material that will give it the required tensile strength. In addition, the reinforcing material must be properly bonded to the concrete so that the beam acts as a single unit. At this time we shall only consider the bending stresses in reinforced concrete beams and limit ourselves to *simple reinforced rectangular beams*, i.e., beams having reinforcement only on the tension side of the neutral axis.

Figure 7.12 shows a cross section of a rectangular reinforced concrete beam with steel reinforcing on the tension side. The deformation [Fig.

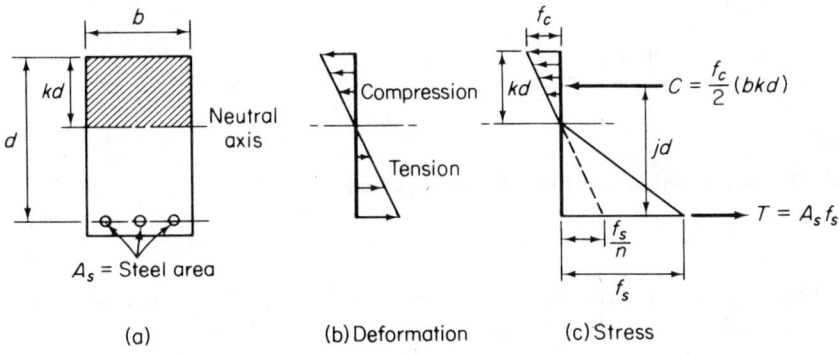

Fig 7.12 Rectangular reinforced concrete beam

7.12(b)] is assumed to be linear, with zero deformation at the neutral axis of the beam. The steel portion of the beam is assumed to have deflected the same amount as it would have if it were concrete, but the stresses in the steel will exceed the stresses in the concrete by a factor n, where $n = E_{steel}/E_{concrete} = E_s/E_c$. The commonly used nomenclature for reinforced concrete differs from that which we have been using. We shall use the following nomenclature:

E_c = modulus of elasticity of concrete in compression

E_s = modulus of elasticity of steel

$n = E_s/E_c$

A_s = area of tension steel

A_t = transformed steel area = nA_s

b = width of beam
C = total compressive force in concrete
d = depth of beam from top to the center of steel reinforcement
f_c = maximum compressive stress in concrete
f_s = maximum tensile stress in steel
j = ratio of moment arm of C-T couple to the depth d
k = ratio of location of neutral axis to the depth d
$p = A_s/bd$
T = total tensile force in concrete

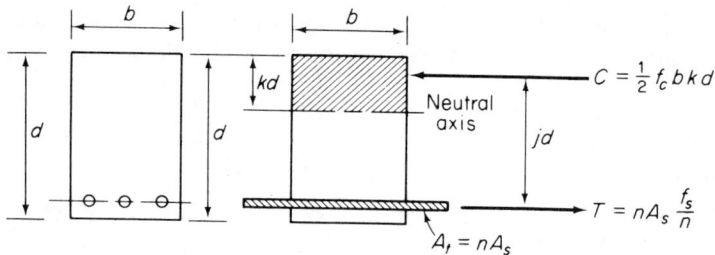

Fig 7.13 Rectangular reinforced concrete beam

The first step in designing a reinforced concrete beam is to transform the steel area to equivalent concrete area, which gives us

$$A_t = nA_s \tag{7.18}$$

This area is assumed to be located at the centerline of the steel and to be so narrow as to have a uniform stress distribution. Therefore the total tensile force T is given by

$$T = f_s A_s = nA_s \frac{f_s}{n} = A_t f_t \tag{7.19}$$

The total compression force in the beam (using only the portion above the neutral axis) is given by the average compressive stress multiplied by the area in compression

$$C = \frac{f_c}{2} bkd \tag{7.20}$$

However, the condition for equilibrium requires $C = T$, and we therefore have a couple with a moment arm of jd.

At this point we will make the assumption (which will be further discussed later in this section) that both the steel and concrete reach their maximum stresses simultaneously. This condition is known as *balanced reinforcement*. The moment of the concrete force is

$$C(jd) = \frac{f_c}{2}(bkd)jd \qquad (7.21)$$

and the moment of the steel is

$$T(jd) = f_s A_s (jd) \qquad (7.22)$$

For balanced reinforcement, the resisting moment (M) must equal either, or both, of the moments given by equations (7.21) and (7.22). Thus

$$M = \frac{f_c}{2}(bkjd^2) = f_s A_s(jd) \qquad (7.23)$$

From Fig. 7.12(c) we can write

$$\frac{kd}{d} = \frac{f_c}{f_c + \dfrac{f_s}{n}}; \quad k = \frac{1}{1 + \dfrac{f_s}{nf_c}} \qquad (7.24)$$

and

$$jd = d - \frac{kd}{3}; \quad j = 1 - \frac{k}{3} \qquad (7.25)$$

Equations (7.23), (7.24), and (7.25) are sufficient for the *design* of a beam for bending when balanced reinforcement is assumed. We can obtain two useful relations from equation (7.23) for this case

$$d = \sqrt{\frac{M}{(f_c/2)\,bkj}} \qquad (7.26)$$

and

$$A_s = \frac{M}{f_s jd} \qquad (7.27)$$

276   Strength of Materials for Engineering Technology

For initial design purposes, k is often taken to be $\frac{3}{8}$, giving a preliminary value for j of $\frac{7}{8}$.

Illustrative problem 7.9

Design a reinforced concrete beam with balanced reinforcement to carry a bending moment of 100 000 N·m. The beam is to be 0.6 m wide, with $n = 15$, $f_s = 140$ MPa and $f_c = 5.25$ MPa.

Solution:

The first step in the solution is to obtain k from equation (7.24). Thus

$$k = \frac{1}{1 + (f_s/nf_c)} = \frac{1}{1 + [140 \times 10^6/(15 \times 5.25 \times 10^6)]} = 0.36$$

$$j = 1 - \frac{k}{3} = 1 - \frac{0.36}{3} = 0.88$$

From equation (7.26)

$$d = \sqrt{\frac{100\,000}{(5.25 \times 10^6/2)(0.36)(0.88) \times 0.6}} = 0.448 \text{ m}$$

The required steel area is obtained from equation (7.27)

$$A_s = \frac{M}{f_s j d} = \frac{100\,000}{140 \times 10^6 \times 0.88 \times 0.448} = 1.81 \times 10^{-3} \text{ m}^2$$

Equations (7.23) through (7.27) are design equations for the balanced reinforcement required when the external moments and allowable internal stresses are given. In those cases in which the reinforcement is not balanced, or in which all of the physical dimensions are known and we desire to calculate either the maximum stresses (or maximum allowable moment when the allowable stresses are given), we cannot use some of the foregoing equations. For example, we cannot use equation (7.24) since the *actual* values of f_c and f_s are not known. We can, however, solve for the location of the neutral axis of a given beam using the concept of the transformed section of the previous article. Referring to Fig. 7.13, we locate the neutral axis from the requirement that the first moment of the area above the neutral axis must equal the first moment of the area below the neutral axis. Thus

$$(bkd)\frac{kd}{2} = nA_s(d - kd) \qquad (7.28)$$

If we denote

$$p = A_s/bd \qquad (7.29)$$

we can solve equation (7.28) for k

$$k = \sqrt{2pn + (pn)^2} - pn \qquad (7.30)$$

Once k is known, we can obtain j from equation (7.25) and the *actual* value of f_s and f_c by applying equation (7.23) twice.

Illustrative problem 7.10

Determine the maximum allowable bending moment in a rectangular reinforced concrete beam if $n = 15$, and the allowable maximum stresses in concrete and steel are $f_c = 5.25$ MPa and $f_s = 140$ MPa. Assume $A_s = 0.0016$ m² and the beam dimensions are $b = 250$ mm and $d = 500$ mm.

Solution:

This is an investigation problem and it is first necessary to find the location of the neutral axis. From equation (7.29)

$$p = \frac{A_s}{bd} = \frac{0.0016}{(0.250)(0.500)} = 0.0128$$

From equation (7.30)

$$k = \sqrt{2 \times 15 \times 0.0128 + (0.0128 \times 15)^2} - 0.0128(15) = 0.457$$

and

$$j = 1 - \frac{k}{3} = 1 - \frac{0.457}{3} = 0.848$$

for concrete:

$$M = \frac{f_c}{2} b(kd)(jd) = \frac{5.25 \times 10^6}{2} (0.250)(0.457)(0.500)(0.848)(0.500)$$
$$= 63\,580 \text{ N} \cdot \text{m}$$

for steel:

$$M = A_s f_s jd = 0.0016 \times 140 \times 10^6 \times 0.848 \times 0.500 = 94\,976 \text{ N} \cdot \text{m}$$

This beam is obviously not balanced, and the stress in the concrete governs giving a maximum allowable moment of 63 580 N·m.

The design and investigation of a reinforced concrete beam is not totally a process of examining the flexural stresses in the beam. It is equally important that shear, bond, diagonal tension, and anchorage be investigated.

The design of reinforced beams of shapes other than rectangular can be analyzed by the method we have used for rectangular sections. However, this becomes cumbersome, and the use of the transformed section method of the previous section is recommended for these cases. These topics are beyond the scope of this text, and the interested student is referred to specialized texts on structural engineering and reinforced concrete.

7.5 SHEAR STRESSES IN BEAMS

In Section 5.6, considerable attention was given to the shear stresses occurring in the faces of a structural element. We shall shortly utilize some of this material in the subsequent study of shearing stresses in beams. Our first concern, however, is to establish the conditions in a beam that give rise to shearing stresses. When a transverse load is placed on a beam, the load tends to displace a portion of the beam as shown in Fig. 7.14.

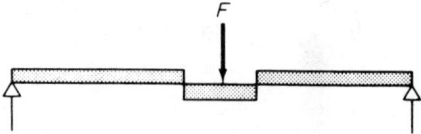

Fig 7.14 **Transverse shear of a beam**

We have already discussed this action in Chapter 6 when we reviewed bending of beams. The action indicated is a pure shearing action and

Chapter 7, Stresses in Beams 279

occurs even where there is no bending of the beam. However, beams do bend and, when they bend, fibers on one side of the neutral axis are placed in compression and those on the other side are placed in tension. In effect, the fibers on either side of the neutral plane tend to slip in directions opposite to one another. If we consider a beam that is composed of two planks that are placed one atop the other and are free to move longitudinally, the planks would bend as in Fig. 7.15. Each

Fig 7.15 Shear due to bending

will move independently of the other so that the lower fibers of the upper plank move with respect to the upper fibers of the lower plank. If the member were solid, internal resistance to this sliding would act to prevent this sliding from occurring.

Before considering this resistance to sliding, let us first review the shearing stresses on an element of volume (see Chapter 5). Figure 7.16, which

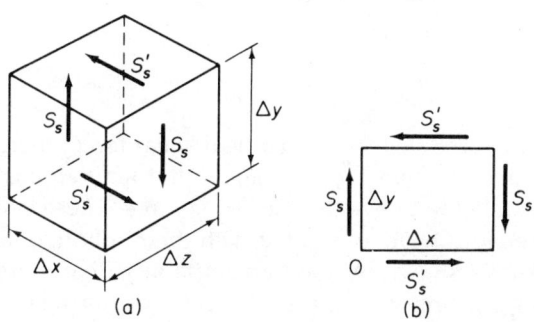

Fig 7.16 Stresses on an element (Fig 5.16 repeated)

is a repeat of Fig. 5.16, shows an elemental cube with shearing stresses S_s on the yz faces. Due to these stresses, the shearing stresses S_s' are induced on the xz faces. Since the area of a yz face is $(\Delta y)(\Delta z)$, the force on each such face is $S_s(\Delta y)(\Delta z)$. For the xz faces, the area is $(\Delta x)(\Delta z)$, and the force on each of these faces is $S'_s(\Delta x)(\Delta z)$. Taking moments about O for these force systems yields

$$S_s(\Delta y)(\Delta z)(\Delta x) = S'_s(\Delta x)(\Delta z)(\Delta y) \qquad (7.31)$$

from which we obtain

$$S_s = S'_s$$

From the above we conclude that a shearing stress cannot occur alone; it must always induce a numerically equal shearing stress on a perpendicular plane.

Figure 7.17 shows the transverse face of a beam that is subjected to

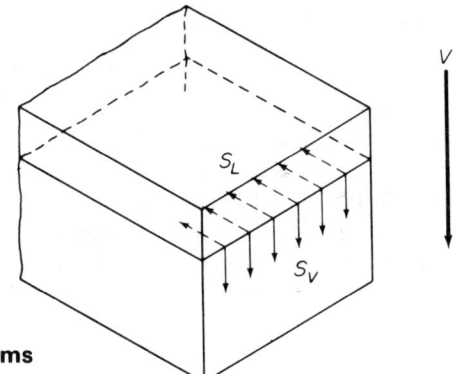

Fig 7.17 Longitudinal shear in beams

a vertical shearing force. Due to the vertical shear force, there are shearing stresses on the face at some location above the neutral axis, stresses that are essentially parallel to the vertical shearing force. These vertical shearing stresses must therefore have associated with them a horizontal (or longitudinal) shear stress as shown. A good assumption is that these longitudinal shearing stresses are uniform across the width of the beam. Thus at any plane, $S_v = S_L$.

It will be recalled from Chapter 6 that the difference in the moment at two positions along a beam can be obtained as the area under the shear diagram between these positions. Keeping this in mind, let us now consider a beam that is subject to any type of transverse loading. From this beam (Fig. 7.18), we will cut out a portion and consider it to be

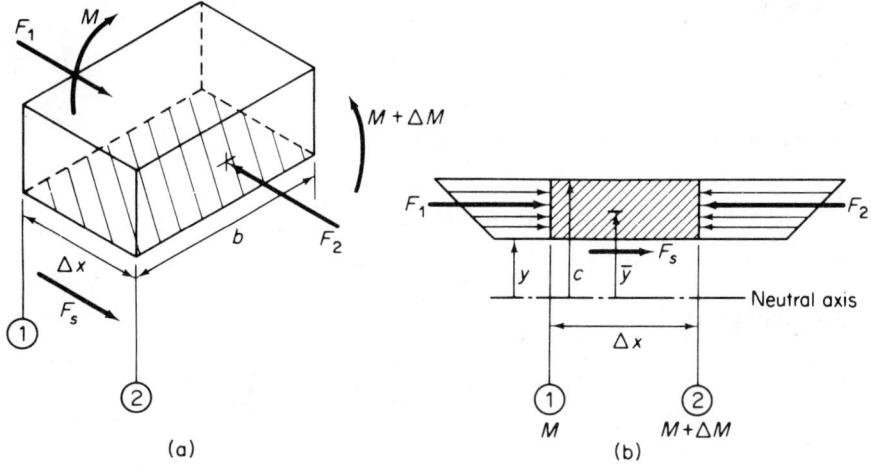

Fig 7.18 Shear in a beam

a free-body diagram in equilibrium. The left face of the element will, in general, be subject to a bending moment M. Due to this moment, there must be a horizontal force F, acting to the right as shown. At the right face, a distance Δx away, we shall assume the moment increases from that on the left face by an amount ΔM and is thus equal to $M + \Delta M$. Due to this moment, $M + \Delta M$, there must be a larger horizontal force F_2 acting in a direction opposite to that of F_1. As Fig. 7.18 shows, a shear force F_s must be present on the horizontal face [cross hatched in Fig. 7.18(a)] such that

$$F_1 + F_s = F_2 \tag{7.32}$$

Force F_1 can be evaluated using the Flexure Formula. Thus

$$F_1 = \sum \frac{M_1 y}{I} (A) \tag{7.33}$$

and similarly

$$F_2 = \sum \frac{M_2 y}{I} (A) \tag{7.34}$$

where Σ denotes the summation of all such terms over the area A and where A is the area of each face [(b)(c − y_1)]. The resisting shearing force F_s is

$$F_s = S_s b (\Delta x) \tag{7.35}$$

If equations (7.33), (7.34), and (7.35) are substituted in equation (7.32),

$$\sum \frac{M_1 y}{I}(A) + S_s b (\Delta x) = \sum \frac{M_2 y}{I}(A) \tag{7.36}$$

Rearranging:

$$S_s = \left(\frac{M_2 - M_1}{\Delta x}\right) \sum \frac{yA}{Ib} = \left(\frac{\Delta M}{\Delta x}\right) \frac{\Sigma yA}{Ib} \tag{7.37}$$

However, the vertical shear $V = \Delta M / \Delta x$; therefore

$$S_s = \frac{V \Sigma yA}{Ib} \tag{7.38}$$

Equation (7.38) can be further simplified by noting that ΣyA is the first moment of area about the neutral axis, i.e., $\bar{y}A$. Denoting

$$Q = \bar{y}A \tag{7.39}$$

the horizontal shearing stress in terms of the vertical shear can be expressed as

$$S_s = \frac{VQ}{Ib} \tag{7.40}$$

The student should note that Q is the first moment about the neutral axis of that part of the cross-sectional area of the beam above the horizontal plane at which the shearing stress is desired, and b is the width of the beam at the section in question. For shapes such as a rectangle, circle, "I," "T," and others, equation (7.40) yields the fact that the maximum shearing stress occurs at the neutral axis. While this is usually the case, it is not always true. (See Problems 7.36 and 7.38).

For the rectangular section shown in Fig. 7.19, we can derive both the maximum shear stress and the shear distribution quite readily. From Fig. 7.19(a), we can calculate the shear stress at the neutral axis.

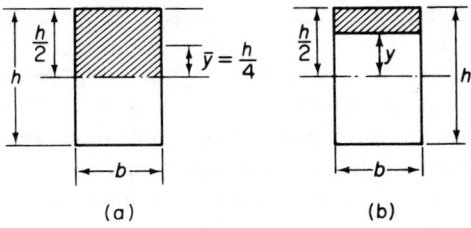

Fig 7.19   Rectangular beam

For this section

$$Q = \left(\frac{bh}{2}\right)\frac{h}{4}, \quad I = \frac{bh^3}{12}, \quad b = b$$

Therefore

$$S_{s_{max}} = \frac{VQ}{Ib} = \frac{V(bh/2) \times (h/4)}{(bh^3/12)\,b} = \frac{3V}{2bh} \qquad (7.41)$$

In other words, the maximum shear stress in a rectangular beam is 50% greater than the average shear stress across the section. The distribution of the stress can be obtained by considering Fig. 7.19(b). In this case consider b, V, and I to be constants.

Then $Q = b(h/2 - y)[y + \frac{1}{2}(h/2 - y)]$, which when simplified yields

$$S_s = \frac{V}{2I}\left(\frac{h^2}{4} - y^2\right) \qquad (7.42)$$

Equation (7.42) is the equation of a parabola that has a maximum value at the neutral axis.

For a circular cross section, the maximum shearing stress also occurs at the neutral axis and is given by

$$S_{s_{max}} = \frac{4V}{3\pi r^2} \qquad (7.43)$$

Illustrative problem 7.11

A rectangular beam L m long and having a concentrated load of F N at its center is to be designed so that the maximum allowable flexure stress and maximum allowable shear stress occur simultaneously. Determine the maximum length of the beam in terms of the allowable shear stress S_s, the allowable flexure stress S_t, and the dimensions of the beam, b and h.

Solution:

The maximum moment in the beam is at the center and equal to $FL/4$. Therefore the maximum flexural stress is

$$S_t = \frac{MC}{I} = \frac{(FL/4) \times (h/2)}{bh^3/12} = \frac{4\,FL}{3\,bh^2}$$

For maximum shear

$$S_s = \frac{3V}{2bh} = \frac{3}{2}\left(\frac{F}{2}\right)\frac{1}{bh} = \frac{3F}{4bh}$$

From flexure

$$F = \frac{S_t bh^2}{L}\left(\frac{4}{3}\right)$$

From the shear stress

$$F = \frac{S_s 4bh}{3}$$

Equating these terms

$$\left(\frac{4}{3}\right)\frac{S_t bh^2}{L} = \frac{S_s 4bh}{3}$$

and

$$L = \left(\frac{S_t}{S_s}\right)(h)$$

The variation of shearing stresses in beams of different cross sections is shown in Fig. 7.20. As noted earlier, the distribution of shear stresses

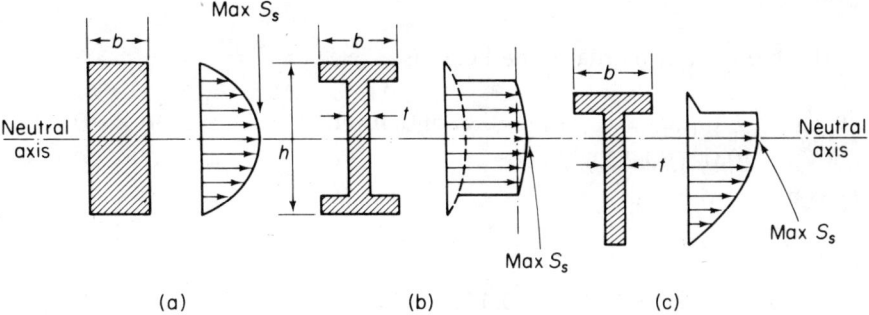

Fig 7.20 Variation of shearing stress in beams. Reprinted with permission from *Applied Strength of Materials* By A. Jensen and H. H. Chenoweth, McGraw-Hill Book Co., Inc., New York, 1967.

in a rectangular beam gives rise to a parabolic curve with a maximum value at the center. For the "I" and "T" sections it is interesting to note that most of the shear is taken by the web of the beam; the flanges are not effective in resisting the vertical shear. Therefore, as a good approximation to the maximum shear stress in an "I" beam, a channel, or a wide flange section, we can use the shear stress obtained by dividing the shearing force V by the web area. Thus

$$S_{s_{max}} = \frac{V}{bt} \qquad (7.44)$$

Illustrative problem 7.12

Determine the maximum shear stress in the "I" beam shown in Fig. 7.21. Compare the results with equation (7.44).

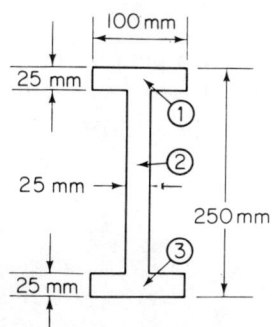

Fig 7.21 Illustrative problem 7.12

Solution:

The moment of inertia of the beam is

For ① $0.100(0.025)^3/12 + (0.025)(0.100)(0.1125)^2 = 3.18 \times 10^{-5} \text{ m}^4$
For ② $(0.025)(0.200)^3/12 \qquad\qquad\qquad\qquad\quad = 1.67 \times 10^{-5} \text{ m}^4$
For ③ $\qquad\qquad\qquad\qquad\qquad\qquad\qquad\qquad = 3.18 \times 10^{-5} \text{ m}^4$

$$I = 8.03 \times 10^{-5} \text{ m}^4$$

$Q = 0.100 \times 0.025 \times 0.1125 + 0.100 \times 0.025 \times 0.050$
$\quad = 4.063 \times 10^{-4}$

$$S_{s_{max}} = \frac{VQ}{Ib} = \frac{V \times 4.063 \times 10^{-4}}{8.03 \times 10^{-5} \times 0.025} = 202.4 \, V$$

Equation (7.44) yields $S_{s_{max}} = V/bh = V/0.75 \times 0.025 = 160 \, V$. Had we used the web area, not including the flanges, we would have obtained $S_{s_{max}} = V/0.200 \times 0.025 = 200 \, V$, which agrees quite well with the more exact value of 202.6 V. In any case, the procedure given by the applicable code should be used.

As a final application of shearing action in beams, let us consider the case of a beam composed of elements that are either riveted or bolted together. For the case shown in Fig. 7.22, it is necessary for the bolt

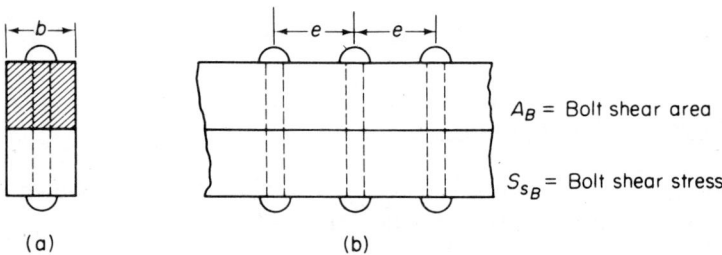

(a) (b)

A_B = Bolt shear area
S_{s_B} = Bolt shear stress

Fig 7.22 Composite beam

to resist the shearing force tending to slide the two portions of the beam relative to each other. If the bolts are assumed to take the shearing force, then each bolt must take the shearing stress multiplied by the shear area between bolts

$$F_B = S_s \times e \times b \qquad (7.45)$$

However, we can use equation (7.40) for S_s. Thus

$$F_B = \frac{VQ}{Ib} \times e \times b = \frac{VeQ}{I} = A_B(S_{s_B}) \qquad (7.46)$$

Equation (7.46) is not restricted to rectangular beams and can be used for other sections. For the beam shown in Fig. 7.23, the value of Q is

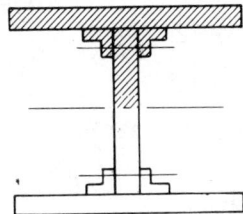

Fig 7.23 Built-up beam

the moment of the shaded cross section with respect to the neutral axis and I is the moment of inertia of the entire cross section of the beam without any deduction for rivet holes. If the vertical shear varies in the interval e, it is usual to take the maximum value in the interval, although the average value is more nearly correct. The use of the maximum value introduces an error on the safe side.

Illustrative problem 7.13

Two steel plates 200 mm × 25 mm are bolted together to form the composite beam shown in Fig. 7.24. If 25-mm-diameter bolts are used

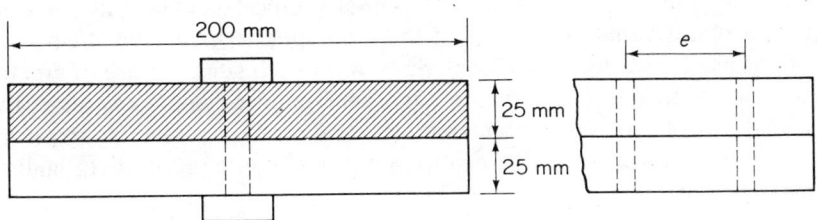

Fig 7.24 Illustrative problem 7.13

to bolt these plates together, determine the spacing of the bolts where the plates are subjected to a shear load of 4.5 kN. Use an allowable shear stress of 56 MPa for the bolts.

Solution:

Q above the neutral axis is

$$Q = (0.200)(0.025)\left(\frac{0.025}{2}\right), \quad I = \frac{bh^3}{12} = \frac{(0.200)(0.050)^3}{12}$$

$$= 2.083 \times 10^{-6} \text{ m}^4$$

$$Q = 6.25 \times 10^{-5} \text{ m}^3$$

$$\frac{VeQ}{I} = (A_B)(S_{s_B}); \quad e = \frac{(A_B)(S_{s_B})I}{VQ}$$

$$= \frac{\frac{\pi}{4}(0.025)^2 \times 56 \times 10^6 \times 2.083 \times 10^{-6}}{4\,500 \times 6.25 \times 10^{-5}} = 0.204 \text{ m}$$

7.6 CLOSURE

The study of stresses in beams leads to formulas such as the Flexure Formula, the shear formula, formulas for reinforced concrete beams, etc. Recognizing that a student has limited time to devote to a given subject, that he must "do" a given number of homework problems, and that on tests he must satisfactorily apply this material under conditions of duress, one cannot blame most students for adopting as an operating procedure the memorization of formulas. However, this is not really a learning process and all too often leads to the misapplication of formulas to situations that are beyond their scope. It is of primary importance (and the author cannot overstress this concept) that the assumptions made, the free-body diagrams used, and the limitations imposed on each situation are of greater importance than the end formula. One very good example of this exists in the study of stresses in rectangular reinforced concrete beams—the formulation used for designing beams is not applicable to the investigation of these beams.

Study each topic, study the derivations, and do not become dependent upon the memorization of formulas. The dividends reaped from this approach will exceed the investment by many orders of magnitude. It will also be invaluable for other professional applications (such as machine design) and for more advanced studies in this field.

PROBLEMS

The weight of the beam is to be neglected in all problems in this chapter.

7.1 A simply supported rectangular beam is 3 m long and has cross-sectional dimensions of 7.5 mm × 150 mm. Determine the maximum concentrated load that can be placed on the center of the beam if the maximum allowable stress is 140 MPa.

7.2 A cantilever beam carries a concentrated load as shown in Fig. P7.2. Determine the maximum bending stress in the beam.

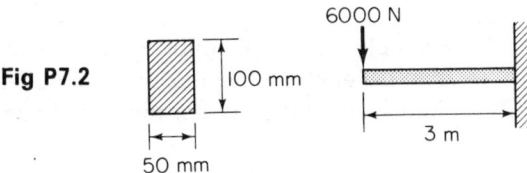

Fig P7.2

7.3 If a rectangular beam is twice as high as it is wide, determine its dimensions if the allowable bending stress is 140 MPa and the maximum bending moment is 12 000 N·m.

7.4 A solid circular bar, 50 mm in diameter, is used as a cantilever beam. The bar is 2 m long and must carry a concentrated load of 1 kN at its free end. Calculate the maximum bending stress in the bar.

7.5 If a square cross section is used to replace a circular cross section in a beam, determine the size of the square. Consider bending only.

7.6 A simply supported steel beam, 3 m long and having a cross-sectional area corresponding to an S5 × 10 beam is uniformly loaded. Determine the loading per m of beam if the maximum bending stress is 154 MPa.

7.7 A cantilever beam carries a uniform load over its entire span as shown in Fig. P7.7. Determine the maximum bending stress in the beam.

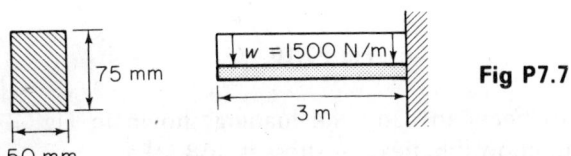

Fig P7.7

7.8 A 4-in. schedule 40 steel pipe is used to support a load as shown in Fig. P7.8. If the maximum bending stress is 154 MPa, determine the maximum load.

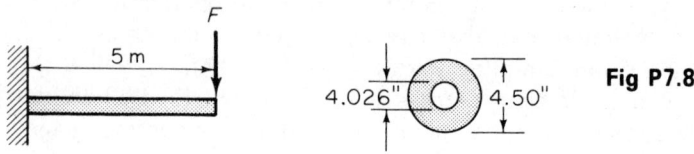

Fig P7.8

7.9 A simply supported hollow square beam has a uniform load distributed over its length. If the maximum bending stress is 126 MPa, determine the permissible load per metre of beam for a beam 3 m long.

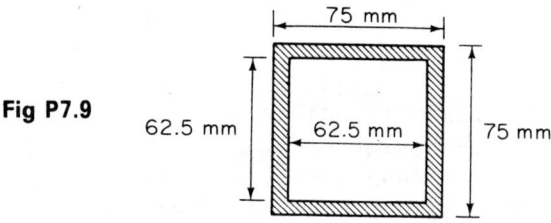

Fig P7.9

7.10 A cantilever beam carries the loads shown in Fig. P7.10. Calculate the maximum bending stress in the beam.

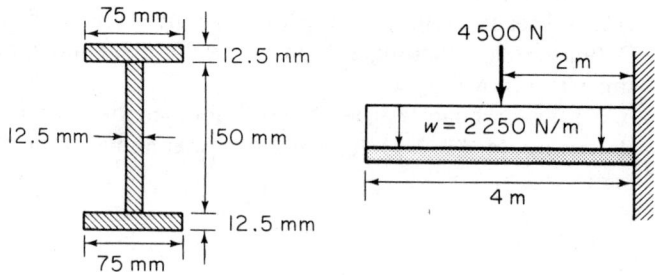

Fig P7.10

7.11 Select a W10 beam to carry the loading shown in Fig. P7.11 if the maximum allowable flexural stress is 168 MPa.

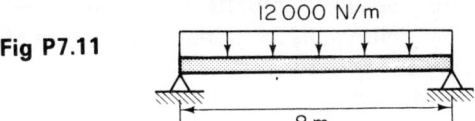

Fig P7.11

7.12 A cantilever beam carries a uniformly varying load over its span as shown in Fig. P7.12. Determine the maximum bending stress in the beam.

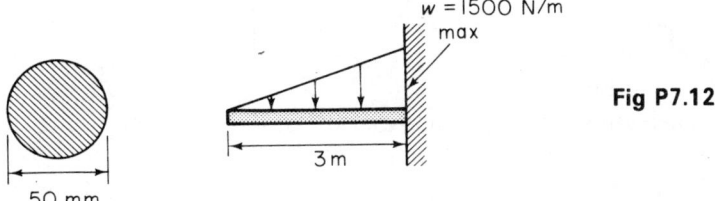

Fig P7.12

7.13 A simply supported beam is loaded as shown in Fig. P7.13. Determine the maximum bending stress in the beam.

Fig P7.13

7.14 If the beam in Problem 7.13 has an additional load of 10 kN placed at its center, what will the maximum bending stress be in the beam?

7.15 Select an S8 beam to carry the loading shown if the maximum allowable bending stress is 140 MPa.

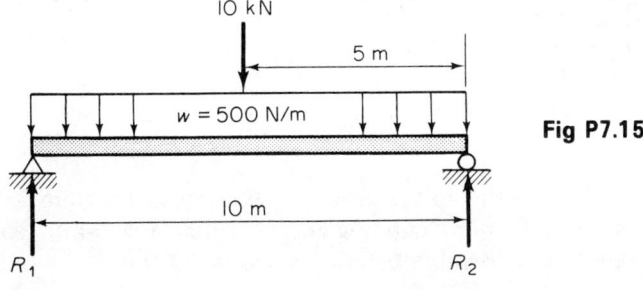

Fig P7.15

7.16 Select a W8 beam to carry the loading shown if the maximum allowable bending stress is 200 MPa.

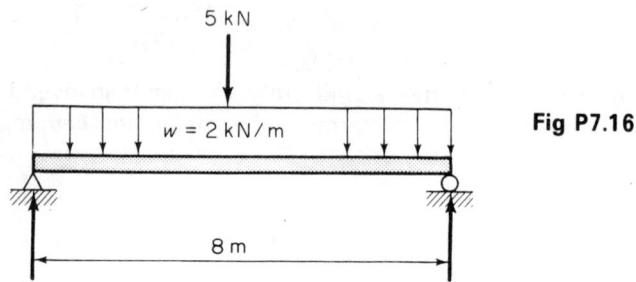

Fig P7.16

7.17 A wooden beam having exact dimensions of 100 mm × 200 mm is used as a floor joist and carries a loading of 1050 N/m. What is the maximum unsupported span if the allowable bending stress is 8.4 MPa?

7.18 A simply supported beam carries the loading shown in Fig. P7.18. Determine the maximum bending stress in the beam.

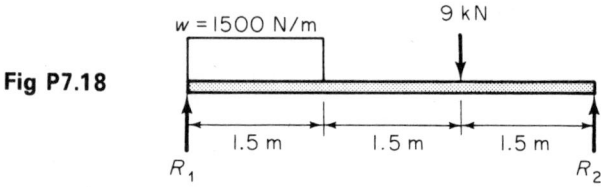

Fig P7.18

7.19 Solve Problem 7.18 if the beam has the cross section shown in Fig. P7.19.

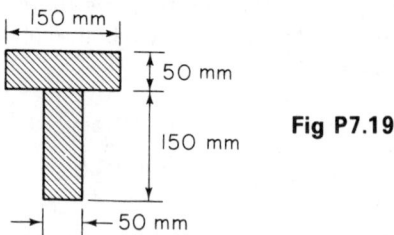

Fig P7.19

7.20 Calculate the stresses in the girders of Illustrative Problem 7.4.

7.21 What is the maximum bending moment that a W14 × 30 beam can carry if the allowable bending stress is 200 MPa?

7.22 A C9 × 20 channel is used as a beam. What is the maximum moment the beam can carry if the allowable bending stress is 164 MPa? Bending is about the X-X axis.

7.23 If an S12 × 35 beam is stressed to 140 MPa, determine the moment at the section of the beam where this occurs.

7.24 A "T" beam is used to carry a moment that places its upper portion in compression and its lower portion in tension. If the allowable stresses are 84 MPa in compression and 42 MPa in tension, determine w when the maximum compressive and tensile stresses occur simultaneously.

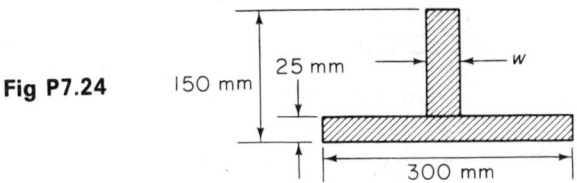

Fig P7.24

7.25 If the cross section of a beam is as shown in Fig. P7.25 and the tensile bending stress at plane A-A is 7 MPa, determine the stresses at the top and bottom of the section. The bottom is in tension.

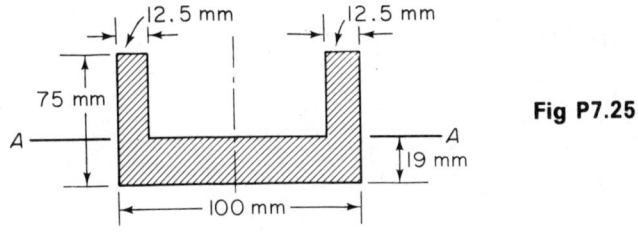

Fig P7.25

7.26 Two 100-mm × 100-mm × 12.5-mm angles are riveted together to form a "T" section. What is the maximum moment this beam can support if the maximum allowable flexural stress is 140 MPa?

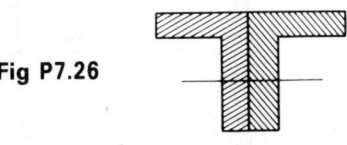

Fig P7.26

294 Strength of Materials for Engineering Technology

7.27 If two $\angle 4 \times 4 \times \frac{1}{2}$ equal angles are arranged in a manner similar to those in Problem 7.26, what maximum moment can such an arrangement support if the maximum allowable bending stress is 210 MPa?

7.28 Two channels are welded together to form a beam as shown. What is the maximum moment that this beam can support if the maximum allowable flexural stress is 140 MPa?

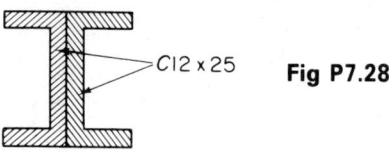

Fig P7.28

7.29 If the channels in Problem 7.28 are rotated 90° to form the letter "H," what maximum moment can the beam support? I_{yy} is 4.47 in.^{4} and A is 7.35 in.^{2}

7.30 If 25-mm cover plates are welded to the beam of Problem 7.28, what increase in maximum moment will occur?

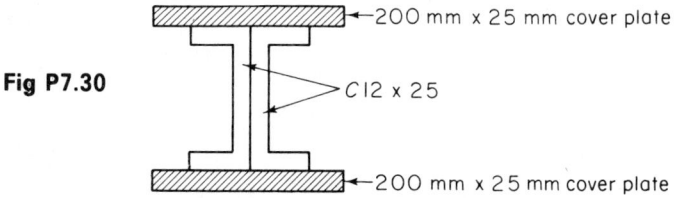

Fig P7.30

7.31 Two equal loads are placed on a beam 6 m long. If the maximum allowable bending stress is 154 MPa, determine the maximum value of these loads.

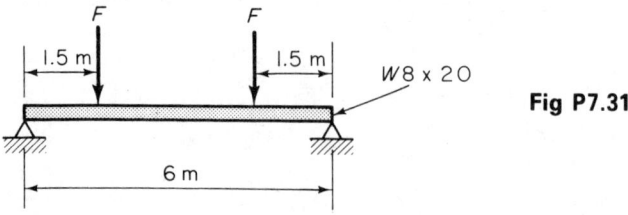

Fig P7.31

7.32 Two S3 × 7.5 beams are riveted together as shown in Fig. P7.32. Determine the maximum uniform load per foot that this beam can carry if the maximum allowable bending stress is 168 MPa.

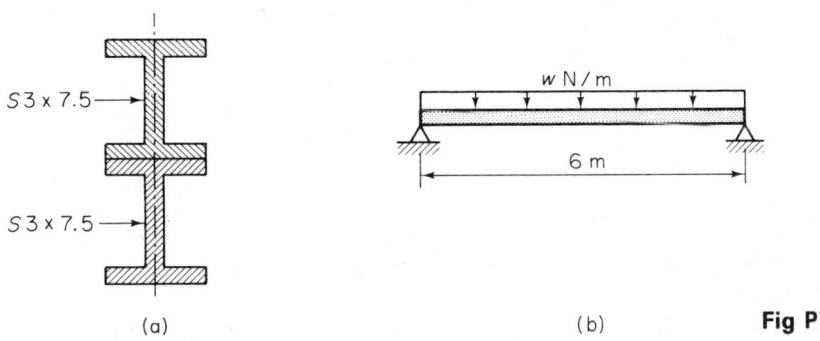

Fig P7.32

7.33 If the two "I" beams of Problem 7.32 are arranged as shown in Fig. P7.33, determine the maximum uniform load per metre that this structure can carry. Neglect the cover plates. Compare the result to that of Problem 7.32.

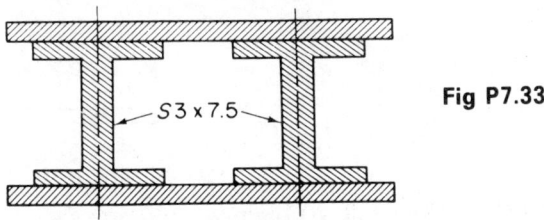

Fig P7.33

7.34 If the cover plates in Problem 7.33 are made 12.5 mm × 200 mm, what percentage increase in allowable load is obtained?

7.35 A beam is made of a full 200-mm × 200-mm wooden beam, reinforced on top and bottom by 19-mm steel plates as shown in Fig. P7.35. If it is to support a maximum moment of 50 000 N·m, with $E_w = 7$ GPa, $E_s = 210$ GPa, determine the maximum stress in each material.

Fig P7.35

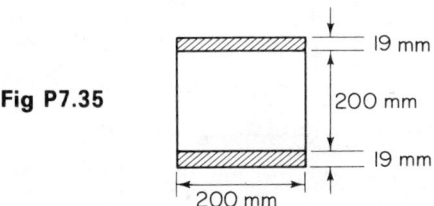

7.36 If the reinforcing plates of Problem 7.35 are rotated through 90°, determine the maximum stress in each material.

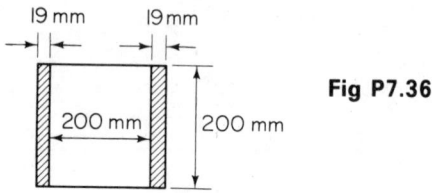

Fig P7.36

7.37 Solve Problem 7.35 if only the top steel plate is used for reinforcement.

7.38 Solve Problem 7.35 if the reinforcing plates are aluminum and E_{al} = 70 GPa.

7.39 A composite beam is made of wood and steel as shown in Fig. P7.39. If $E_s = 210 \times 10^9$ Pa and $E_w = 10.5 \times 10^9$ Pa, determine the maximum stress when the beam is subjected to a 50 000 N·m bending moment.

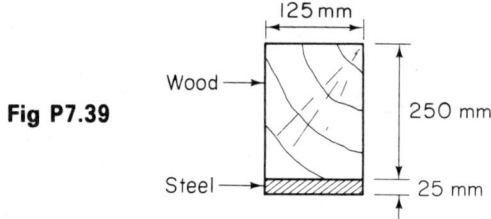

Fig P7.39

7.40 A wooden beam having a full 200-mm × 250-mm cross section is reinforced by a 25-mm steel plate on top and a 25-mm aluminum plate on the bottom. Determine the maximum allowable bending moment on this beam if E_w = 7 GPa, E_s = 210 GPa, E_{al} = 70 GPa, S_{wood} = 14 MPa, S_{steel} = 140 MPa and S_{al} = 105 MPa.

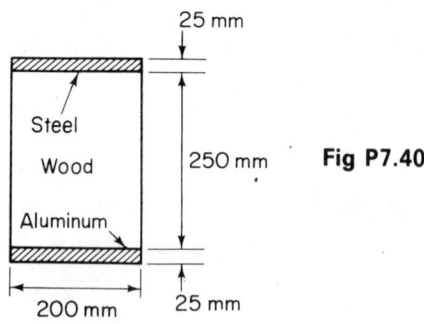

Fig P7.40

7.41 A reinforced beam is made of wood and steel elements as shown. Determine the maximum stress in the wood and steel if $E_w = 7$ GPa and $E_s = 210$ GPa. Is this construction an economical use of material?

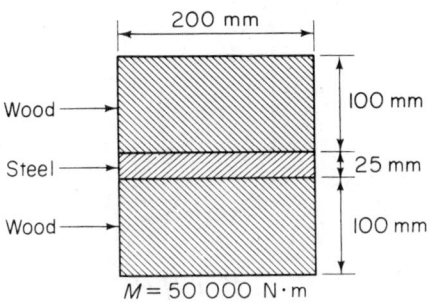

Fig P7.41

7.42 Design a reinforced concrete beam with balanced reinforcement to carry a bending moment of 80 000 N·m. The beam is to be 0.5 m wide, $n = 15$, $f_s = 126$ MPa, $f_c = 4.2$ MPa.

7.43 Design a reinforced concrete beam to resist a moment of 60 kN·m if $n = 15$. Assume the beam to be 200 mm wide and the reinforcement to be balanced. Use $f_c = 5.25$ MPa and $f_s = 105$ MPa.

7.44 A reinforced concrete beam is to resist a bending moment of 40 kN·m. If it is made 450 mm deep and has 1.5×10^{-3} m² of steel reinforcement, determine the maximum stress in the steel and concrete. $E_s = 210$ GPa and $n = 15$; $f_c = 5.25$ MPa and $f_s = 105$ MPa.

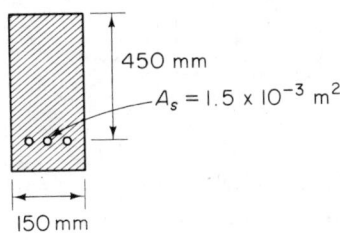

Fig P7.44

7.45 Determine the maximum allowable bending moment in the reinforced concrete beam shown in Fig. P7.45.

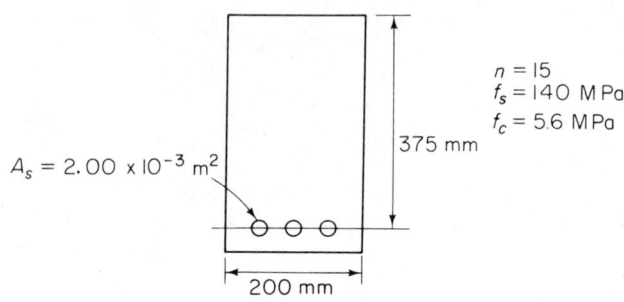

Fig P7.45

7.46 Design a rectangular reinforced concrete beam to carry a bending moment of 75 000 N·m if $d = 2b$. Assume balanced reinforcement and $n = 12$, $f_s = 126$ MPa, $f_c = 4.9$ MPa.

7.47 Determine the area of steel required for balanced reinforcement of a rectangular reinforced concrete beam if $b = 300$ mm, $d = 0.5$ m, $n = 15$, $f_s = 140$ MPa and $f_c = 5.6$ MPa.

7.48 Determine the maximum allowable bending moment that the reinforced concrete beam shown in Fig. P7.48 can support. Take $n = 14$, $f_c = 4.9$ MPa, $f_s = 126$ MPa.

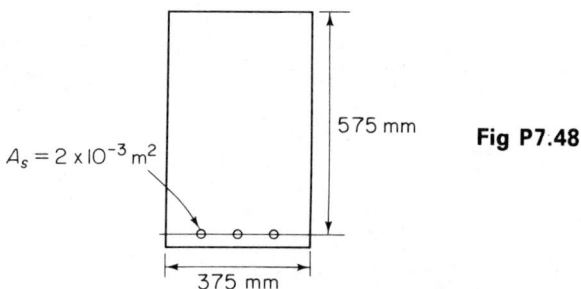

Fig P7.48

7.49 Investigate the reinforced concrete beam shown in Fig. P7.49 to determine the maximum allowable moment if $n = 20$, $f_c = 3.5$ MPa, $f_s = 105$ MPa, $b = 350$ mm, $d = 625$ mm, $A_s = 2.0 \times 10^{-3}$ m^2.

Fig P7.49

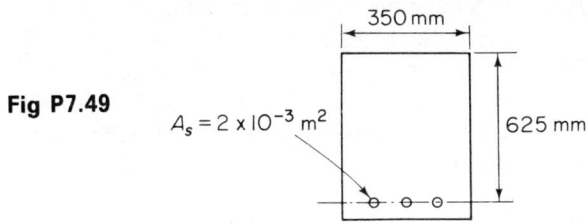

7.50 Calculate the maximum shear stress in Problem 7.2.
7.51 Calculate the maximum shear stress in Problem 7.10.
7.52 Calculate the maximum shear stress in Problem 7.12.
7.53 A simply supported rectangular wooden beam has a full 150-mm × 200-mm cross section. It is to support a concentrated load at the center of its span of 3 m; the allowable bending stress is 5.6 MPa and the allowable shear stress is 1.75 MPa. Determine the maximum value of the load.
7.54 Derive equation (7.43) in the text.
7.55 Determine the maximum value of F for the wooden beam shown in Fig. P7.55.

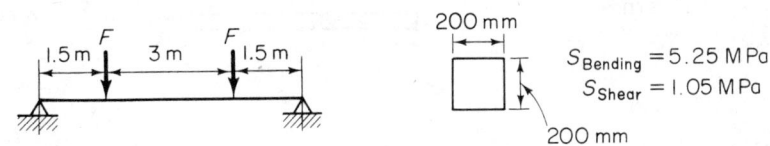

Fig P7.55

7.56 A $W10 \times 33$ beam has an allowable bending stress of 140 MPa and an allowable shear stress of 56 MPa. Determine the maximum value of F.

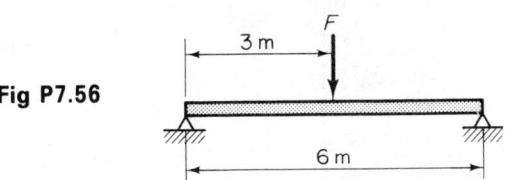

Fig P7.56

7.57 If the beam in Problem 7.56 is a W24 × 100 beam, determine the value of F.

7.58 Determine the maximum allowable uniform load per foot of beam for an S4 × 9.5 beam loaded as shown in Fig. P7.58. Use an allowable tensile stress of 154 MPa and an allowable shear stress of 84 MPa.

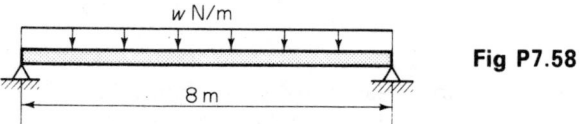

Fig P7.58

7.59 Solve Problem 7.58 for an allowable shear stress of 70 MPa.

7.60 A solid circular bar having a 100-mm diameter is to support a concentrated load at its center. If the allowable bending stress is 168 MPa and the allowable shear stress is 70 MPa, determine the maximum load the bar can carry.

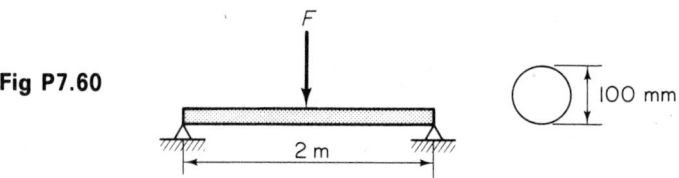

Fig P7.60

7.61 Solve Problem 7.60 for an allowable bending stress of 5.25 MPa and an allowable shear stress of 1.05 MPa.

7.62 For the shape shown in Fig. P7.62, where does the maximum shear stress occur?

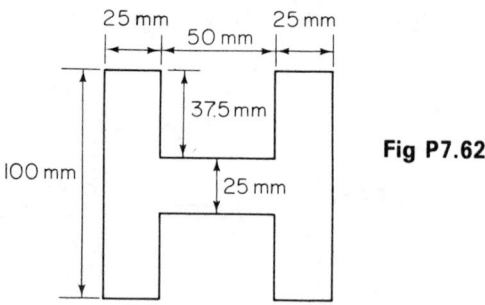

Fig P7.62

7.63 If the vertical shear force on the beam shown in Fig. P7.63 is 35 kN, determine the maximum shear.

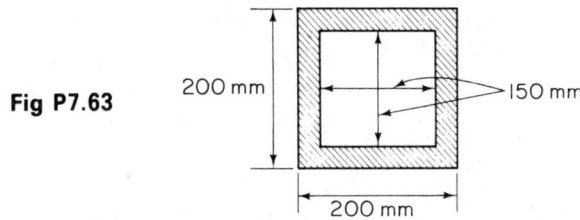

Fig P7.63

7.64 If the figure in Problem 7.63 is a rectangle that is 400 mm long, determine the maximum shear.

7.65 Does the maximum shear occur at the neutral axis of the beam shown in Fig. P7.65?

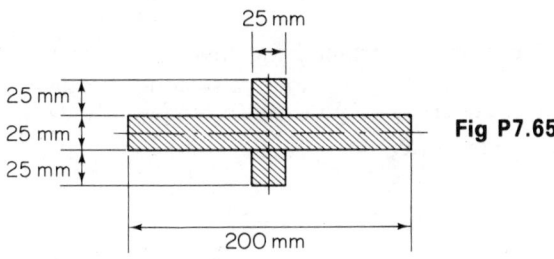

Fig P7.65

7.66 Determine the maximum shear force that a beam having the section shown in Fig. P7.66 can support. Use an allowable shear stress of 56 MPa.

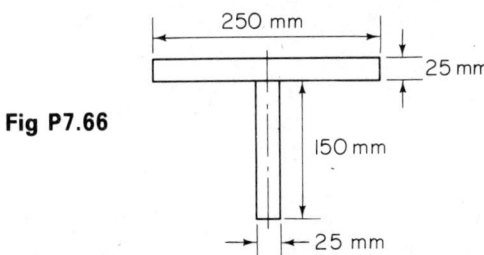

Fig P7.66

302 Strength of Materials for Engineering Technology

7.67 If two 100-mm × 100-mm bars are bolted together to form a beam as shown in Fig. 7.22, assuming a shear load of 22 kN, bolts of 19-mm diameter, and an allowable bolt shear stress of 56 MPa, determine the necessary bolt spacing.

7.68 Three 50-mm × 100-mm full planks are bolted together as shown in Fig. P7.68. Determine the maximum shear force if the bolts are 19 mm in diameter and are spaced 200 mm apart. The allowable shear stress in the bolts can be taken as 70 MPa.

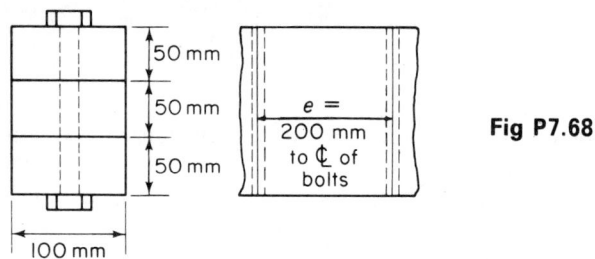

Fig P7.68

7.69 Two S3 × 7.5 beams are fastened together as shown by pairs of 25-mm rivets spaced at 150-mm intervals along the beam. If the allowable rivet shear stress is 224 MPa, determine the maximum shear force the rivets can carry.

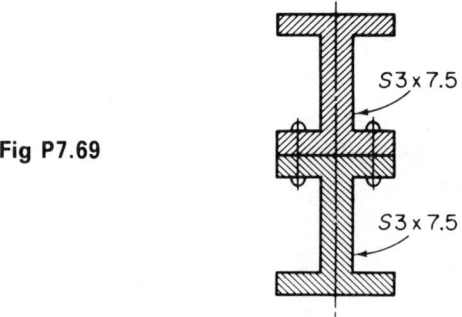

Fig P7.69

REFERENCES

American Institute of Steel Construction. *Manual of Steel Construction.* 7th ed., 1970.

Arges, K. P. and A. E. Palmer, *Mechanics of Materials.* New York: McGraw-Hill Book, Co., 1963.

Higdon, A., E. E. Ohlsen, and W. B. Stiles. *Mechanics of Materials*. 3rd ed. New York: John Wiley & Sons, 1976.

Jensen, A. and H. H. Chenoweth. *Applied Strength of Materials*. 3rd ed. New York: McGraw-Hill Book Co., 1971.

Levinson, Irving J. *Mechanics of Materials*. 2nd ed. Englewood Cliffs, NJ: Prentice-Hall, Inc., 1970.

Olsen, G. A. *Elements of Mechanics of Materials*. 3rd ed. Englewood Cliffs, NJ: Prentice-Hall, Inc., 1974.

Sheiry, E. S. *Elements of Structural Engineering*. International Textbook Co.

Singer, F. L. *Strength of Materials*. 2nd ed. New York: Harper and Row, 1962.

Timoshenko, S. and Donovan H. Young. *Elements of Strength of Materials*. 5th ed. New York: D. Van Nostrand Co., 1968.

8

The deflection of beams

8.1 INTRODUCTION

We have seen that when an elastic member is subjected to a load it will deflect and set up internal stresses as it resists the applied loading. In Chapter 7 our concern was to calculate these stresses; in this chapter we shall concern ourselves with the deflections of beams as they deform under the applied loading. Quite frequently, a restriction on the maximum allowable deflection of a beam is the governing design factor and limits the load that the beam can be designed for. One typical example is the limitation that the maximum deflection of a ceiling beam shall be less than $\frac{1}{360}$th of its span to prevent the cracking of plaster ceilings. The deformation of shafts under loads is also of considerable importance. The calculation of beam deflections is often approached from a mathematical viewpoint that requires the solution of a second order differential equation subject to the loading and the type of end supports of the beam. While this approach is mathematically straightforward, it presents formidable problems associated with the evaluation of the proper boundary conditions as well as in the algebra required to obtain the solution. Fortunately, an equivalent procedure can be carried out by relatively simple arithmetic computations based upon general rules derived from the conditions for equilibrium of the deformed beam. There are available several such arithmetic procedures, and in subsequent portions of this chapter we shall consider the *moment area method*, the procedure most generally in use. In each case we shall derive the necessary principles and then apply them to specific problems. It should be noted at this time that the deflection

Chapter 8, The Deflection of Beams   305

of a beam causes its ends to rotate and that this angular rotation must be considered in the design of structures. Frequently the angular deflection of the ends of a beam and/or its deflection will yield the conditions necessary for the solution of the category of beams that we have previously denoted as being statically indeterminate.

8.2 THE ELASTIC CURVE

In Chapter 7 we studied the action of a beam as it flexed under the action of the applied loading and subsequently derived the Flexure Formula. The assumptions made concerning the beam action are equally applicable to the present discussion and are thus repeated here:

1. The beam is straight before bending and has a uniform cross section throughout its length.
2. The material of the beam is homogeneous and its stress-strain curve is linear, i.e., stress is proportional to strain. It obeys Hooke's Law.
3. The beam has the same modulus of elasticity in compression and tension.
4. A transverse plane remains plane after bending.
5. The neutral axis is curved due to the bending, and the curved neutral plane is known as the *elastic curve* of the beam.

With the foregoing assumptions in mind, let us now consider the section of a beam shown in Fig. 8.1. On the left side of this section [Fig. 8.1(b)]

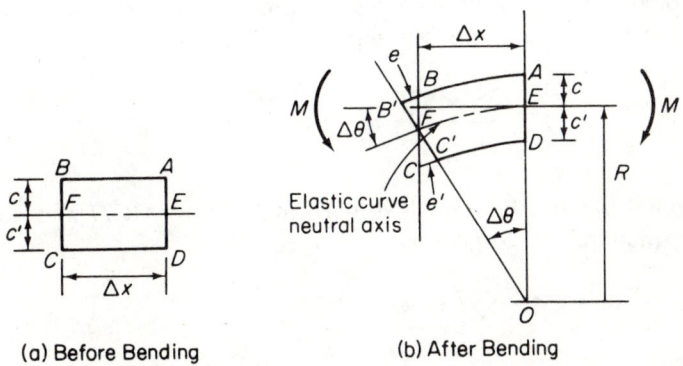

(a) Before Bending   (b) After Bending

Fig 8.1 The bending of a beam

a counter-clockwise couple M is shown; on the right side there is a clockwise couple shown. Thus two equal but opposite couples, M, act on the section in question. Originally the segment, Δx long, is straight as shown in Fig. 8.1(a). It bends as shown in Fig. 8.1(b). Due to the bending action, the upper fibers of the beam elongate by an amount e and the lower fibers contract by an amount e'. This action causes B to move to B' and C to move to C' while the neutral axis (the neutral plane is also the elastic curve of the beam) does not change in length although it does become curved. Due to the curvature, the plane BC rotates through the angle $\Delta\theta$ to the position $B'C'$, and it remains plane. The center of curvature of the beam can be found as the intersection of AD and $B'C'$ extended to meet at point O; the radius of curvature is denoted as R.

Let us now consider the sectors OEF and FBB', which we can see are geometrically similar. The following relationship must therefore hold

$$\frac{EF}{EO} = \frac{\Delta x}{R} = \frac{BB'}{FB} = \frac{e}{C} \tag{8.1}$$

therefore,

$$\frac{\Delta x}{R} = \frac{e}{C} \tag{8.2}$$

The symbol e has been used to denote the total deformation of the top fiber. The unit strain ε is $e/\Delta x$. Thus

$$\varepsilon = \frac{C}{R} \tag{8.3}$$

From Hooke's Law we know stress is proportional to strain, i.e., $S = E\varepsilon$ or $\varepsilon = S/E$. Inserting this into equation (8.3) yields

$$\frac{S}{E} = \frac{C}{R} \quad \text{or} \quad S = \frac{EC}{R} \tag{8.4}$$

At this point we invoke the Flexure Formula: $S = MC/I$ and substitute for S in equation (8.4) to give us

$$\frac{MC}{EI} = \frac{C}{R} \tag{8.5}$$

simplifying

$$\frac{1}{R} = \frac{M}{EI} \tag{8.6}$$

or

Chapter 8, The Deflection of Beams

$$M = \frac{EI}{R} \tag{8.6a}$$

Equations (8.6) and (8.6a) are the basic equations for the elastic curve of the bent beam. It is interesting to note that these equations were derived for the case of a beam in pure bending, but since they express the relation between bending moment and radius of curvature, they are applicable to all bent members.

One other relation will be derived at this time since it will be used in the following sections of this chapter—an expression for *the angle of curvature* (also known as *the deflection angle*). The tangent of the angle $\Delta\theta$ is $\Delta x/R$. However, since $\Delta\theta$ is a small angle, the tangent can be replaced by the angle $\Delta\theta$ in radians. Thus

$$\Delta\theta = \frac{\Delta x}{R} \tag{8.7}$$

Substituting for R from equation (8.6)

$$\Delta\theta = \frac{M}{EI}(\Delta x) \tag{8.8}$$

Illustrative problem 8.1

Determine the radius of curvature at the center of the beam shown in Fig. 8.2.

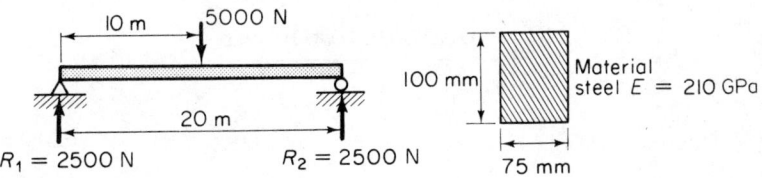

Fig 8.2 Illustrative problem 8.1

Solution:

The moment at the center of the beam = $R_1 \times L/2$ = 2500 × 10 = 25 000 N·m. The moment of inertia I is $bh^3/12$ = $\dfrac{0.075(0.100)^3}{12}$

$= 6.25 \times 10^{-6}\,\text{m}^4$. Using equation (8.6)

$$R = \frac{EI}{M} = \frac{210 \times 10^9 \times 6.25 \times 10^{-6}}{25\,000} = 52.5\,\text{m}$$

The radius of curvature is quite large, indicating that the curvature of the beam is relatively small, i.e., the deformed beam is not curved a great deal.

Illustrative problem 8.2

A steel band saw, 1 mm thick, 18 mm wide, is bent over a pulley wheel whose diameter is 1 m. Determine the stress in the saw blade if $E = 210$ GPa.

Solution:

This problem can be solved in two ways. The first is simply to use equation (8.4) directly as follows

$$S = \frac{CE}{R} = \frac{(0.001/2) \times 210 \times 10^9}{1/2} = 210^4\,\text{MPa}$$

Notice that the width of the saw blade (18 mm) did not enter into the solution.

The alternate approach is to use the Flexure Formula. The procedure is as follows

$$M = \frac{EI}{R} = \frac{210 \times 10^9 (0.018)(0.001)^3/12}{1/2} = 0.630\,\text{N} \cdot \text{m}$$

By the Flexure Formula

$$S = \frac{MC}{I} = \frac{0.630 \times 0.001/2}{0.018(0.001)^3/12} = 210\,\text{MPa}$$

Both methods yield equivalent results, but the second procedure shows that the direct load on the saw is $0.63/(\frac{1}{2}) = 1.26\,\text{N}$ ($F \times R/R$). The direct stress in the blade is F/A, which equals $1.26/0.001 \times 0.018 = 70$ kPa, which obviously is not the critical design stress in the saw blade.

8.3 THE MOMENT AREA PROPOSITIONS

Consider a beam that has been bent by the loading applied to it. Such a case is shown in Fig. 8.3 with the curvature of the bent beam greatly

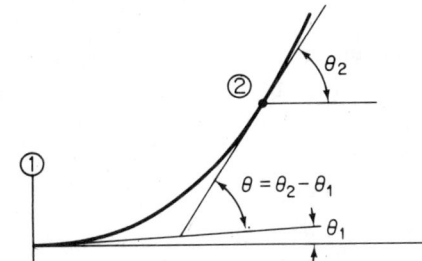

Fig 8.3 Beam curvature

exaggerated. Let us draw tangents to the elastic curve at points ① and ②. The angle that the tangent to the elastic curve makes with the horizontal (undeformed curve) is denoted as θ, the angle at ① as θ_1 and the angle at ② as θ_2. We can see from Fig. 8.3 that the difference in angle between the tangents drawn at points ① and ② is

$$\theta = \theta_2 - \theta_1 \tag{8.9}$$

and we may define the angle of curvature of a segment of a bent beam to be the angle formed by the two end tangents to the elastic curve.

Let us now study the situation shown in Fig. 8.4. Due to some applied

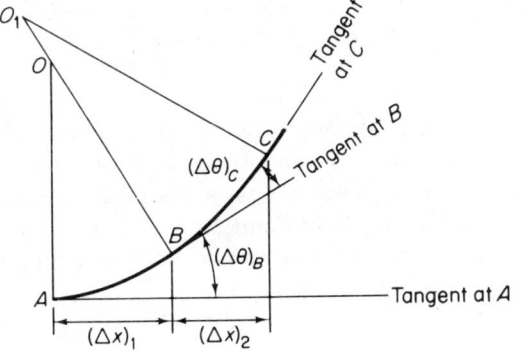

Fig 8.4 The elastic curve

constant moment M_1, the segment of the neutral axis $(\Delta x)_1$ deflects to form the elastic curve. We shall assume that the length of the elastic curve AB is very nearly equal to the undeformed section of the neutral axis. This conclusion is based upon the developments of Sec. 8.2. Thus

AB and $(\Delta x)_1$ are taken to be essentially equal. If an adjacent section of the beam $(\Delta x)_2$ is subjected to a constant moment M_2 the elastic curve will undergo further curvature indicated by point C. The change in curvature from A to C equals $(\Delta\theta)_B + (\Delta\theta)_C$, i.e.,

$$\Delta\theta = (\Delta\theta)_B + (\Delta\theta)_C \tag{8.10}$$

However, in equation (8.8) we had already derived an expression relating the change in angle to the applied moment, the modulus of elasticity of the material, and the properties of the cross section of the beam. Using this relation with equation (8.10) yields

$$\Delta\theta = \frac{M_1}{EI}(\Delta x)_1 + \frac{M_2}{EI}(\Delta x)_2 \tag{8.11}$$

For the present, E and I will be taken to be constants. If we continue this summation to some point on the beam, D

$$\theta \Big|_A^D = \theta_D - \theta_A = \sum_A^D \frac{M}{EI}(\Delta x) \tag{8.12}$$

where $\sum_A^D$ is used to denote the sum of $M/EI(\Delta x)$ terms from A to D.

Equation (8.12) is the mathematical statement of the first moment area principle. The product term EI represents *the flexural rigidity* of the beam.

Principle 1: The difference in the slope angles (in radians) between tangents drawn to any two points on the elastic curve of a beam is equal to the area of the moment curve between the two points divided by EI, the flexural rigidity of the beam.

In order to maintain the conventions already adopted for moments in beams, we will take areas due to positive moments to be positive and areas due to negative moments to be negative. Thus, a net positive area of the M/EI summation indicates that the right-hand tangent (point C in Fig. 8.4) makes a positive (or counter-clockwise) angle with respect to the left-hand tangent (point A in Fig. 8.4).

Illustrative problem 8.3

A cantilever beam is fixed at one end and is loaded by a concentrated force at the free end. Determine the angular deflection of the beam if E and I are constant.

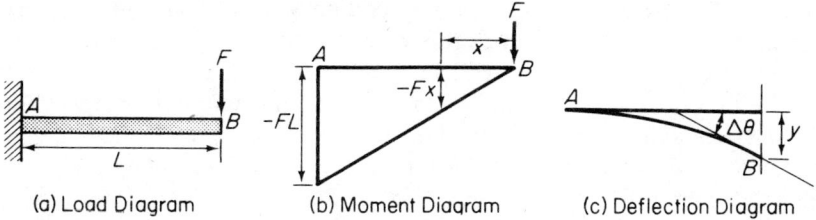

(a) Load Diagram   (b) Moment Diagram   (c) Deflection Diagram

Fig 8.5 Illustrative problem 8.3

Solution:

As shown in Fig. 8.5(b), the moment is negative and increases (in the absolute sense) from zero at the free end to $-FL$ at the fixed end. The area between A and B of the moment curve is the area of the triangle whose base is L and whose altitude is $-FL$. This area is $-FL^2/2$, and the angle between tangents to the elastic curve at points A and B is

$$\theta = \frac{-FL^2}{2EI} \text{ radians}$$

with the negative value indicating a positive rotation of point B with respect to point A. Since the beam is built in at the fixed end the tangent at point A is horizontal.

The next step is to develop a procedure that will enable us to determine the deflection of a beam at any point along the beam. By definition we shall take the deflection of a beam at a given section to be equal to the distance between the elastic curve and the neutral axis measured normal to the neutral axis. In the development of Principle 1, it was considered that the elastic curve of the beam was composed of a series of small arcs having varying curvature; we shall now proceed to extend this concept to determine beam deflections. For this purpose, refer to Fig. 8.6, which

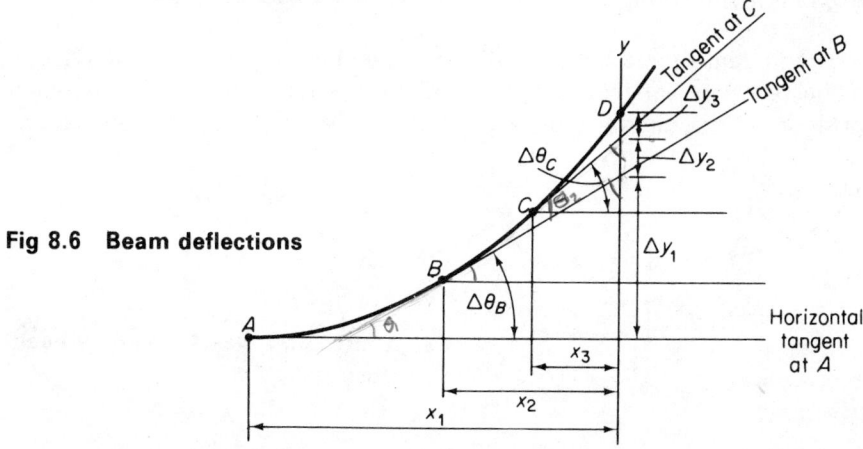

Fig 8.6 Beam deflections

is essentially the same as Fig. 8.4 with certain additions, specifically, $(\Delta y)_1$ and $(\Delta y)_2$ are shown as well as the x_1 and x_2 distances. For convenience we will take the tangent at A to be horizontal and to coincide with the undeformed neutral axis of the beam. Due to this assumption, the chords AB and BC will coincide with the tangents at B and C since the arcs are taken to be very small. If the line Y-Y is made normal to the neutral axis (and therefore vertical in this case) at D, the intercept of the tangents at B and C on Y-Y can be taken to be the tangent of the arc corresponding to the angle made by the tangents. Thus for the intercept due to the tangent at B

$$(\Delta y)_1 = (x_1)(\Delta\theta)_1 \tag{8.13}$$

and due to the tangent at C

$$(y)_2 = (x_2)(\Delta\theta)_2 \tag{8.14}$$

The total displacement of D from the tangent at A is denoted as y. Thus

$$y = (\Delta y)_1 + (\Delta y)_2 = x_1(\Delta\theta)_1 + x_2(\Delta\theta)_2 \tag{8.15}$$

Using equation (8.8) with equation (8.15) yields

$$y = x_1 \frac{M_1 \Delta x}{EI} + \frac{x_2 M_2 \Delta x}{EI} \tag{8.16}$$

Equation (8.16) allows us to state Principle 2 of the moment area method as follows:

Principle 2: *The displacement of a point on the elastic curve of a beam from the tangent at another point on the curve measured normal to the unstrained neutral axis is equal to the moment of the area of the M/EI diagram included between the two points taken about the first point.*

The student should note that displacement only becomes true deflection if the reference tangent is parallel to the unstrained (usually horizontal) position of the neutral axis. Figure 8.7 shows the tangential deviations

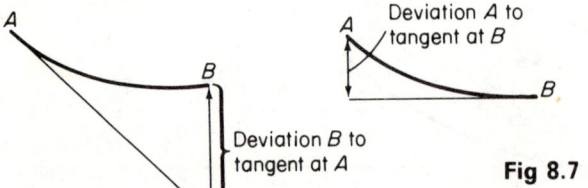

Fig 8.7 **Deviations on a beam**

of point A with respect to a tangent at B and of point B with respect to a tangent at point A. Note that these are not equal, and care must be exercised to keep the proper order in mind when setting up a problem. Also bear in mind the convention that moments of areas of positive moment diagrams are considered to be positive and produce positive deflections; moment arms are therefore positive when they extend into the bending moment diagram lying between the points in question. The following illustrative problems will serve to exemplify the foregoing discussion.

8.3a Cantilever beams

Illustrative problem 8.4

Determine the deflection of the free end of the cantilever beam shown in Fig. 8.5.

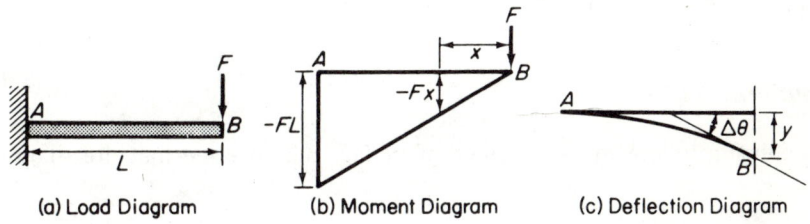

(a) Load Diagram   (b) Moment Diagram   (c) Deflection Diagram

Fig 8.5   repeated

Solution:

We have chosen the cantilever beam as the first illustration of both Principles 1 and 2 since the built-in end of the beam does not rotate, and therefore, the tangent to the elastic curve at this end is horizontal. This condition simplifies the calculation considerably. Applying the second principle, we can state that the deflection at B (the free end) from the tangent to the elastic curve at A (the built-in end) is the moment of the area of the M/EI diagram between points A and B taken about point B. The area of the M/EI diagram (since E and I are constants) is $(-FL)(L/2)(1/EI) = -FL^2/2EI$. In order to obtain the moment of this area about point B, it is necessary to determine the distance of the centroid

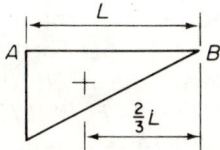

Fig 8.5(d)   Centroid of a triangle

of this area from point B. This distance is the moment arm to use in determining the desired moment. For a triangle whose base is L, this distance is $\frac{2}{3}L$. Thus the moment is

$$\left(\frac{-FL^2}{2EI}\right)(\tfrac{2}{3}L) = \frac{-FL^3}{3EI}$$

The moment arm is taken as positive since it lies between A and B. The displacement $(-FL^3/3EI)$ of B from the tangent at A is therefore the deflection of B from its unloaded position since the tangent at A is horizontal.

Illustrative problem 8.5

Determine the deflection of the cantilever beam of Illustrative Problems 8.3 and 8.4 at the center of the beam.

Solution:

Referring to the moment diagram of Fig. 8.8, we see that the diagram

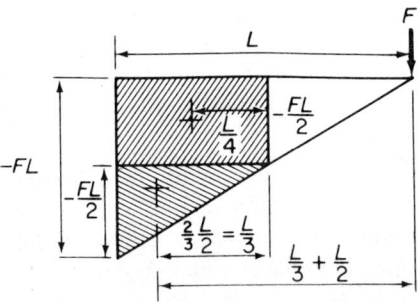

Fig 8.8 Moment diagram for illustrative problem 8.5

can be assumed to consist of a triangle and a rectangle. Since it is simpler to treat simple geometric figures such as rectangles and triangles, we will use this procedure wherever possible to simplify problems. For the rectangular portion, the area is $(1/EI)(-FL/2)(L/2)$ and the moment arm is $L/4$. The moment is therefore $(-FL/2)(L/2EI)(L/4) = -FL^3/16EI$. For the triangle, the area is $(1/EI)(-FL/2)(L/2)\tfrac{1}{2}$ and the moment arm is $\tfrac{2}{3}(L/2) = L/3$. The moment of the triangle about the center is $(1/EI)(-FL/2)(L/2)(\tfrac{1}{2})(L/3)$ $= -FL^3/24EI$. The total deflection is the sum of these two terms, namely,

$$y_{center} = \left(\frac{-FL^3}{16EI}\right) + \left(\frac{-FL^3}{24EI}\right) = \frac{-5FL^3}{48EI}$$

Illustrative problem 8.6

A concentrated load F is placed at the mid-span of a uniform cantilever beam. Determine the deflection of the beam at the midpoint and at the free end.

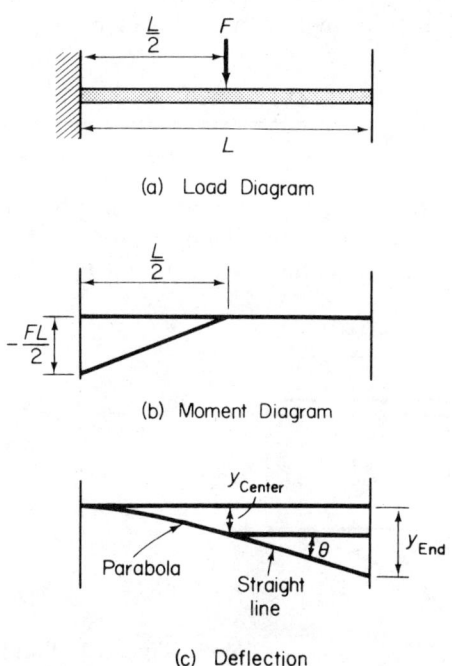

Fig 8.9 Illustrative problem 8.6

Solution:

The deflection at the mid-span is obtained exactly as it was in Illustrative Problem 8.4. Thus

$$y_{center} = \left(\frac{-FL}{2}\right)\left(\frac{L}{2}\right)\left(\frac{1}{2}\right)\left(\frac{L}{3}\right)\left(\frac{1}{EI}\right) = \frac{-FL^3}{24EI}$$

From the moment diagram, we note that there is zero additional moment contributed at any section of the beam to the right of the load. From Principle 1, this is to be interpreted to mean that there is no additional angular change beyond the midpoint as one proceeds to the right of the midpoint. This is shown schematically in Fig. 8.9(c). We see that the deflection at the free end is $y_{center} + (L/2)\theta$ where we have substituted θ for tan θ. Using Principle 1, $\theta = (-FL/2)(L/2)(\frac{1}{2})(1/EI) = -FL^2/8EI$. $(L/2)\theta$ therefore $= (L/2)(-FL^2/8EI) = -FL^3/16EI$. The total free end deflection is finally given as $y_{end} = y_{center} + (L/2)\theta = (-FL^3/24EI) + (-FL^3/16EI) = -5FL^3/48EI$. The student will note that this problem differs from the two previous problems in that here the elastic curve consists of two parts rather than a single continuous element.

Illustrative problem 8.7

A cantilever beam supports a uniformly distributed load of w N/m over its full length. Determine the deflection of the free end if the beam has a constant cross section and has a constant E.

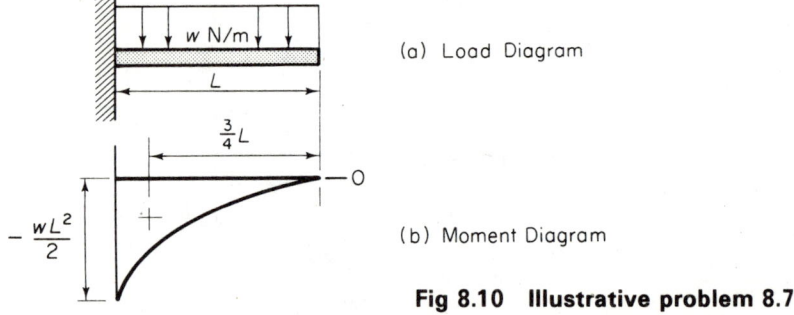

Fig 8.10   Illustrative problem 8.7

Solution:

As shown in Fig. 8.10(b), the moment diagram due to the load is a parabola with a maximum negative value of $-wL^2/2$. The properties of various areas were discussed in Chapter 4 and for convenience Table 4.1 is repeated at this time. The spandrel noted in the table is also a parabola of the nth degree. Since the properties of the parabola are needed quite often in our present study, the properties of various parabolas are presented in Table 8.1. Using either Table 4.1 or Table 8.1 for the problem under discussion yields an area of $(1/3EI)(-wL^2/2)(L)$ for the moment diagram and a moment arm of $\frac{3}{4}L$. The deviation of the free end from

Chapter 8, The Deflection of Beams

Table 4.1 (*repeated*)

Section	Area	$\bar{x}$	$\bar{y}$
Square	b^2	$\dfrac{b}{2}$	$\dfrac{b}{2}$
Rectangle	bh	$\dfrac{b}{2}$	$\dfrac{h}{2}$
Triangle	$\dfrac{1}{2} bh$	$\dfrac{b}{3}$	$\dfrac{h}{3}$
Semi-circle	$\dfrac{\pi R^2}{2}$	0	$\dfrac{4R}{3\pi}$ or $0.4244R$
Quarter-circle	$\dfrac{\pi R^2}{4}$	$\dfrac{4R}{3\pi}$ or $0.4244R$	$\dfrac{4R}{3\pi}$ or $0.4244R$
Quadrant of ellipse	$\dfrac{\pi ab}{4}$	$\dfrac{4a}{3\pi}$ or $0.4244a$	$\dfrac{4b}{3\pi}$ or $0.4244b$
Spandrel $y=kx^n$, Vertex	$\dfrac{bh}{n+1}$	$\dfrac{b}{n+2}$	$\left\{\dfrac{n+1}{4n+2}\right\}h$

Table 8.1 Properties of Parabolas

	Shape	Area	Centroid x
①	Vertex, h, b, Parabola – $y = ax^2$	$\frac{1}{3}bh$	$\frac{b}{4}$
②	Vertex, h, b, Cubic Parabola $y = ax^3$	$\frac{1}{4}bh$	$\frac{b}{5}$
③	Vertex, h, b, Parabola	$\frac{2}{3}bh$	$\frac{3}{8}b$
④	h, b, Parabola m^{th} Degree (Spandrel) $y = ax^m$	$\frac{bh}{m+1}$	$\frac{b}{m+2}$

the tangent at the fixed end (also the deflection of the free end in this case) is therefore

$$y_{\text{free end}} = \frac{1}{3EI}\left(\frac{-wL^2}{2}\right) L\left(\tfrac{3}{4}L\right) = \frac{-wL^4}{8EI} = \frac{-WL^3}{8EI}$$

where W, the total load, is equal to wL.

Illustrative problem 8.8

A cantilever beam supports a uniformly distributed load of 400 N/m over its length of 3 m. If the beam is an S7 × 20 beam having a modulus of elasticity of 210 GPa, determine the deflection of the free end.

Solution:

Using the results of Illustrative Problem 8.7, and $I = 42.4$ in.^{4}

$$y_{\text{free end}} = \frac{-(wL)L^3}{8EI} = \frac{(-400)(3)(3)^3}{8 \times 210 \times 10^9 \times 42.4 \times (0.0254)^4}$$
$$= 1.09 \times 10^{-3} \text{ m} = 1.09 \text{ mm}$$

Since the data for the S7 × 20 beam is given in in.^{4}, the value of 42.4 in.^{4} has been multiplied by $(0.0254)^4$ to obtain consistent SI units.

Illustrative problem 8.9

As a final example involving the simple cantilever beam, let us determine the deflection of the free end of a uniform cantilever beam that is subjected to an end moment M.

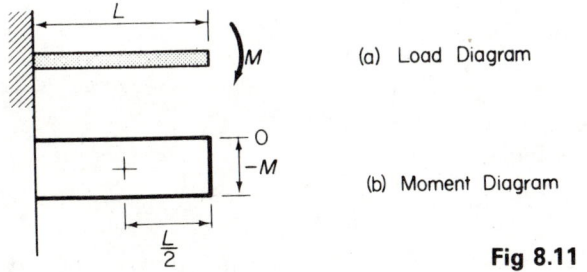

Fig 8.11   Illustrative problem 8.9

Solution:

Since the moment diagram is a rectangle, the area of the M/EI diagram is $(1/EI)(-ML)$ and its moment arm is $L/2$. Using these facts and noting that a tangent to the deflection curve at the built-in end is horizontal yield

$$y_{\text{free end}} = -\frac{1}{EI}\left(\frac{ML^2}{2}\right) = \frac{-ML^2}{2EI}$$

8.3b Simply supported beams

Later in this chapter we shall return to the cantilever beam as we develop methods of treating statically indeterminate beams. For the present, we shall leave the cantilever beam and direct our discussion to the simply supported beam. We shall first develop a general procedure for handling this type of problem and then illustrate it with several examples. For this

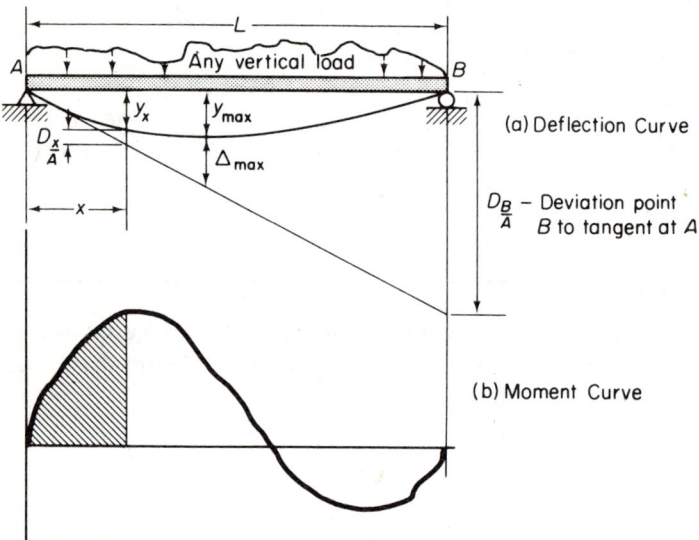

Fig 8.12 Simply supported beam

discussion refer to Fig. 8.12, which depicts a uniform beam, simply supported at its ends and carrying any arbitrary vertical load.

Referring to Fig. 8.12(a), we have from similar triangles at section x

$$\frac{y_x + Dx/a}{x} = \frac{D_{B/A}}{L} \tag{8.17}$$

where $D_{B/A}$ is used to denote the deviation of point B to the tangent to the elastic curve at point A. Our concern at the moment is to evaluate the deflection y_x at x. Thus

$$y_x + D_{x/A} = \frac{x}{L} D_{B/A} \tag{8.18}$$

The deviation $D_{B/A}$ is obtained as the moment of the M/EI diagram between points A and B about point B. Since $D_{x/A}$ is the deviation at x from a tangent to the elastic curve at A, it is evaluated as the moment of the M/EI curve between x and A [shaded in Fig. 8.12(b)] about x. Thus we can calculate all of the terms needed in equation (8.18) to determine y_x. We now note that the location of the section at which the deflection is maximum can be determined by two independent methods. Both methods use the fact that the tangent to the beam at the section where the deflection is maximum is horizontal. Thus, in Fig. 8.13 we note that the angles

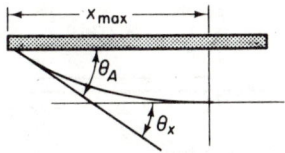

Fig 8.13 Simply supported beam—maximum deflection

θ_A and θ_x must be equal. This requirement can be evaluated by applying Principle 1 to yield

$$\theta_A = \theta_x; \left(\text{Area } \frac{M}{EI}\right)_{A-B} = \left(\text{Area } \frac{M}{EI}\right)_{A-x_{max}} \tag{8.19}$$

or in words: the area of the M/EI diagram between the ends of the beam must equal the area between the left support and the section at which the maximum deflection occurs. This locates the section at which the maximum deflection occurs and its value y_{max} is determined as previously outlined for any section x.

The second alternate (but equivalent) method of determining the location of the section at which the maximum deflection occurs, is based upon the geometry in Fig. 8.14. The deviation of A from the tangent to the

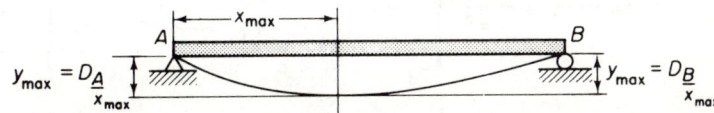

Fig 8.14 The maximum deflection of simple beams

deflection curve at the section where the maximum deflection occurs must also equal the deviation of B from this same tangent since both deviations equal the maximum deflection. Both deviations are found from the second moment area principle enabling us to determine the location of section x_{max}.

Illustrative problem 8.10

A simply supported beam has a concentrated load F at its center. If the length of the beam is L and the beam has a constant cross section, determine the maximum deflection of the beam.

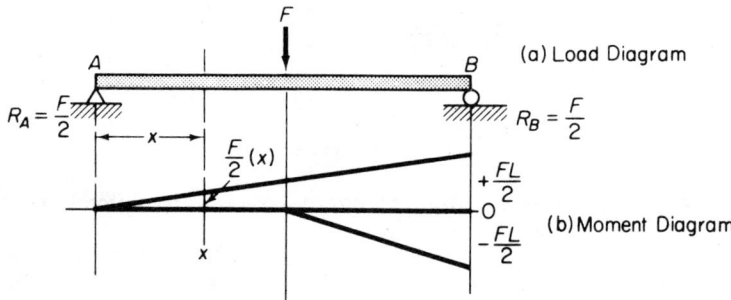

Fig 8.15   Illustrative problem 8.10

Solution:

The beam is symmetrically loaded so that the maximum deflection occurs at its center. Taking advantage of this fact, we note that the deflection at the center can be obtained by our considering the center to be fixed and either end loaded by concentrated loads of $F/2$, as shown in Fig. 8.15(c). The end deflection of the cantilever is equal to the central deflection

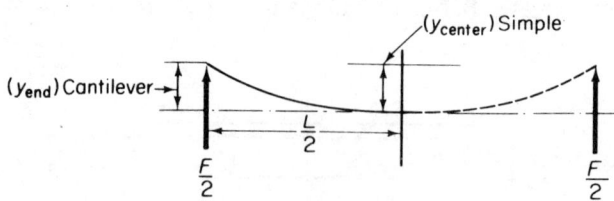

Fig 8.15(c)   Illustrative problem 8.10

of this simple beam. The end load is $F/2$ and the length of beam is $L/2$ for the cantilever, yielding

$$(y_{end}) \text{ cantilever} = (y_{center}) \text{ simple} = \frac{(F/2)(L/2)^3}{3EI} = \frac{-FL^3}{48EI} = y_{max}$$

Had we not noted and used symmetry, we would first have had to calculate the location of the section of maximum deflection x, which is also the section at which the tangent to the deflection curve is horizontal. In order to simplify the work, the moment diagram of Fig. 8.15(b) is shown in parts. We have for θ_A

$$\theta_A = \frac{1}{EI}\left[\frac{(FL)}{2}\frac{(L)}{2} - \frac{(FL)}{2}\left(\frac{L}{4}\right)\right] = \frac{1}{EI}\left(\frac{FL^2}{8}\right)$$

If we assume that x is to the left of the center, as shown in Fig. 8.15(b), $\theta_x = [(F/2)x](x)(1/EI) = (1/EI)(Fx^2/2)$. Therefore, for $\theta_A = \theta_x$; $(1/EI)(FL^2/8) = (1/EI)(Fx^2/2)$ and solving, $x = L/2$, which we anticipated. We now utilize the fact that at this section a tangent is horizontal. Thus, $y_{max} = -D_{A/x}$ (with the sense of the deviation reversed giving rise to the negative sign before the right-hand term). The deviation $-D_{A/x} = -1/EI$ $[(FL/4)(L/2)(\frac{1}{2})(\frac{2}{3})(L/2)] = -FL^3/48EI$. Thus, $y_{max} = -FL^3/48EI$, which agrees with our earlier result.

Illustrative problem 8.11

Determine the maximum deflection of a simply supported beam of constant cross section when loaded over its entire length by a uniform load of w N/m.

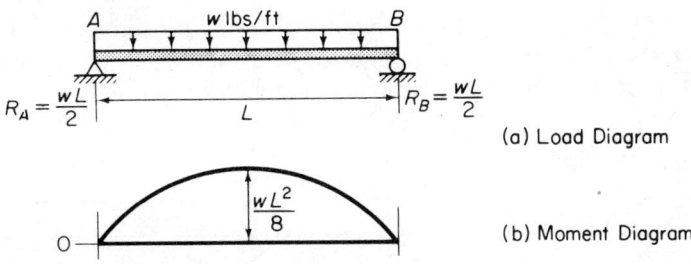

(a) Load Diagram

(b) Moment Diagram

Fig 8.16 Illustrative problem 8.11

Solution:

In the solution of this problem we shall utilize symmetry. Thus, we conclude that the beam is symmetrical about its center and the maximum deflection occurs at the center. From Chapter 6, we can conclude that the bending moment curve is a parabola with its vertex at the center and a maximum height of $wL^2/8$. From Table 8.1, case 3, the area of such a parabola is $\frac{2}{3}bh$, and the centroid of the area is located $\frac{3}{8}b$ from the right side ($\frac{5}{8}b$ from the left support). The moment of the M/EI curve between the center and left support about the left support yields $-y_{max}$.

$$y_{max} = -\frac{1}{EI}\left(\frac{2}{3}\frac{L}{2}\right)\frac{wL^2}{8}\left(\frac{5}{8}\frac{L}{2}\right) = \frac{-5wL^4}{384EI} = \frac{-5WL^3}{384EI}$$

where $W = wL$

Illustrative problem 8.12

A standard steel beam is used as a simply supported beam. If the beam is 5 m between supports, the load is 6000 N/m and the deflection is limited to $\frac{1}{360}$ th of its span, will a $W8 \times 31$ beam be suitable? Use $E = 203$ GPa.

Solution:

From Illustrative Problem 8.11, and $I = 110$ in.^{4} for a $W8 \times 31$ beam

$$y_{max} = \frac{-5wL^4}{384EI} = -\frac{5(6000)(5)^4}{384 \times 203 \times 10^9 \times 110 \times (0.0254)^4}$$
$$= -5.25 \times 10^{-3} \text{ m}$$

Note that $(0.0254)^4$ is used to convert in.^{4} to m^4 for dimensional consistency. The requirement is for y_{max} to be less than $\frac{1}{360}$ th of the span, i.e., ($\frac{5}{360} = 1.39 \times 10^{-2}$ metres). The beam is therefore satisfactory.

Illustrative problem 8.13

If the beam in Illustrative Problem 8.12 is also required to have a bending stress that is not to exceed 140 MPa, is the $W8 \times 31$ beam suitable?

Solution

We have already checked that the deflection is within the desired limits, i.e., $\frac{1}{360}$th of the span. Using the flexure formula, $S = \dfrac{MC}{I}$ and the data for the $W8 \times 31$ beam gives us

$$S = \frac{(6000)/4 \times 5 \times 4 \times 0.0254}{110 \times (0.0254)^4} = 16.6 \text{ MPa}$$

Since this stress is well below the 140 MPa allowed, the beam is satisfactory. Note that it is necessary to check *both* the deflection and the stress to see that both are within the allowable limits when designing beams.

8.4 THE SUPERPOSITION METHOD AND ITS APPLICATION

At this point in our study let us note a very important fact: Both the slope and the deflection have been found to be directly proportional to the magnitude of the assumed loads for every case that we have studied. In addition, the moment area method was predicated on the assumption that the deflections (both angular and linear) are small. Based upon the linearity between load and deflection and for the case of small beam deflections, it can be shown that the deflection and slope of a beam at a point where it is subjected to several loads can be obtained by determining the slope and deflection due to each load and then summing each item at the section in question. The particular importance of the foregoing is that it makes it possible to solve many deflection problems utilizing the results of a relatively few simple cases. Table 8.2 summarizes the cases studied in the previous sections of this chapter as well as several others that will prove useful. This table is based upon the extensive tabulations of beam diagrams and formulas given in the seventh edition of the *Steel Construction Manual of the American Institute of Steel Construction*. In addition to the shear and moment curves, general expressions are given for the deflection, the maximum deflection, and the section at which the maximum deflection occurs. It should be borne in mind that the superposition method is not an independent method for the determination of beam deflections, but rather a convenient, useful procedure that can be used in conjunction with other methods to give us solutions for complicated beam load problems.

Table 8.2 Slope and Deflection Formulas for Statically Loaded Uniform Beams

1. Cantilever Beam – Concentrated Load at Free End

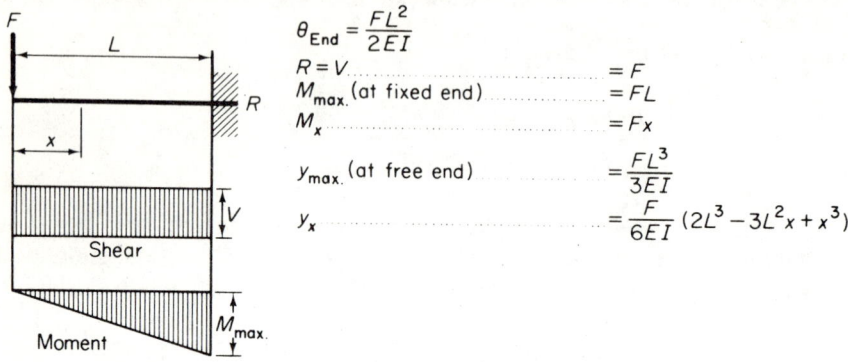

$$\theta_{End} = \frac{FL^2}{2EI}$$

$R = V$ .... $= F$
$M_{max.}$ (at fixed end) .... $= FL$
M_x .... $= Fx$

$y_{max.}$ (at free end) .... $= \dfrac{FL^3}{3EI}$

y_x .... $= \dfrac{F}{6EI}(2L^3 - 3L^2 x + x^3)$

2. Cantilever Beam – Concentrated Load at Any Point

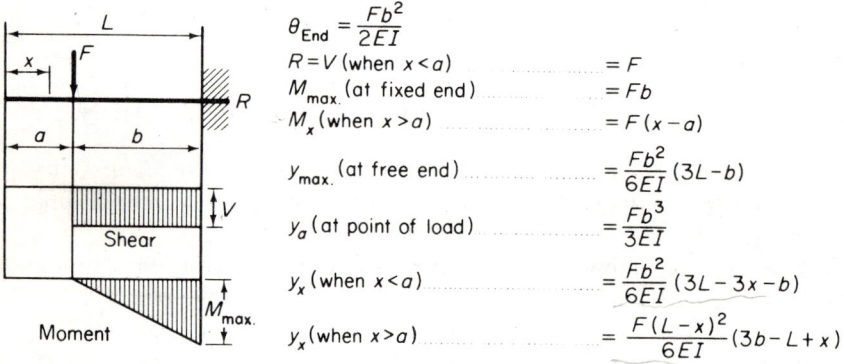

$$\theta_{End} = \frac{Fb^2}{2EI}$$

$R = V$ (when $x < a$) .... $= F$
$M_{max.}$ (at fixed end) .... $= Fb$
M_x (when $x > a$) .... $= F(x-a)$

$y_{max.}$ (at free end) .... $= \dfrac{Fb^2}{6EI}(3L-b)$

y_a (at point of load) .... $= \dfrac{Fb^3}{3EI}$

y_x (when $x < a$) .... $= \dfrac{Fb^2}{6EI}(3L - 3x - b)$

y_x (when $x > a$) .... $= \dfrac{F(L-x)^2}{6EI}(3b - L + x)$

3. Cantilever Beam – Uniformly Distributed Load

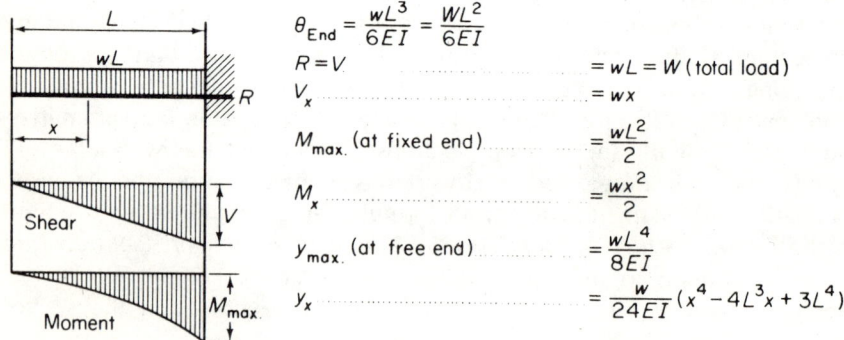

$$\theta_{End} = \frac{wL^3}{6EI} = \frac{WL^2}{6EI}$$

$R = V$ .... $= wL = W$ (total load)
V_x .... $= wx$

$M_{max.}$ (at fixed end) .... $= \dfrac{wL^2}{2}$

M_x .... $= \dfrac{wx^2}{2}$

$y_{max.}$ (at free end) .... $= \dfrac{wL^4}{8EI}$

y_x .... $= \dfrac{w}{24EI}(x^4 - 4L^3 x + 3L^4)$

Table 8.2 (continued)

4. Cantilever Beam – Load Increasing Uniformly to Fixed End

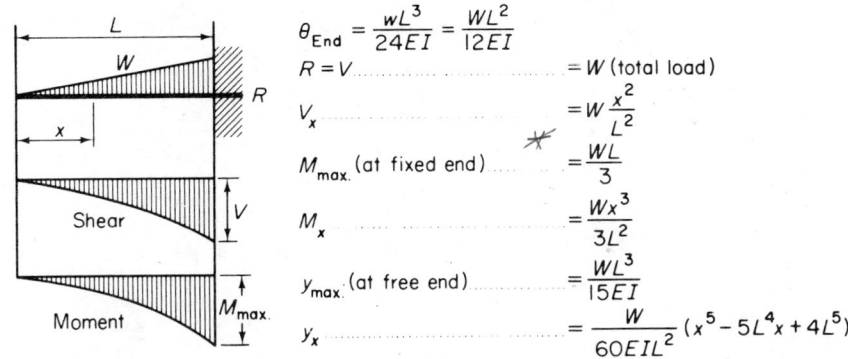

$$\theta_{End} = \frac{wL^3}{24EI} = \frac{WL^2}{12EI}$$

$R = V$ .... $= W$ (total load)

V_x .... $= W\frac{x^2}{L^2}$

$M_{max.}$ (at fixed end) .... $= \frac{WL}{3}$

M_x .... $= \frac{Wx^3}{3L^2}$

$y_{max.}$ (at free end) .... $= \frac{WL^3}{15EI}$

y_x .... $= \frac{W}{60EIL^2}(x^5 - 5L^4x + 4L^5)$

5. Cantilever Beam – Moment at Free End

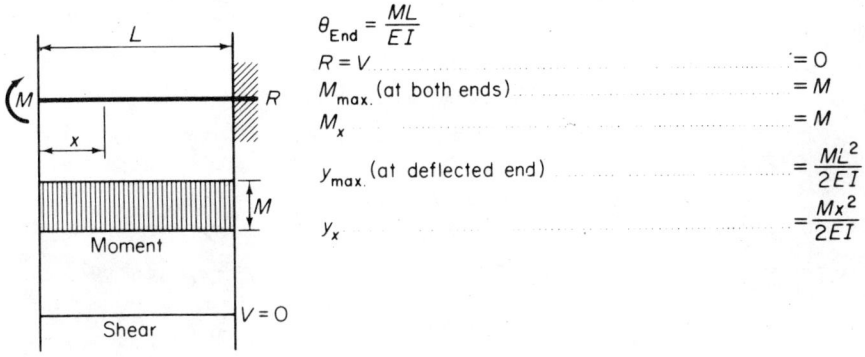

$$\theta_{End} = \frac{ML}{EI}$$

$R = V$ .... $= 0$

$M_{max.}$ (at both ends) .... $= M$

M_x .... $= M$

$y_{max.}$ (at deflected end) .... $= \frac{ML^2}{2EI}$

y_x .... $= \frac{Mx^2}{2EI}$

6. Simple Beam – Concentrated Load at Center

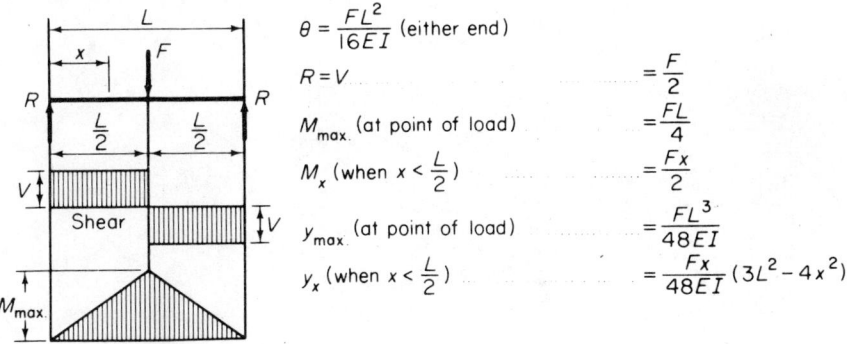

$$\theta = \frac{FL^2}{16EI} \text{ (either end)}$$

$R = V$ .... $= \frac{F}{2}$

$M_{max.}$ (at point of load) .... $= \frac{FL}{4}$

M_x (when $x < \frac{L}{2}$) .... $= \frac{Fx}{2}$

$y_{max.}$ (at point of load) .... $= \frac{FL^3}{48EI}$

y_x (when $x < \frac{L}{2}$) .... $= \frac{Fx}{48EI}(3L^2 - 4x^2)$

Table 8.2 (continued)

7. Simple Beam — Concentrated Load at Any Point

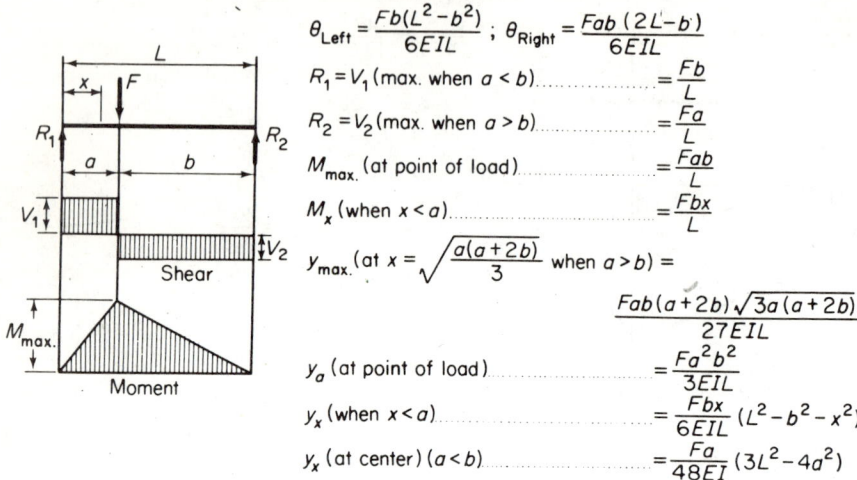

$\theta_{Left} = \dfrac{Fb(L^2 - b^2)}{6EIL}$; $\theta_{Right} = \dfrac{Fab(2L-b)}{6EIL}$

$R_1 = V_1$ (max. when $a < b$) $= \dfrac{Fb}{L}$

$R_2 = V_2$ (max. when $a > b$) $= \dfrac{Fa}{L}$

M_{max} (at point of load) $= \dfrac{Fab}{L}$

M_x (when $x < a$) $= \dfrac{Fbx}{L}$

y_{max} (at $x = \sqrt{\dfrac{a(a+2b)}{3}}$ when $a > b$) $= \dfrac{Fab(a+2b)\sqrt{3a(a+2b)}}{27EIL}$

y_a (at point of load) $= \dfrac{Fa^2 b^2}{3EIL}$

y_x (when $x < a$) $= \dfrac{Fbx}{6EIL}(L^2 - b^2 - x^2)$

y_x (at center) ($a < b$) $= \dfrac{Fa}{48EI}(3L^2 - 4a^2)$

8. Simple Beam — Two Equal Concentrated Loads Symmetrically Placed

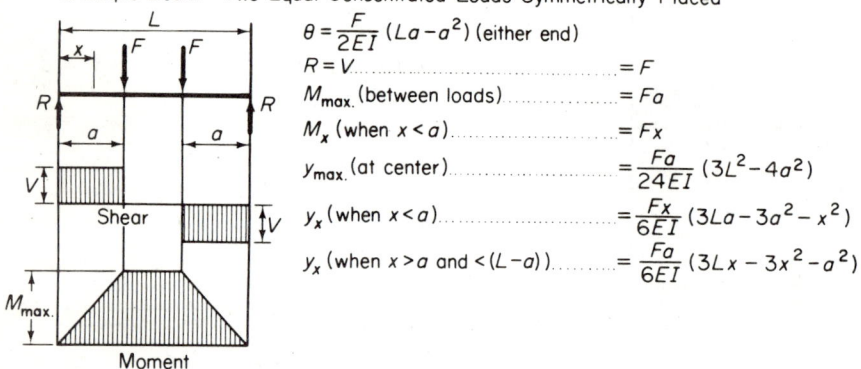

$\theta = \dfrac{F}{2EI}(La - a^2)$ (either end)

$R = V$ $= F$

M_{max} (between loads) $= Fa$

M_x (when $x < a$) $= Fx$

y_{max} (at center) $= \dfrac{Fa}{24EI}(3L^2 - 4a^2)$

y_x (when $x < a$) $= \dfrac{Fx}{6EI}(3La - 3a^2 - x^2)$

y_x (when $x > a$ and $< (L-a)$) $= \dfrac{Fa}{6EI}(3Lx - 3x^2 - a^2)$

9. Simple Beam — Uniformly Distributed Load

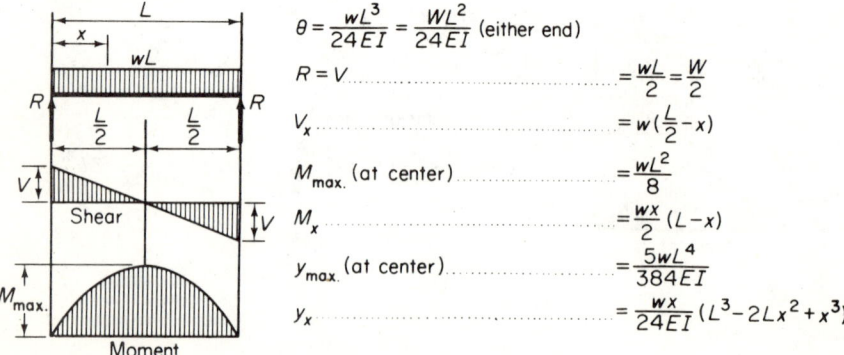

$\theta = \dfrac{wL^3}{24EI} = \dfrac{WL^2}{24EI}$ (either end)

$R = V$ $= \dfrac{wL}{2} = \dfrac{W}{2}$

V_x $= w\left(\dfrac{L}{2} - x\right)$

M_{max} (at center) $= \dfrac{wL^2}{8}$

M_x $= \dfrac{wx}{2}(L - x)$

y_{max} (at center) $= \dfrac{5wL^4}{384EI}$

y_x $= \dfrac{wx}{24EI}(L^3 - 2Lx^2 + x^3)$

Chapter 8, The Deflection of Beams 329

Table 8.2 *(continued)*

10. Simple Beam – Load Increasing Uniformly to One End

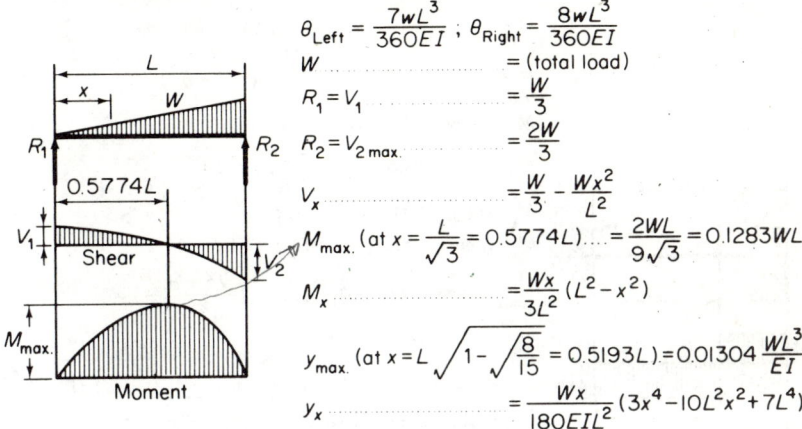

$\theta_{Left} = \dfrac{7wL^3}{360EI}$; $\theta_{Right} = \dfrac{8wL^3}{360EI}$

W = (total load)

$R_1 = V_1$ = $\dfrac{W}{3}$

$R_2 = V_{2\,max}$ = $\dfrac{2W}{3}$

V_x = $\dfrac{W}{3} - \dfrac{Wx^2}{L^2}$

M_{max} (at $x = \dfrac{L}{\sqrt{3}} = 0.5774L$) = $\dfrac{2WL}{9\sqrt{3}} = 0.1283WL$

M_x = $\dfrac{Wx}{3L^2}(L^2 - x^2)$

y_{max} (at $x = L\sqrt{1 - \sqrt{\dfrac{8}{15}}} = 0.5193L$) = $0.01304\dfrac{WL^3}{EI}$

y_x = $\dfrac{Wx}{180EIL^2}(3x^4 - 10L^2x^2 + 7L^4)$

11. Simple Beam – Load Increasing Uniformly to Center

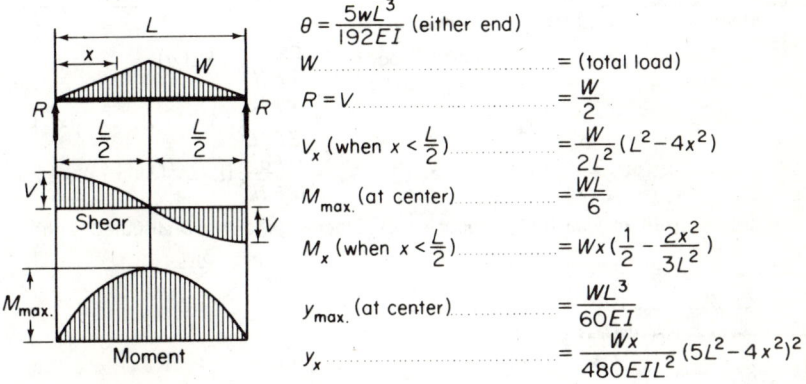

$\theta = \dfrac{5wL^3}{192EI}$ (either end)

W = (total load)

$R = V$ = $\dfrac{W}{2}$

V_x (when $x < \dfrac{L}{2}$) = $\dfrac{W}{2L^2}(L^2 - 4x^2)$

M_{max} (at center) = $\dfrac{WL}{6}$

M_x (when $x < \dfrac{L}{2}$) = $Wx\left(\dfrac{1}{2} - \dfrac{2x^2}{3L^2}\right)$

y_{max} (at center) = $\dfrac{WL^3}{60EI}$

y_x = $\dfrac{Wx}{480EIL^2}(5L^2 - 4x^2)^2$

12. Simple Beam – Moment at One End

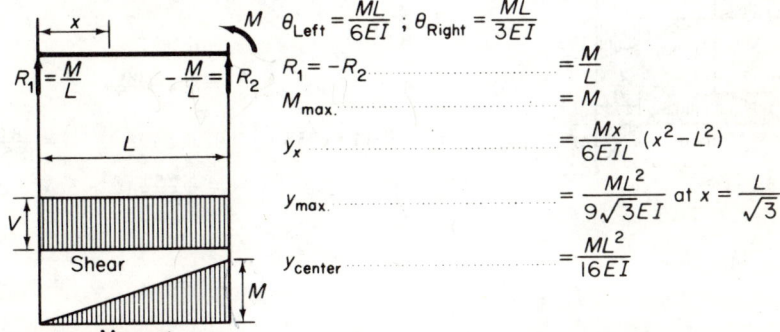

$\theta_{Left} = \dfrac{ML}{6EI}$; $\theta_{Right} = \dfrac{ML}{3EI}$

$R_1 = -R_2$ = $\dfrac{M}{L}$

M_{max} = M

y_x = $\dfrac{Mx}{6EIL}(x^2 - L^2)$

y_{max} = $\dfrac{ML^2}{9\sqrt{3}EI}$ at $x = \dfrac{L}{\sqrt{3}}$

y_{center} = $\dfrac{ML^2}{16EI}$

Table 8.2 (continued)

13. Beam Fixed at One End, Supported at Other – Concentrated Load at Any Point

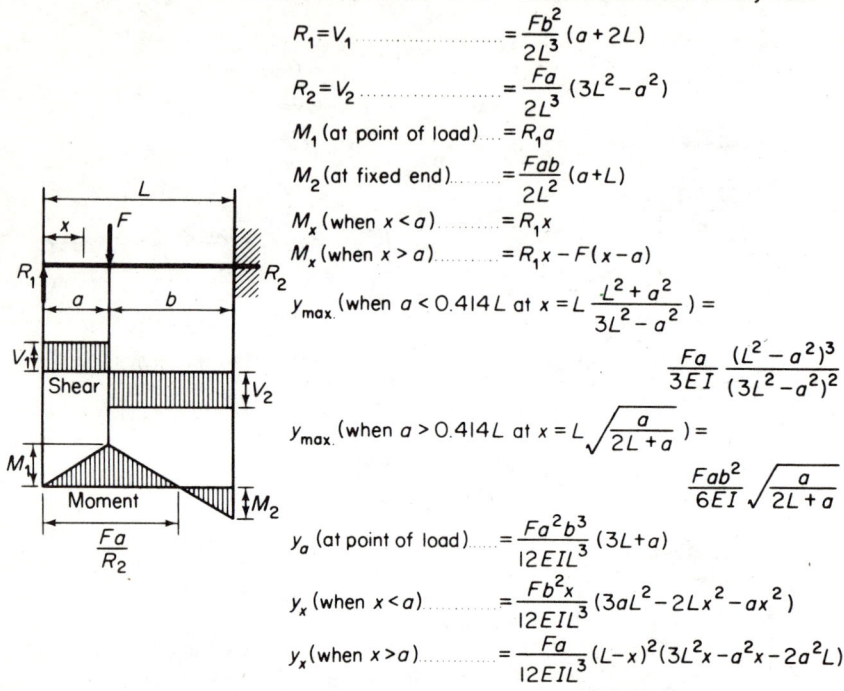

$$R_1 = V_1 = \frac{Fb^2}{2L^3}(a+2L)$$

$$R_2 = V_2 = \frac{Fa}{2L^3}(3L^2 - a^2)$$

$$M_1 \text{ (at point of load)} = R_1 a$$

$$M_2 \text{ (at fixed end)} = \frac{Fab}{2L^2}(a+L)$$

$$M_x \text{ (when } x < a) = R_1 x$$

$$M_x \text{ (when } x > a) = R_1 x - F(x-a)$$

$$y_{max} \text{ (when } a < 0.414L \text{ at } x = L\frac{L^2+a^2}{3L^2-a^2}) = \frac{Fa}{3EI}\frac{(L^2-a^2)^3}{(3L^2-a^2)^2}$$

$$y_{max} \text{ (when } a > 0.414L \text{ at } x = L\sqrt{\frac{a}{2L+a}}) = \frac{Fab^2}{6EI}\sqrt{\frac{a}{2L+a}}$$

$$y_a \text{ (at point of load)} = \frac{Fa^2 b^3}{12EIL^3}(3L+a)$$

$$y_x \text{ (when } x < a) = \frac{Fb^2 x}{12EIL^3}(3aL^2 - 2Lx^2 - ax^2)$$

$$y_x \text{ (when } x > a) = \frac{Fa}{12EIL^3}(L-x)^2(3L^2 x - a^2 x - 2a^2 L)$$

14. Beam Fixed at One End, Supported at Other – Uniformly Distributed Load

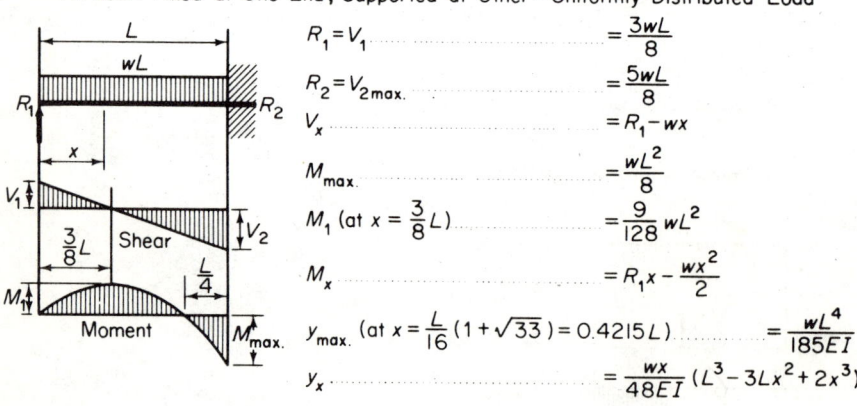

$$R_1 = V_1 = \frac{3wL}{8}$$

$$R_2 = V_{2max} = \frac{5wL}{8}$$

$$V_x = R_1 - wx$$

$$M_{max} = \frac{wL^2}{8}$$

$$M_1 \text{ (at } x = \frac{3}{8}L) = \frac{9}{128}wL^2$$

$$M_x = R_1 x - \frac{wx^2}{2}$$

$$y_{max} \text{ (at } x = \frac{L}{16}(1+\sqrt{33}) = 0.4215L) = \frac{wL^4}{185EI}$$

$$y_x = \frac{wx}{48EI}(L^3 - 3Lx^2 + 2x^3)$$

Table 8.2 (continued)

15. Beam Fixed at Both Ends – Concentrated Load at Any Point

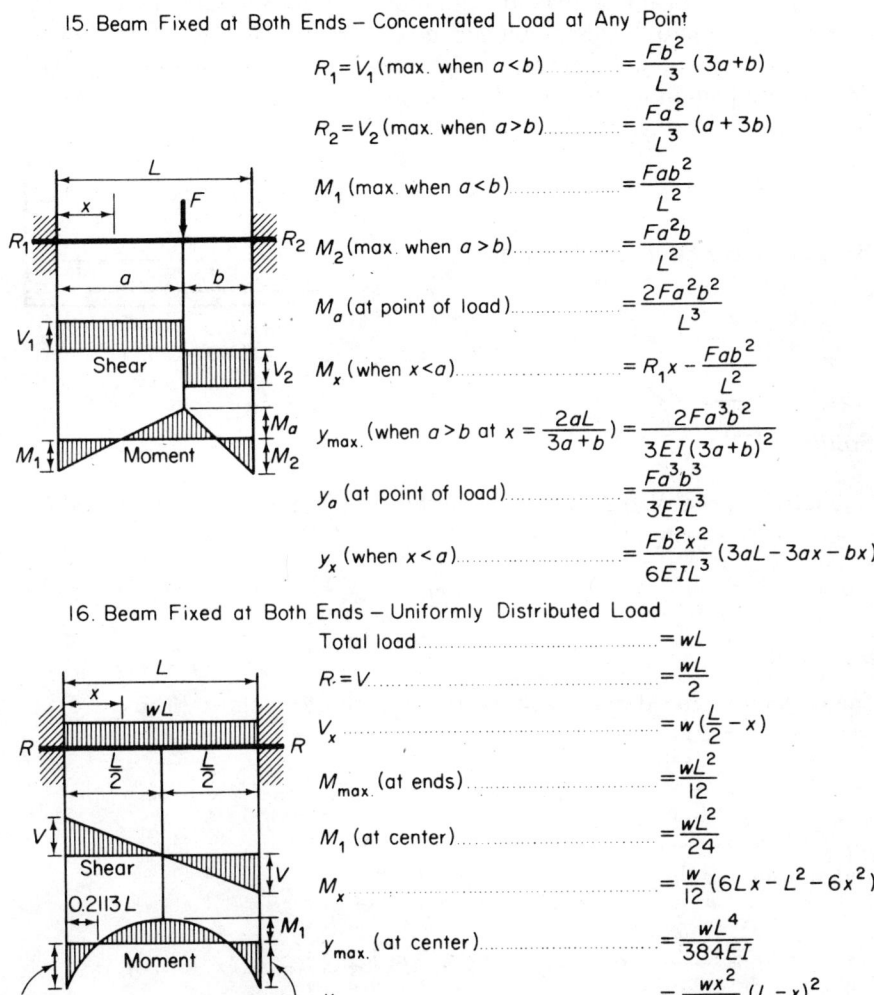

$R_1 = V_1$ (max. when $a<b$) $= \dfrac{Fb^2}{L^3}(3a+b)$

$R_2 = V_2$ (max. when $a>b$) $= \dfrac{Fa^2}{L^3}(a+3b)$

M_1 (max. when $a<b$) $= \dfrac{Fab^2}{L^2}$

M_2 (max. when $a>b$) $= \dfrac{Fa^2b}{L^2}$

M_a (at point of load) $= \dfrac{2Fa^2b^2}{L^3}$

M_x (when $x<a$) $= R_1 x - \dfrac{Fab^2}{L^2}$

y_{max} (when $a>b$ at $x = \dfrac{2aL}{3a+b}$) $= \dfrac{2Fa^3b^2}{3EI(3a+b)^2}$

y_a (at point of load) $= \dfrac{Fa^3b^3}{3EIL^3}$

y_x (when $x<a$) $= \dfrac{Fb^2x^2}{6EIL^3}(3aL - 3ax - bx)$

16. Beam Fixed at Both Ends – Uniformly Distributed Load

Total load $= wL$

$R = V$ $= \dfrac{wL}{2}$

V_x $= w\left(\dfrac{L}{2} - x\right)$

M_{max} (at ends) $= \dfrac{wL^2}{12}$

M_1 (at center) $= \dfrac{wL^2}{24}$

M_x $= \dfrac{w}{12}(6Lx - L^2 - 6x^2)$

y_{max} (at center) $= \dfrac{wL^4}{384EI}$

y_x $= \dfrac{wx^2}{24EI}(L-x)^2$

Illustrative problem 8.14

A cantilever beam is loaded with both a concentrated load F at its free end and a uniformly distributed load, w N/m over its length L. Determine the deflection of the free end using the method of superposition. Also derive an equation for the deflection at any section x.

Fig 8.17 Illustrative problem 8.14

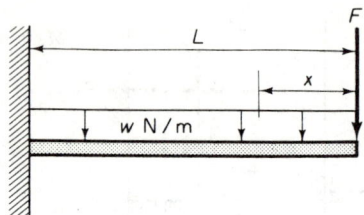

Solution:

Referring to Table 8.2, we note that this problem is a combined loading, cases 1 and 3 being used. Thus

$$y_{\text{free end}} = \frac{FL^3}{3EI} + \frac{wL^4}{8EI}$$

The deflection at x due to each load is summed to yield y_x. Thus

$$y_x = \frac{F}{6EI}(2L^3 - 3L^2x + x^3) + \frac{w}{24EI}(x^4 - 4L^3x + 3L^4)$$

and simplifying

$$y_x = \frac{1}{24EI}(8FL^3 - 12FL^2x + 4Fx^3 + x^4 - 4L^3x + 3L^4)$$

Illustrative problem 8.15

As shown in Fig. 8.18, a simple beam is symmetrically loaded by two equal forces, F (case 8). Using the results of case 7, derive the value of

$$y_{\text{max}} \text{ (at center)} = \frac{Fa}{24EI}(3L^2 - 4a^2)$$

Chapter 8, The Deflection of Beams 333

Fig 8.18 Illustrative problem 8.15

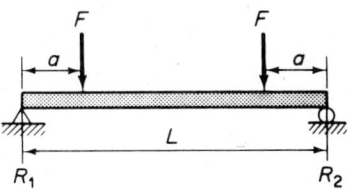

Solution:

Consider the beam to be loaded only with the left force F. The deflection at the center due to this load is (case 7)

$$y_{center} = \frac{Fa}{48EI}(3L^2 - 4a^2)$$

by symmetry, y_{center} due to the right-hand load must equal the deflection of the center due to the left-hand load. Thus

$$y_{max} \text{ (at center)} = \frac{Fa}{24EI}(3L^2 - 4a^2)$$

Illustrative problem 8.16

Determine the deflection at any section between the supports for the beam shown in Fig. 8.19 when loaded at w N/m.

Fig 8.19 Illustrative problem 8.16

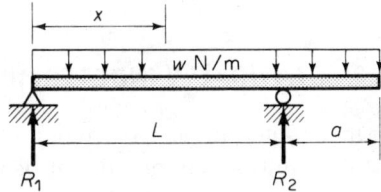

Solution:

Due to the overhang, there is a moment of $wa^2/2$ on the right support, which serves to reduce our problem to one of a uniformly loaded simple beam with an end moment of $wa^2/2$, as shown in Fig. 8.19(b). This

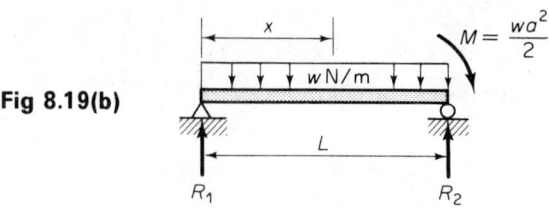

Fig 8.19(b)

reduces the problem to the superposition of cases 9 and 12. For case 9,

$$y_x = \frac{wx}{24EI}(L^3 - 2Lx^2 + x^3)$$

and for case 12,

$$y_x = \frac{-Mx}{6EIL}(x^2 - L^2)$$

where the negative sign is used due to the direction of M and $M = wa^2/2$. The total deflection is then

$$y_x = \frac{wx}{24EI}(L^3 - 2Lx^2 + x^3) - \frac{wa^2 x}{12EIL}(x^2 - L^2)$$

or

$$y_x = \frac{wx}{24EI}\left[L^3 - 2Lx^2 + x^3 - \frac{2a^2}{L}(x^2 - L^2)\right]$$

8.5 STATICALLY INDETERMINATE BEAMS

In chapter 6, various types of beams were discussed and classified. At that time it was noted that beams having either more supports or more conditions at the supports than could be evaluated from the three conditions necessary for static equilibrium are called *statically indeterminate beams*. The three general types of beams falling under this heading are the *built-in beam*, the *propped beam* and the *continuous beam*. The student should review the relevant sections of Chapter 6 that discussed these types of beams. In the following discussion we shall apply the superposition method to evaluate the additional information required to obtain a solution to statically indeterminate situations. We will use any condition available to us such as the deflection at a support being zero, the relation between the slopes at two positions, etc., and a certain amount of ingenuity will be necessary. Most often the necessary condition is found in the statement of the problem.

The foregoing discussion is best illustrated by considering the propped beam. Figure 6.5 (repeated) shows the propped beam as a cantilever beam

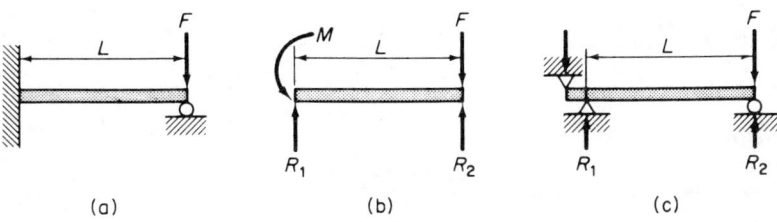

Fig 6.5 Propped beams (repeated)

with a support under the free end as shown in Fig. 6.5(a). The equivalent beams, from a force and moment viewpoint, are shown in Figs. 6.5(b) and (c). It is interesting to note that the equivalent propped beam shown in Fig. 6.5(b) is (in principle) the same as was solved for in Illustrative Problem 8.16. Our problem is to evaluate the unknown moment at the built-in end or the magnitude of each of the reactions at the supports. Two conditions are available to us to perform this evaluation; specifically, the deflection at the supported end is zero, and the slope at the built-in end is zero. The condition that the deflection at the propped end is zero is easier to apply to this case. The procedure consists of first removing the end support and determining the end deflection of the beam as if no support were present. The next step is to calculate the magnitude of the concentrated force applied to this end that is necessary to cause the unloaded beam to return to its undeflected initial position from the deflected position. This must be the value of the reaction at the support at the propped end. Once this is known, the requirements for static equilibrium, $\Sigma F = 0, \Sigma M = 0$, suffice to complete the solution of the problem.

Illustrative problem 8.17

Determine the reactions, end moment, and central deflection for the uniform simple beam shown in Fig. 8.20.

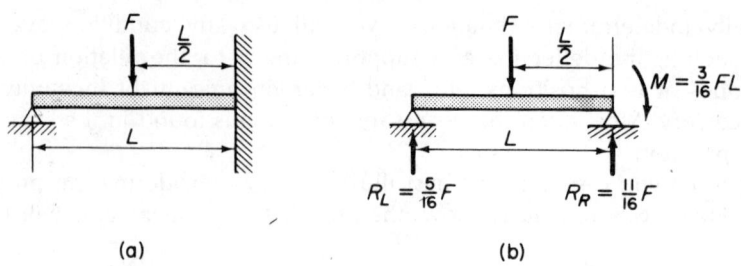

(a)   (b)

Fig 8.20 Illustrative problem 8.17

Solution:

As the first step, let us remove the end support. The deflection of the free end can be found from case 2 of Table 8.2 as

$$y_{\text{free end}} = \frac{Fb^2}{6EI}(3L - b) = \frac{F(L/2)^2}{6EI}\left(3L - \frac{L}{2}\right) = \frac{5FL^3}{48EI}$$

The force required at the end of the beam to return it to its initial undeflected position is found from case 1 to be

$$y = \frac{5FL^3}{48EI} = \frac{R_L L^3}{3EI}; \quad R_L = \tfrac{5}{16}F$$

The reaction at the left support is therefore $\tfrac{5}{16}F$ and at the right support it is $F - \tfrac{5}{16}F$ or $\tfrac{11}{16}F$. If we now invoke the condition that the moment at the left support must be zero and take moments about the right support

$$R_L(L) - F\left(\frac{L}{2}\right) - M = 0$$

$$M = \frac{-FL}{2} + R_L(L) = \frac{-FL}{2} + \frac{5FL}{16} = \frac{-3FL}{16}$$

The deflection at the center is found as in Illustrative Problem 8.16. The deflection at the center due to the central concentrated load is found from case 6 to be

$$y_{\text{center}} = \frac{FL^3}{48EI}$$

The central deflection due to the right end moment is obtained from case 12

$$y_{center} = \frac{ML^2}{16EL} = \frac{-3}{16}\left(\frac{FL^3}{16EI}\right) = \frac{-3}{256}\frac{FL^3}{EI}$$

The net central deflection, $y_{center} = FL^3/48EI - 3FL^3/256EI = 7FL^3/768EI$. This result can be verified from case 13.

Illustrative problem 8.18

If a propped cantilever is loaded with a uniform load of w lb/ft, determine the deflection at its center.

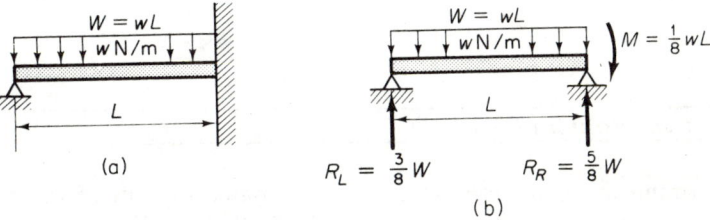

Fig 8.21 Illustrative problem 8.18

Solution:

Proceeding as before, we first calculate the deflection of the beam at the left support with the support removed. From case 3,

$$y_{free\ end} = \frac{wL^4}{8EI}$$

The force required to move the free end to its initial undeformed position is found using case 1. Thus, $y = wL^4/8EI = F'L^3/3EI$: $F' = 3wL/8$. The reaction at the left support is therefore $\frac{3}{8}W$, where $W = wL$, and at the right support the reaction is $\frac{5}{8}W$. Using the condition of zero moment at the left support and taking moments about the right support yields

$$R_L - \frac{(wL)L}{2} - M = 0$$

$$M = \frac{-wL^2}{2} + R_L(L) = \frac{-wL^2}{2} + \tfrac{3}{8}wL = \tfrac{3}{8}wL - \tfrac{1}{2}wL = -\tfrac{1}{8}wL = -\tfrac{1}{8}wL^2$$

The problem has now resolved itself to solving the problem shown in Fig. 8.21(b). From case 9 of Table 8.2, the deflection at the center of a simple beam with a uniform load is $5wL^4/384EI$. From case 12, the central deflection due to the right end moment is $ML^2/16EI = -\tfrac{1}{8}wL^2(L)^2/16EI$. The net central deflection is thus

$$y_{center} = \frac{5wL^4}{384EI} - \frac{wL^4}{128EI} = \frac{1}{192}\left(\frac{wL^4}{EI}\right)$$

Case 14 yields a check on this result.

Illustrative problem 8.19

Determine the end moment at the right-hand support of the beam in Illustrative Problem 8.18 using the condition that the angle at the built-in end is zero.

Solution:

The angular deflection at the built-in end can be thought to consist of two effects, a deflection due to the load w and a counter rotation due to M, thus yielding a net angular rotation of zero. The rotation due to the uniform load on a simple beam is obtained from case 9 as $\theta = wL^3/24EI$. The moment required at the right end to rotate the beam at this end a negative value equal to $wL^3/24EI$ is obtained from case 12 as $ML/3EI$. Therefore $ML/3EI = -wL^3/24EI$ and $M = -\tfrac{1}{8}wL^2$. This is the same result obtained in Illustrative Problem 8.18. The remainder of the solution for obtaining the deflection at the center (if desired) proceeds as in Illustrative Problem 8.18 after the end reactions are evaluated.

The propped beams that we have studied thus far can be considered to be a special case of beams that have both ends completely restrained or built-in. As noted in Chapter 6, built-in beams can be thought of as double-ended cantilever beams, or simple beams with moments at both ends. These conditions are shown diagrammatically in Fig. 6.4 (repeated).

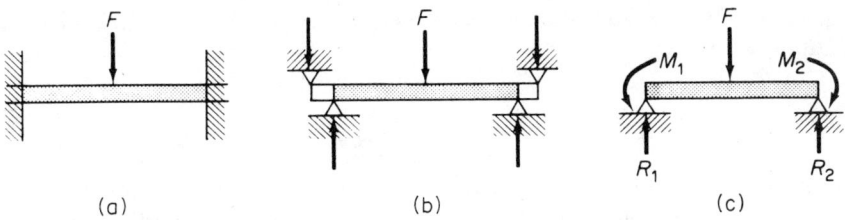

(a)　　　　　　　　(b)　　　　　　　　(c)

Fig 6.4 Built-in beams (repeated)

The method that we shall use to treat this category of beams is the same as that used for propped beams in the previous sections. However where we required one extra condition to reach a solution for propped beams, we shall require two conditions to solve built-in beams. We can utilize both the shear and moment at one end of the beam as unknowns and by solving for them, obtain the required conditions. It is also possible to consider a symmetrical beam as shown in Fig. 6.4(c) and by using the condition that $\theta_1 = \theta_2$ to solve for the end moments at each end of the beam. The following illustrative examples will serve to demonstrate the application of the superposition method to this type of structure.

Illustrative problem 8.20

A uniform built-in beam has a concentrated load placed on the center of the beam. Determine the deflection of the beam.

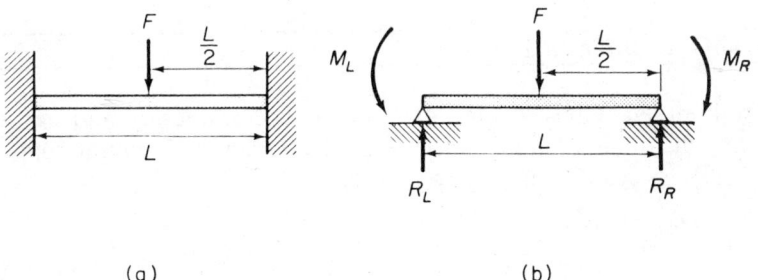

(a)　　　　　　　　(b)

Fig 8.22 Illustrative problem 8.20

Solution:

As our first observation, we note that the beam in question has complete symmetry and that $R_L = R_R = F/2$; $M_L = M_R$ and $\theta_L = \theta_R = 0$. From case 6 of Table 8.2, the angular deflection due to the central load only is given as

$$\theta_L = \theta_R = \frac{FL^2}{16EI}$$

From case 12, a moment applied on the right side of the beam will give rise to an angular deflection of $ML/3EI$ *on the right* and an angular deflection of half of this value *on the left* or $ML/6EI$. Equal moments M at both ends of the beam cause an angular deflection of $ML/3EI + ML/6EI$ or $ML/2EI$ at each end. Thus, the moment required at each end is found by the condition that $\theta_L = \theta_R = 0$ on the beam. The angular deflection due to the central load must be equal and opposite to that produced by the end moments

$$\frac{FL^2}{16EI} = \frac{ML}{2EI} \quad \text{and} \quad M = \frac{FL}{8}$$

The deflection at the center due to the concentrated load less the deflection at the center due to the end moments gives us the desired result. For the concentrated load $y_{center} = FL^3/48EI$ (from case 6), and the central deflection due to each moment is found from case 12 to be $ML^2/16EI$ and, when M is substituted for, is $(FL/8)(L^2/16EI)$ or $FL^3/128EI$. Performing the required subtraction yields

$$y_{center} = \frac{FL^3}{48EI} - 2\left(\frac{FL^3}{128EI}\right) = \frac{FL^3}{192EI}$$

The student can verify this result using case 15 of Table 8.2.

Illustrative problem 8.21

A built-in beam has a span of 5 m between supports and carries a uniformly distributed load of 3000 N/m. Determine the central deflection if an S5 × 10 beam is used. Use $E = 210 \times 10^9$ Pa.

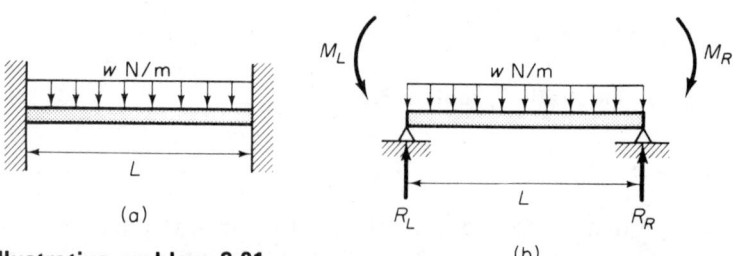

Fig 8.23 Illustrative problem 8.21

Solution:

We can go directly to case 16 of Table 8.2 to obtain the value of y_{center} as $wL^4/384EI$, or we can derive this result using the procedure of Illustrative Problem 8.20. Using this latter procedure, we first obtain the end angular deflections of the equivalent beam shown in Fig. 8.23(b) from case 9 as: $\theta_L = \theta_R = wL^3/24EI$. The required end moments are obtained as in Illustrative Problem 8.20 from the angular deflection due to the moments of $ML/2EI$. Thus, $wL^3/24EI = ML/2EI$ and $M = wL^2/12$. Case 9 gives us the central deflection due to the uniform load as $5wL^4/384EI$, and case 12 gives us the central deflection due to the end moments as $ML^2/16EI$ or $(wL^2/12)(L^2/16EI)$. The total central deflection is therefore

$$-y_{center} = \frac{5wL^4}{384EI} - 2\left[\left(\frac{wL^2}{12}\right)\left(\frac{L^2}{16EI}\right)\right] = \frac{wL^4}{384EI}$$

the same result we had from case 16 of Table 8.2. Since $I = 12.3$ in.^{4} for the $S5 \times 10$ beam

$$y_{center} = \frac{3000 \times (5)^4}{384 \times 210 \times 10^9 \times 12.3 \times (0.0254)^4} = 4.54 \times 10^{-3} \text{ m}$$

If the deflection were limited to $\frac{1}{360}$th of the span, the maximum allowable deflection would be $5/360 = 1.39 \times 10^{-2}$ m, and we see that this deflection is within this limit.

8.6 CONTINUOUS BEAMS—THE THREE-MOMENT EQUATION

A *continuous beam* is a beam having three or more supports. This type of beam has more supports than are necessary for static equilibrium and is the last type of indeterminate beam that we shall study. Figure 6.6 (repeated) shows an example of this type of beam. When the beam has three supports, we can use the procedure of the previous section of this

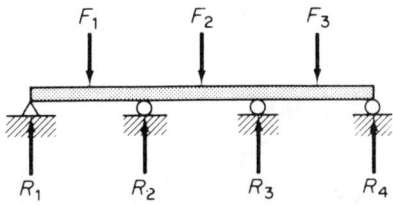

Fig 6.6 Continuous beam (repeated)

chapter; when more than three supports exist, it is necessary to determine the bending moments at the supports using a general relation among the bending moments at any two sections of the beam known as the three-moment equation. We shall first consider a beam having three supports and apply the methods already developed in this chapter. Then the three-moment equation will be used to solve the same problem.

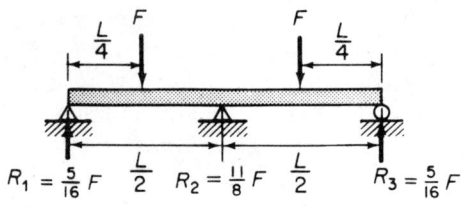

(a) Beam Loading

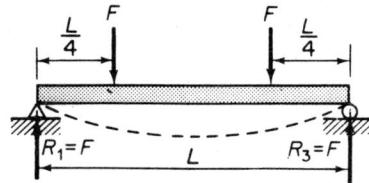

$$y_{centre} = \frac{\frac{FL}{4}}{29EI}\left[3(L)^2 - 4\left(\frac{L}{4}\right)^2\right] = \frac{11FL^3}{314EI}$$

(b) Beam Deflection — Central Support Removed

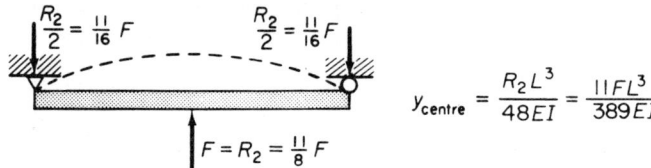

$$y_{centre} = \frac{R_2 L^3}{48EI} = \frac{11FL^3}{389EI}$$

(c) Beam Deflection — Central Support Only

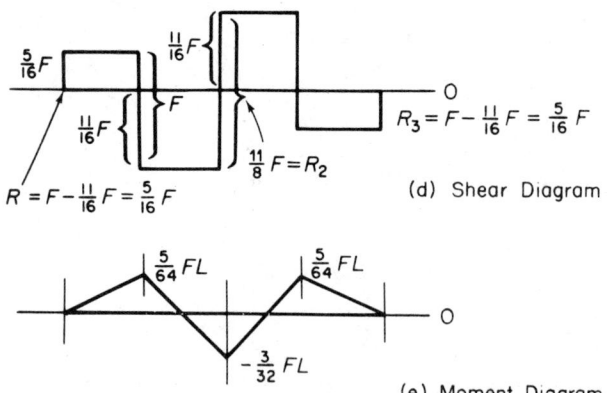

(d) Shear Diagram

(e) Moment Diagram

Fig 8.24 Two-span, three-support continuous beam

The illustration that we shall use is in Fig. 8.24. It consists of a two-span beam having equal spans and equal centrally-placed, concentrated loads on each span. Using the method of superposition, we can consider the beam to be loaded as in Fig. 8.24(a) with the central support removed. For this beam, the end reactions are equal to F and the central deflection is found from case 8 of Table 8.2 to be $(Fa/24EI)(3L^2 - 4a^2)$ (see also Illustrative Problem 8.15). The condition that the middle support imposes is that the deflection at the middle support must be zero. Thus, we have a simply supported beam with a central force acting upward to cause the net deflection at this force to be zero. From case 6 of Table 8.2

$$y_{center} = \frac{F'L^3}{48EI}$$

where the F' in this relation is the required force at support R_2. Thus, $F'L^3/48EI = 11FL^3/384EI$ and $F' = \frac{11}{8}F$. Notice that half of this load, acting as shown in Fig. 8.24(c), must act at each support. The net force at each support is the sum of the terms at each support, yielding a support reaction on each of the outer supports of $\frac{5}{16}F$. The complete shear diagram for the beam is shown in Fig. 8.24(d). With the beam's shear diagram known and the moment at the center span known, the complete moment diagram for the beam can be drawn (and checked) using the principles developed in Chap. 6. We can now calculate the deflection of the beam using the superposition method applied to each span. The loading on each span consists of the applied loads and the end moments due to the supports.

A general relation known as the *three-moment equation* can be derived from the moment area method for the bending moments at any two sections (three supports) on a beam. Referring to Fig. 8.25(a) and (b), we can write

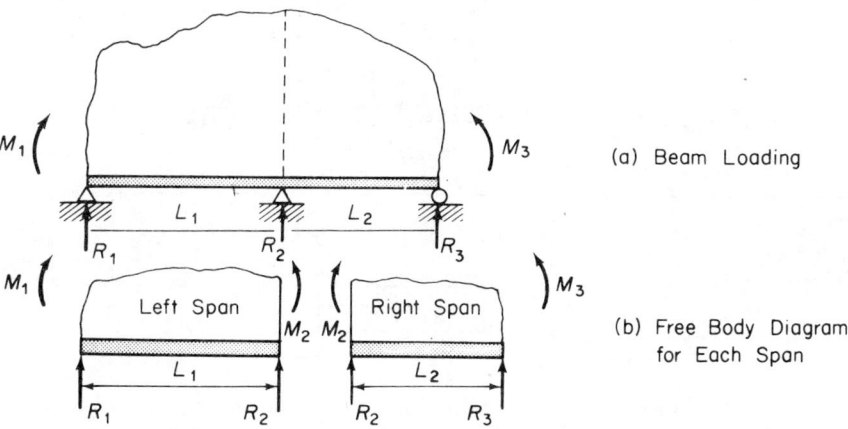

Fig 8.25 Generalized loads on a continuous beam

the three moment equation as follows, if we assume all of the supports to be at the same level:

$$M_1 L_1 + 2M_2(L_1 + L_2) + M_3 L_2 + [(\text{Load Term}) \text{ Left span}]$$
$$+ [(\text{Load Term}) \text{ Right span}] = 0 \qquad (8.17)$$

Table 8.3 Load Terms for Use with the Three-Moment Equation*

w = uniform load, lbs/ft W = total load on span = wL

Left span (load term)	no.	Type of loading	Right span (load term)
$\frac{Fa}{L}(L^2 - a^2)$	1	concentrated load F at distance a from left, b from right	$\frac{Fb}{L}(L^2 - b^2)$
$\frac{wL^3}{4} = \frac{WL^2}{4}$	2	uniform load w over full span L	$\frac{wL^3}{4} = \frac{WL^2}{4}$
$\frac{8}{60}wL^3 = \frac{8}{30}WL^2$	3	triangular load increasing to w at right	$\frac{7}{60}wL^3 = \frac{7}{30}WL^2$
$\frac{7}{60}wL^3 = \frac{7}{30}WL^2$	4	triangular load decreasing from w at left	$\frac{8}{60}wL^3 = \frac{8}{30}WL^2$
$\frac{5}{32}wL^3 = \frac{5}{16}WL^2$	5	triangular load peak w at center, $L/2$ each side	$\frac{5}{32}wL^3 = \frac{5}{16}WL^2$
$\frac{w(b^2 - a^2)}{4L}[2L^2 - (b^2 + a^2)]$	6	partial uniform load w over segment between a and d, with b, c	$\frac{w(d^2 - c^2)}{4L}[2L^2 - (d^2 + c^2)]$

*Reproduced with permission from *Applied Strength of Materials* by Alfred Jensen and H. H. Chenoweth, McGraw-Hill Book Co., Inc. p. 251, 1967.

Equation (8.17) is based upon the assumptions that all supports are on the same level, the beam is initially straight, E and I are constant throughout the beam length, and the supports remain in a straight line after loading is applied. If the moment at any point is negative, it must be used as negative quantity in equation (8.17), and a negative result indicates a negative moment at a given point. The load terms to be used for various types of loading can be derived from the moment area method. When this is done, the results are those shown in Table 8.3. This table can be used for other loadings by noting that the load terms for any combination of loads in Table 8.3 are additive.

Let us now solve the previous problem by applying the three-moment equation as follows:

$$M_1 L_1 + 2M_2(L_1 + L_2) + M_3 L_2 + [\text{(Load Term) Left span}]$$
$$+ [\text{(Load Term) Right span}] = 0$$

$M_1 = 0$, $L_1 = \dfrac{L}{2}$, $L_2 = \dfrac{L}{2}$, $M_3 = 0$, [(Load Term) Left span]

$$= [\text{(Load Term) Right span}]$$

Therefore

$$2M_2(L) + 2\left\{\frac{F(L/4)}{L/2}\left[\left(\frac{L}{2}\right)^2 - \left(\frac{L}{4}\right)^2\right]\right\} = 0 \quad \text{and} \quad M_2 = -\frac{3FL}{32}$$

the same result as was obtained by the earlier approach. Knowing M_2, we can calculate the deflection on each span using the superposition method.

Where there are more than two spans, it is necessary to apply the three-moment equation successively to each pair of spans, i.e., span 1 and 2, span 2 and 3, span 3 and 4, etc., and then solve the resulting simultaneous equations for the unknown moments at the supports.

8.7 CLOSURE

It is not possible for any text to adequately cover all of the combinations of methods that can be applied to the solution of beam deflection problems. In this chapter we have considered the moment area method and the superposition method to obtain solutions to typical beam deflection situations. There is a great deal of arithmetic computation involved in these problems, and too often these computations mask the principle being used. The student should always try to qualitatively set up each problem, drawing the necessary sketches of beam shear, bending moment, and deflection as well as clearly noting all quantities given in each problem on these diagrams. Attention must also be given to the proper use of dimensions since multiple powers of length occur in all of these problems.

The student may have noted that there are more illustrative problems in this chapter than in any other previous chapter. The reason for this is simply that there are more principles to be illustrated to more situations than before. Understand each problem and the situation illustrated before proceeding to the next one, and in this manner you will develop an understanding of the material as well as confidence in applying it to different situations.

PROBLEMS

Use $E = 210 \times 10^9$ Pa in all problems.

8.1 A simple beam having a span of 5 m is loaded with a uniform load of 3000 N/m over its entire length. If the beam is an S10 × 35 section, determine its radius of curvature at the center.

8.2 A simple beam has a span of 6 m and has a uniform load of 6000 N/m over its entire length. If the beam is a W8 × 31 section, what is its radius of curvature?

8.3 A cantilever beam has a span of 10 m and a concentrated load of 1000 N is placed at its free end. What is the radius of curvature at the built-in end if the beam is a W12 × 14 beam?

8.4 A cantilever beam has a span of 12 m and a concentrated load of 2 kN is placed at its free end. What is the radius of curvature of the beam at its built-in end if the beam has a moment of inertia of a W8 × 31 beam?

8.5 A cantilever beam has a span of 6 m and has a uniform load of 6500 N/m over its entire length. What is the radius of curvature of the beam at its built-in end if the beam has a moment of inertia of 100 in.^{4}?

8.6 A band saw has a blade 25 mm wide and 1 mm thick. If it wraps around a pulley that has a radius of 600 mm, determine the bending stresses in the blade.

8.7 A flagpole may be considered to be a cantilever beam fastened at the ground and subjected to a uniform wind load. At 120 miles per hour, the loading is estimated to be 1000 N/m on a 75-mm-diameter rod. If the flagpole is 6 m long, determine the deflection of the free end.

8.8 Calculate the maximum deflection for the beam shown in Fig. P8.8.

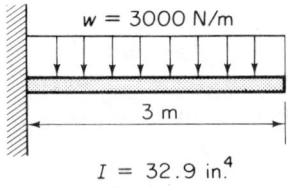

Fig P8.8

8.9 Calculate the angular deflection of the free end of the beam in Fig. P8.8.
8.10 Determine the maximum deflection for the beam shown in Fig. P8.10.

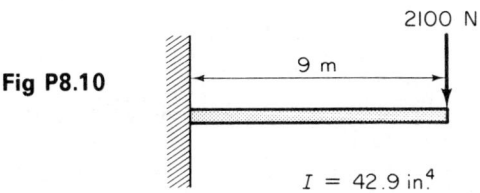

Fig P8.10

8.11 Determine the angular deflection of the free end of the beam in Fig. P8.10.
8.12 An S10 × 35 beam is used as a cantilever having a span of 18 m and carrying a load of 1500 N at its free end. What is the slope and deflection at the free end?
8.13 If instead of the 1500-N load, the beam in Problem 8.12 had a 1500-N·m moment at its free end, determine the deflection and slope its free end would have.
8.14 If instead of the 1500-N load in Problem 8.12, the beam had a uniform load of 1500 N/m over its entire length, determine the deflection of its free end.
8.15 What concentrated load F is necessary, acting upward on the free end of a cantilever beam, to keep the deflection of the free end zero if there is a uniform downward load of w N/m acting on the beam?
8.16 Derive the expressions for the maximum deflection and slope at the free end of the cantilever beam of case 4 of Table 8.2 using the moment area method.
8.17 A cantilever beam carries a uniform load of 100 N/m as shown in Fig. P8.17. Using the moment area method, determine the slope and deflection of the free end.

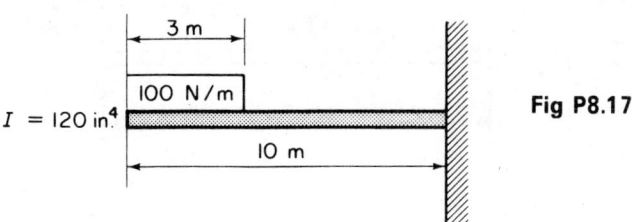

Fig P8.17

8.18 Calculate the central deflection of the simple beam shown in Fig. P8.18.

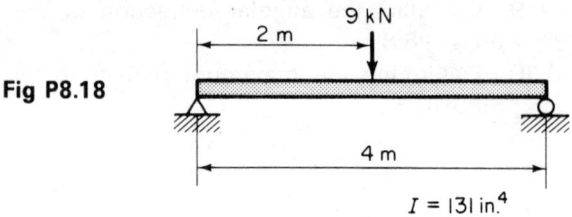

Fig P8.18

$I = 131 \text{ in.}^4$

8.19 Calculate the angular deflection at either end of the beam in Fig. P8.18.

8.20 A simple beam of length L is loaded by a concentrated load F at the center of the beam. If the maximum deflection is limited to $\frac{1}{360}$ th of the span, determine an expression for the maximum stress in the beam in terms of E, L, and C.

8.21 Select a W beam to support a uniform load of 2250 N/m over the 6 m span of a simple beam. Assume the maximum deflection of the beam is limited to $\frac{1}{360}$ th of the span and the allowable bending stress is 140 MPa.

8.22 Calculate the central deflection for the simple beam shown in Fig. P8.22.

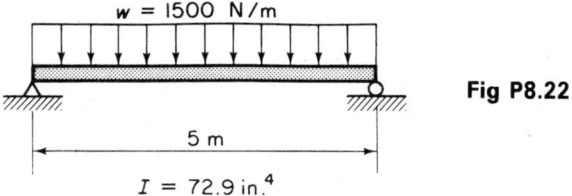

Fig P8.22

$I = 72.9 \text{ in.}^4$

8.23 Calculate the angular deflection at either end of the beam in Fig. P8.22.

8.24 Calculate the central deflection for the simple beam shown.

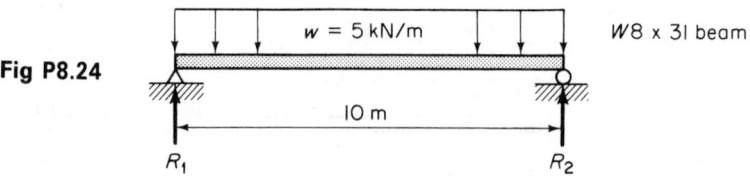

Fig P8.24

8.25 Calculate the angular deflection at either end of the beam in Problem 8.24.

Chapter 8, The Deflection of Beams   349

8.26 Using the moment area method, determine the slope at each end of the beam shown in Fig. P8.26 and the maximum deflection.

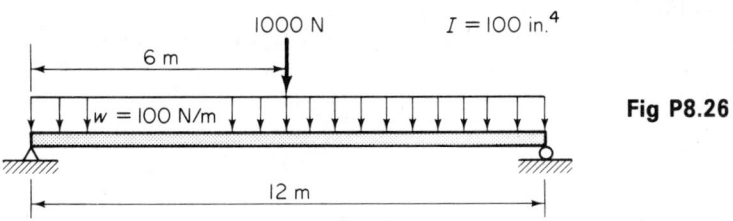

Fig P8.26

8.27 Solve Problem 8.26 using Table 8.2.
8.28 Using Table 8.2, determine the central deflection of the beam shown.

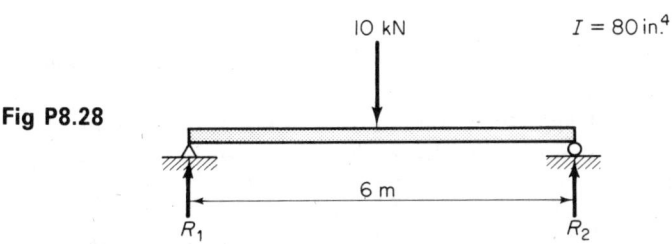

Fig P8.28

8.29 Using Table 8.2, determine the slope at each end of the beam shown in Problem 8.28.
8.30 Determine the deflection of the center of the beam shown in Fig. P8.30.

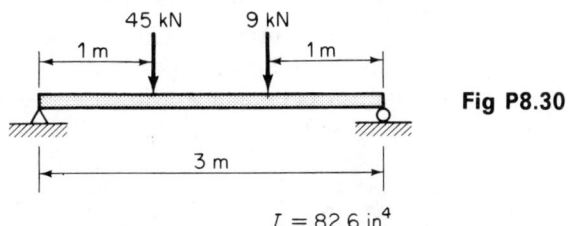

Fig P8.30

8.31 Determine the angular deflection at the left support for the beam in Fig. P8.30.
8.32 A shaft having a 50-mm diameter is supported by bearings 3.5 m apart. Consider the bearings to be equivalent to simple supports and calculate the maximum deflection of the shaft due to its own weight. Steel weighs 78 540 N/m^3.

8.33 A simply supported beam is 16 m long between supports and has a concentrated load of 1000 N placed 4 m to the right of the left support. If $I = 119$ in.^{4}, determine the deflection at the load and the slope at each end of the beam.

8.34 Using Table 8.2, solve for the deflection at the center of the beam shown.

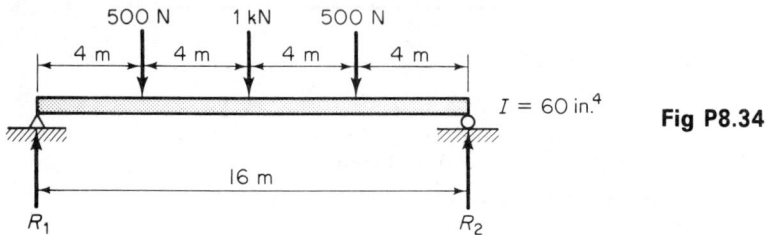

Fig P8.34

8.35 Determine the slope at each end of the beam of Problem 8.34 using Table 8.2.

8.36 Using the moment area method, determine the deflection at the center of the beam shown in Fig. P8.36 if $I = 49$ in.^{4}

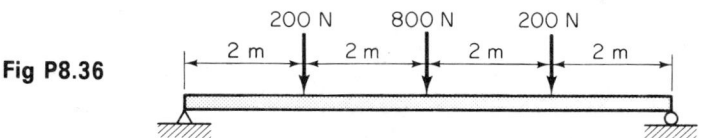

Fig P8.36

8.37 Solve Problem 8.36 using Table 8.2.

8.38 A cantilever beam carries a load of w N/m. If a support is placed under the free end to limit the deflection to one-half of that which would occur if no support were present, determine the moment at the built-in end. The length of the beam is L.

8.39 A simply supported beam has an overhang as shown in Fig. P8.39. Determine the deflection at the 4000-N load.

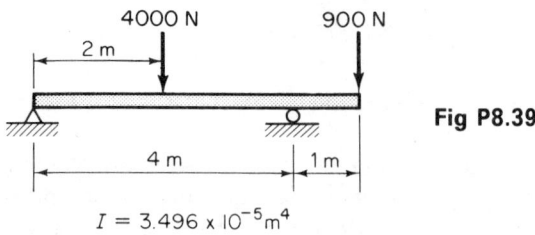

Fig P8.39

8.40 Determine the deflection at the midpoint of the span for the beam shown.

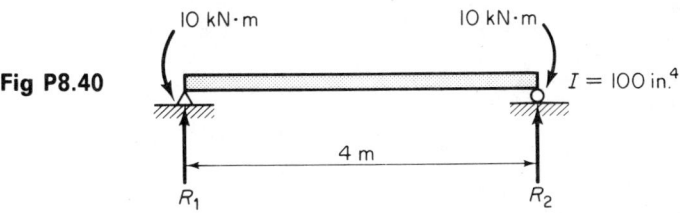

Fig P8.40

8.41 Determine the deflection at the midpoint of the central span for the beam shown in Fig. P8.41.

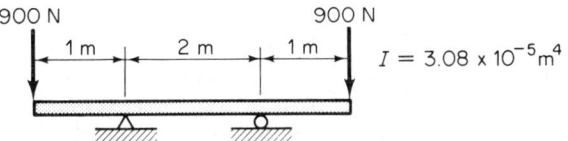

Fig P8.41

8.42 Determine the deflection at the midpoint of the center span of the beam shown if $I = 3.0 \times 10^{-5}$ m^4.

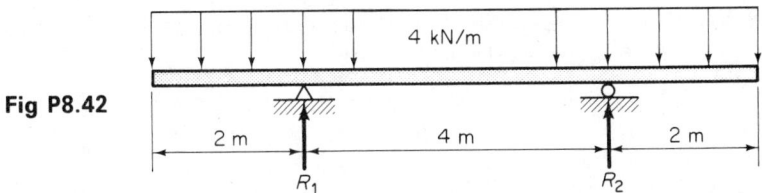

Fig P8.42

8.43 Determine the deflection at the center of the span of the beam shown in Fig. P8.43 if $I = 3.183 \times 10^{-5}$ m^4.

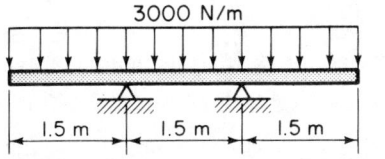

Fig P8.43

8.44 If the allowable stress is 140 MPa and an S7 × 20 beam is used as shown in Fig. P8.44 with the maximum deflection limited to $\frac{1}{360}$ th of the span, is it suitable for this application?

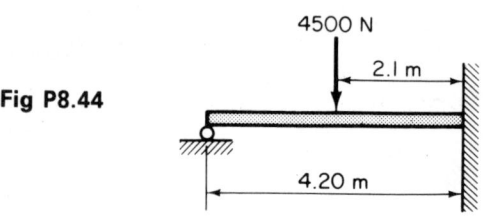

Fig P8.44

8.45 Determine the maximum deflection for the beam shown in Fig. P8.45. $I = 2.302 \times 10^{-5}$ m^4.

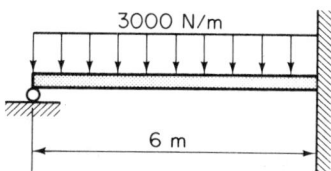

Fig P8.45

8.46 What is the deflection at the center of the beam shown in Fig. P8.46 if $I = 4.745 \times 10^{-5}$ m^4?

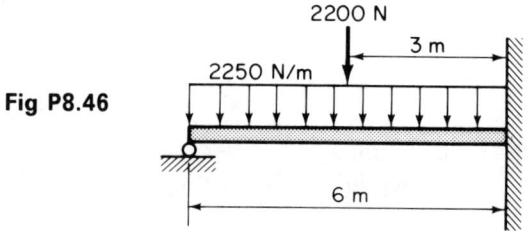

Fig P8.46

8.47 Determine the deflection at section x for the beam shown if $I = 6.140 \times 10^{-5}$ m^4.

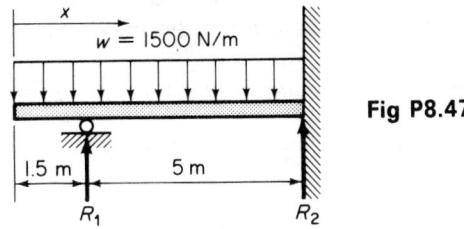

Fig P8.47

8.48 Determine the deflection at the midpoint of the beam shown.

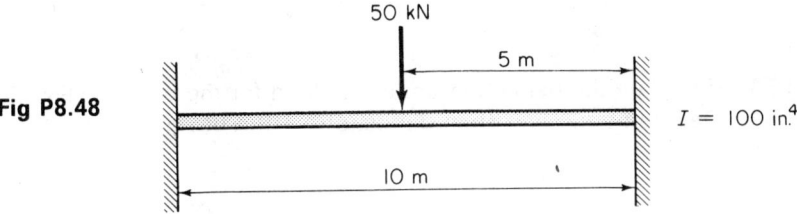

Fig P8.48

8.49 Determine the deflection at the middle of the span of the beam shown.

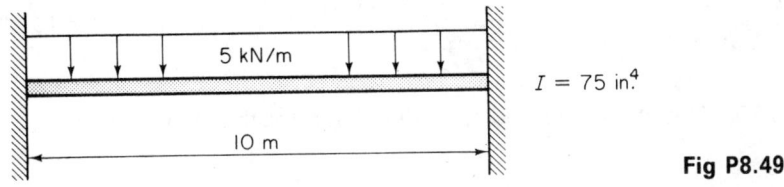

Fig P8.49

8.50 What moment of inertia is required for the beam loaded as shown in Fig. P8.50? Assume the deflection is limited to $\frac{1}{360}$ th of the span.

Fig P8.50

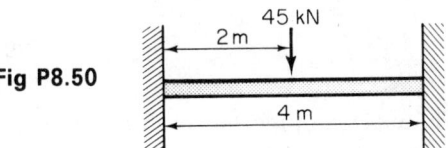

8.51 If the moment of inertia determined in Problem P8.50 is for a rectangle whose height is twice its width, determine the maximum stress in the beam.

8.52 Calculate the deflection at the center of the built-in beam shown in Fig. P8.52.

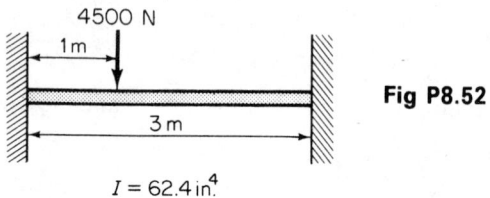

Fig P8.52

$I = 62.4 \, in^4$

8.53 Calculate the deflection under the load for the beam of Fig. P8.52.

8.54 Determine the mid-span deflection for the beam shown.

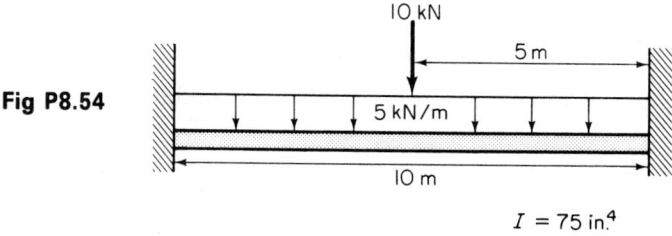

Fig P8.54

$I = 75 \, in^4$

8.55 Determine the mid-span deflection for the beam shown.

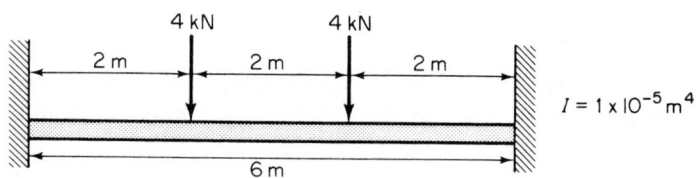

$I = 1 \times 10^{-5} \, m^4$

Fig P8.55

8.56 Determine the moment and the reactions at each support for the beam shown in Fig. P8.56. Also evaluate the mid-span deflection.

Fig P8.56

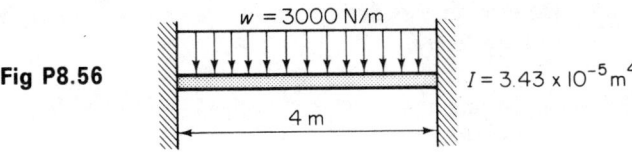

8.57 Determine the mid-span deflection for the beam shown.

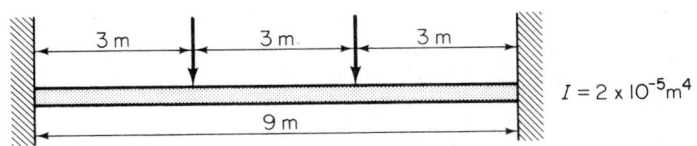

Fig P8.57

8.58 Calculate the mid-span deflection for the beam shown in Fig. P8.58.

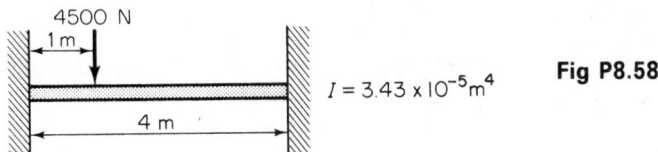

Fig P8.58

8.59 Determine the deflection under either load for the beam in Problem 8.57.

8.60 What is the deflection at the center of the beam shown in Fig. P8.60?

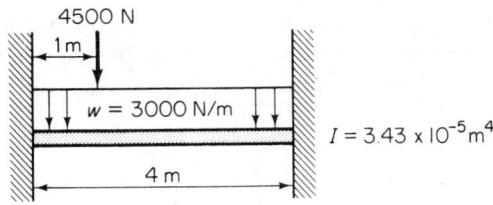

Fig P8.60

356   Strength of Materials for Engineering Technology

8.61   Determine the deflection at the 2200-N load for the beam shown in Fig. P8.61.

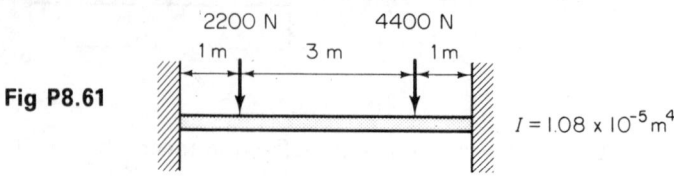

Fig P8.61

8.62   Show that the moment at the central support of the beam shown in Fig. P8.62 equals $-wL^2/8$.

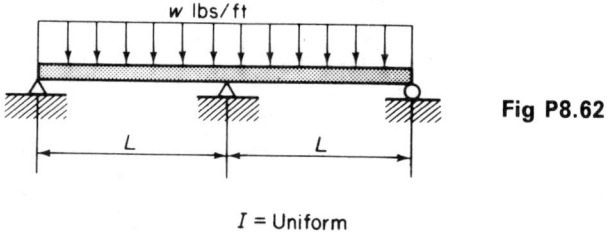

Fig P8.62

8.63   Determine the moment at the central support of the beam shown in Fig. P8.63 if its cross section is constant.

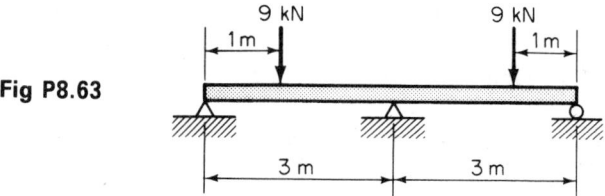

Fig P8.63

8.64   Determine the reaction at each support for the beam shown in Fig. P8.64. What is the moment at the center support? Use either the moment area method or superposition method.

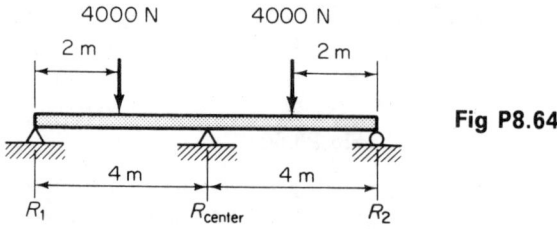

Fig P8.64

8.65 Determine the reaction at each support for the beam shown in Fig. P8.65. Use the superposition method.

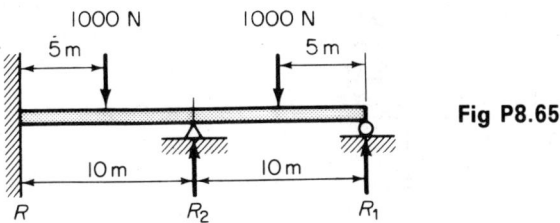

Fig P8.65

8.66 Solve Problem 8.64 using the three-moment equation.
8.67 Determine the moment at the center support of the beam shown in Fig. P8.67 using the superposition method. Also determine R_1, R_2, and R_3.

Fig P8.67

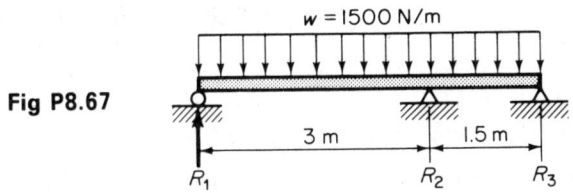

8.68 Use the three-moment equation to determine the moment at the center support of the beam in Problem 8.67.
8.69 Determine the moment at the center support for the beam shown in Fig. P8.69 using the superposition method.

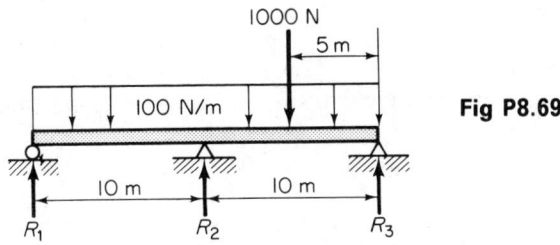

Fig P8.69

8.70 Verify the results obtained in Problem 8.69 using the three-moment equation.

8.71 Use the three-moment equation to determine the moment at each of the supports of the beam shown in Fig. P8.71.

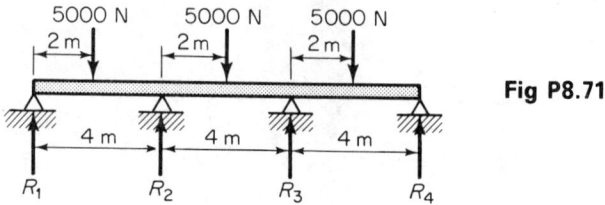

Fig P8.71

8.72 A multi-span beam carries a uniform load of 3000 N/m. Determine the moment at each support.

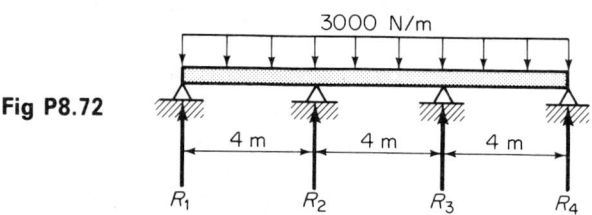

Fig P8.72

8.73 A multi-span beam carries the loads shown in Fig. P8.73. Determine the moment at each support. Compare your result to Problem 8.72.

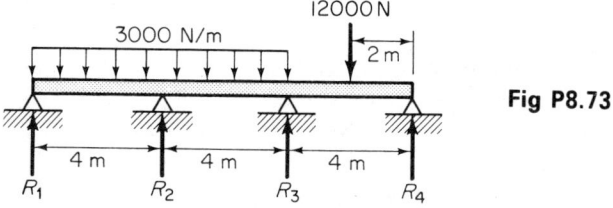

Fig P8.73

8.74 Determine the moment at each support for the beam shown in Fig. P8.74.

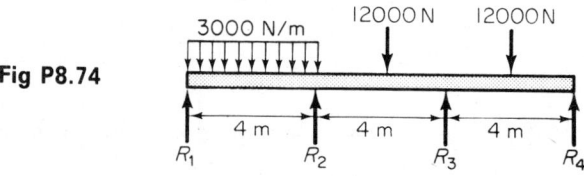

Fig P8.74

8.75 Two beams are placed at right angles to each other and are loaded at their central intersection by a load, F. If both beams have the same moment of inertia and they are made of the same material, determine the amount of the load F that each beam carries.

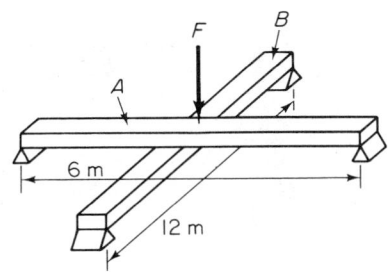

Fig P8.75

REFERENCES

American Institute of Steel Construction. *Manual of Steel Construction.* 7th ed., 1970.
Amirikian, A. *Analysis of Rigid Frames.* U.S. Government Printing Office.
Arges, K. P. and A. E. Palmer. *Mechanics of Materials.* New York: McGraw-Hill Book Co., 1963.
Bassin, M. E. and S. M. Brodsky. *Statics and Strength of Materials.* 2nd ed. New York: McGraw-Hill Book Co., 1969.
Breneman, J. W. *Strength of Materials.* 3rd ed. New York: McGraw-Hill Book Co., 1965.
Conway, H. D. *Mechanics of Materials.* Englewood Cliffs, NJ: Prentice-Hall, Inc.
Jensen, A. and H. H. Chenoweth. *Applied Strength of Materials.* 3rd ed. New York: McGraw-Hill Book Co., 1971.
Levinson, Irving J. *Mechanics of Materials.* 2nd ed. Englewood Cliffs, NJ: Prentice-Hall, Inc., 1970.
Lisarelli, F. R. *Essential Strength of Materials.* New York: McGraw-Hill Book Co.
Singer, F. L. *Strength of Materials.* 2nd ed. New York: Harper and Row, 1962.
Timoshenko, S. and Donovan H. Young. *Elements of Strength of Materials.* 5th ed. New York: D. Van Nostrand Co., 1968.

9
Columns

9.1 INTRODUCTION

When a structural member was subjected to a compressive axial load, we assumed that it could be treated as an elastic member that would fail at a loading dependent upon its elastic properties, i.e., its yield point. However, when the member is long relative to its lateral dimensions, it is called a *column* and columns fail primarily due to lateral bending leading to lateral instability. This type of failure, called *buckling,* does not depend upon the strength of the material, but is found to be a function only of the dimensions of the member and its modulus of elasticity. Thus two identical columns will fail (buckle) at the same load even though one may be made of a high strength alloy steel and the other of carbon steel. Unfortunately, the point at which a compression member becomes a column is not clearly defined. There is a region where the member acts as both a column and a compression member before it can be said to be a true column. In the subsequent sections of this chapter we will classify columns as *short columns, intermediate columns,* and *long columns.* It will also be necessary to discuss the character of the loading of a column (centric or eccentric), and the type and action of the end conditions in the column. All of these variables are of such nature as to be only partly amenable to mathematical analysis. As a consequence we shall resort to empirical relations and rules to a degree only exceeded by our usage of such relations when we dealt with riveted joints in Chapter 3.

9.2 COLUMNS—GENERAL

Before studying columns, let us first discuss the failure of a column of cross-sectional area A when subjected to an axial load F. Assume that the column in question is centrally loaded, is of a given material, and has end conditions that are kept constant. The only variables for our initial consideration will be the load and unsupported length of the column L. If the column is short, it can be loaded until the stress F/A reaches the yield stress of the material. If we make the column longer, failure will still be characterized by the yield stress of the material. Note that a property of the material (its yield stress) is the limiting condition for failure.

If we continue our experiment, we find some unsupported column length where the column becomes unstable (fails) and where the stress F/A at failure is greater than the proportional limit, but does not exceed the yield stress of the material. Continued increase in the length of the column gives the same type of action, but an ever decreasing failure value of F/A is needed to cause it to occur.

Let us consider the action of the column as the load F is applied at some still larger value of unsupported column length. While F/A is still within the elastic range of the material, the column will fail due to buckling or lateral bending. This condition is one of elastic instability and is not a function of the properties of the material. The continued increase in length L results in smaller and smaller values of the applied load F causing buckling to occur.

If the experiment just described is carried out for differing cross sections (all other conditions such as material and restraints being kept constant), it will be found that a single failure curve can be plotted on which the failure stress F/A is plotted as a function of L/r where L is the unsupported length and r is the least radius of gyration of the column.* Because of the universal appearance of the term L/r in column formulas, it is called the *slenderness ratio*.

Figure 9.1 shows a typical plot of F/A against L/r for the conditions we have described. For L/r values up to approximately 60, we can consider the column to be *a compression member;* for L/r values between approximately 60 and 120, we shall call the column an *intermediate column;* L/r values above 120 characterize the column as a *slender column*. The student should again note that L is the effective length of the column, and r is the least radius of gyration of the column about the bending axis.

---

*The use of r for the radius of gyration is common in column design. Note that this is k as defined in Chapter 4.

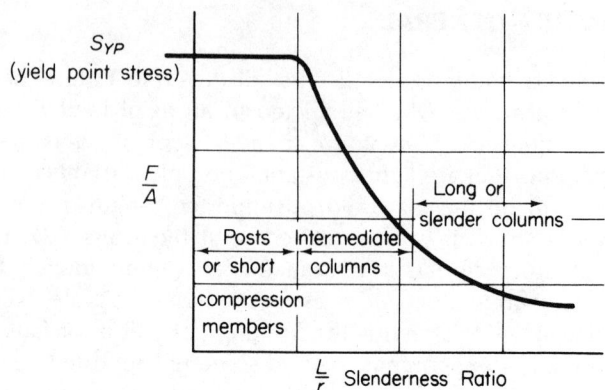

Fig 9.1

Illustrative problem 9.1

A rectangular steel member 50 mm × 100 mm is to be used as a column. If its length is 1.5 m, is it a short, intermediate, or long column?

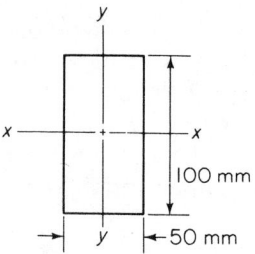

Fig 9.2   Illustrative problem 9.1

Solution:

The moment of inertia of a rectangle is $bh^3/12$. Applying this to both the x-x and y-y axes, and noting the area is $0.050 \times 0.100 = 0.005$ m^2

$$I_{xx} = \frac{0.050(0.100)^3}{12} = 4.17 \times 10^{-6} \text{ m}^4$$

$$I_{yy} = \frac{(0.100)(0.050)^3}{12} = 1.04 \times 10^{-6} \text{ m}^4$$

$$r_{xx} = \sqrt{\frac{I_{xx}}{A}} = \sqrt{\frac{4.17 \times 10^{-6}}{.5 \times 10^{-3}}} = 2.89 \times 10^{-2} \text{ m}$$

$$r_{yy} = \sqrt{\frac{I_{yy}}{A}} = \sqrt{\frac{1.04 \times 10^{-6}}{5 \times 10^{-3}}} = 1.44 \times 10^{-2} \text{ m}$$

r_{yy} is least and limits the design

Therefore $L/r = 1.5/1.44 \times 10^{-2} = 104.2$; the column is intermediate.

Illustrative problem 9.2

What is the longest length of $W12 \times 27$ column if L/r is not to exceed 120?

Solution:

From Appendix B, this section has a least value of r equal to 1.52 in. for the y-y axis. The length L is therefore

$$L = 1.52 \times (0.0254)(120) = 4.63 \text{ m}$$

9.3 LONG COLUMNS

Buckling is defined as that state of loading that will cause the column to undergo excessively large deflections while stressed within the elastic limits of the material. The load that will just cause buckling to occur is known as the *critical load*. We can illustrate this condition by reference to Fig. 9.3. The column shown in Fig. 9.3(a) is loaded by a central force

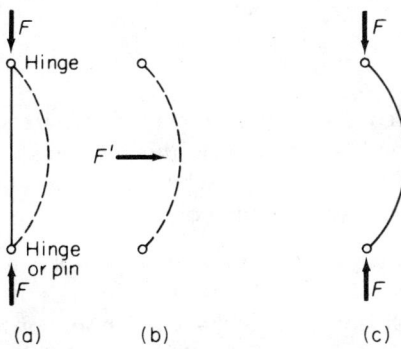

Fig 9.3 Long column behavior under load

F, which is below the critical load. If the load F is initially zero, the column will be straight and the application of the load F causes the column to deflect (exaggerated) as shown. Assuming the load to be below the critical load, we find that the withdrawal of F will return the column to its initial straight condition. A similar action will take place by the horizontal application of the force F'. By keeping F' sufficiently small, we insure that its removal would cause the column to return to its original shape. If the load F were applied vertically and exceeded the critical load, the column would undergo a large deformation as shown in Fig. 9.3(c) and would not return to its initial straight condition upon removal of the load. If the load F were near the critical load, the application of a lateral force [similar to F' in Fig. 9.3(b)] would cause an excessive deflection (buckling or elastic instability), which would in turn cause the column to take on a permanent set.

The action of long columns within the elastic limits of their materials was first investigated in the mid-1700's by a Swiss mathematician named Leonhard Euler. The following are a consequence of his analysis:

1. The *critical stress*, defined as the ratio of the critical load to the cross-sectional area of the column (F_{cr})/A, is independent of the strength of the material of the column. As noted earlier, a column of alloy steel having a given cross-sectional area will have the same critical stress as a column of carbon steel having the same cross-sectional area.
2. The critical stress is directly proportional to E, the modulus of elasticity of the column material.
3. The critical stress varies inversely as the square of the slenderness ratio of the column, i.e., $1/(L/r)^2$.
4. The critical stress is a function of the method by which the ends of the column are restrained. We can express these results mathematically as

$$S_{cr} = \frac{F_{cr}}{A} = \frac{n\pi^2 E}{(L/r)^2} \qquad (9.1)$$

where S_{cr} is the critical stress (not a maximum stress); n is a constant for a given type of end restraint; E is the modulus of elasticity; L/r is the slenderness ratio of the column where L is the effective length; and r is the least radius of gyration of the column.

Equation (9.1) is known as *Euler's Equation,* and it is the rationale behind most column formulas. Explicit in its derivation is the assumption that the critical stress shall not exceed the yield point of the material (more properly the stress at the proportional limit, but the yield point is usually used to determine the critical stress). This enables us to evaluate the maximum value of L/r for any material.

Illustrative problem 9.3

A mild alloy steel ($E = 203$ GPa) has a yield stress of 175 MPa. If $n = 1$ in Euler's Equation, what is the least value of L/r for the member to act as a long column?

Solution:

$$S_{cr} = \frac{n\pi^2 E}{(L/r)^2} \quad \text{or}$$

$$L/r = \sqrt{\frac{n\pi^2 E}{S_{cr}}} = \sqrt{\frac{1 \times \pi^2 \times 203 \times 10^9}{175 \times 10^6}} = 107$$

In Illustrative Problem 9.3 we used a value of $n = 1$ as part of the statement of the problem. Since n (and the effective column length as well) is dependent upon the end conditions of the column, let us briefly consider the end fixity of columns. The column shown in Fig. 9.3 had ends which were pinned or hinged, which permitted them to rotate, but they were not permitted to translate. In effect this condition (used to derive Euler's Equation) determines the value of L to be used in equation (9.1); the effective length L is the length between points of zero moment on the deflection curve of the column. Figure 9.4 shows four conditions of

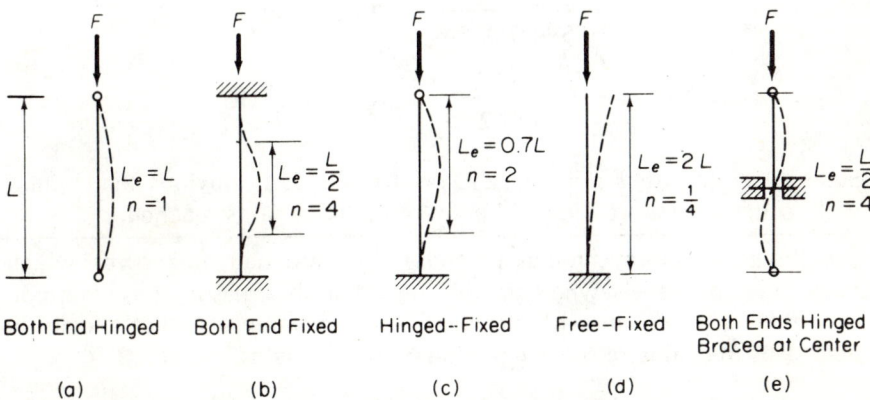

Fig 9.4 End restraint of columns

end restraint for a column:

1. Both ends hinged $L_e = L$ $n = 1$
2. Both ends fixed (built-in) $L_e = L/2$ $n = 4$
3. Hinged at one end, fixed at the other $L_e = 0.7 L$ $n = 2$
4. One end free, one end fixed $L_e = 2L$ $n = \frac{1}{4}$

For end conditions that afford partial restraint, the effective length can be satisfactorily taken to be between $0.5L$ and L, which gives values of n between 1 and 4. Also, where a column is braced [as shown in Fig. 9.4(e)], the effective length is halved if the bracing is at the center; for bracing at the third points, $L_e = L/3$ and $n = 9$. The student should note that since $n = 1$ for both ends hinged, n can be thought of as a factor of comparison that indicates the relative strength of a column of a given end fixity to a column having both ends hinged.

Illustrative problem 9.4

An S5 × 10 beam is to be used as a 6 m long column. If $E = 210$ GPa, and one end is free while the other is fixed, determine the value of S_{cr}. Compare this stress to the 210 MPa yield stress of the material.

Solution:

For this case $n = \frac{1}{4}$; $S_{cr} = n\pi E/(L/r)^2$. For an S5 × 10 beam, $r_{xx} = 2.05$ and $r_{yy} = 0.643$. Thus, the least value of

$$L/r = \frac{6}{2.05 \times 0.0254} = 115.2 \quad \text{and}$$

$$S_{cr} = \frac{\frac{1}{4}\pi^2 \times 210 \times 10^9}{(115.2)^2} = 39.04 \times 10^6 \text{ Pa}$$

Since S at the yield point is 210×10^6 Pa, it is obvious that failure will occur long before the yield point of the material is reached.

In the design of any structural member there are many unknowns, which have been noted elsewhere in this text. For this reason it is common to modify Euler's Equation by introducing a *factor of safety f*, which effectively decreases the value of S_{cr} in equation (9.1) and yields

$$S_{cr} = \frac{n\pi^2 E}{f(L/r)^2} \tag{9.2}$$

The load factor of safety thus defined is the ratio of the critical stress to the applied stress (load). For steady loads, f is commonly taken to be 2.5; its values may vary from 2 to 3.

Illustrative problem 9.5

A 4 in. Schedule 40 steel pipe is used as a column to support a centric load of 45 000 N. If the column is hinged at both ends and is 25 ft long, determine the factor of safety of this design. Use $E = 210$ GPa.

Solution:

From Appendix D, a 4 in. Schedule 40 pipe has $I_x = 7.233$ in.^{4}, $A = 3.173$ in.^{2} and $r = \sqrt{I/A} = \sqrt{7.233/3.173} = 1.51$ in. Therefore, $L/r = (25 \times 12)/1.51 \cong 198.7$.

$$S_{cr} = \frac{n\pi^2 E}{(L/r)^2} = \frac{1 \times \pi^2 \times 210 \times 10^9}{(198.7)^2} = 52.5 \text{ MPa}$$

The applied stress is F/A; A for 4 in. Schedule 40 pipe is 3.173 sq. in. and $F/A = 45\ 000/3.173 \times (0.0254)^2 = 21.98$ MPa·$f = 52.5/21.98 = 2.39$.

A word of caution at this point: For practical reasons, the AISC Code prohibits the use of compression members beyond an L/r value of 200. In the case of main members that are subjected to shock and/or vibration, L/r is further limited to a value of 120 or less. As will be discussed later on in this chapter, main members whose slenderness ratios lie between 120 and 200 have their allowable stress decreased by a factor that is a function of L/r in this range. Also, a column can be braced at unequal lengths along either axis requiring that both axes be investigated.

Illustrative problem 9.6

An S12 × 35 steel beam is to be used as a hinged column 16 m long. A safety factor of 3 is specified. If the column is braced at its midpoint in its weakest direction and is unbraced along its strongest direction, determine the maximum load it can carry. Use $E = 210$ GPa.

Solution:

The properties of an S12 × 35 beam are: $r_{xx} = 4.72$ in., $r_{yy} = 0.98$ in., $A = 10.3$ sq. in. Along the x-x axis

$$S_{cr} = \frac{n\pi^2 E}{3(L/r)^2}; \quad \frac{L}{r} = \frac{16}{4.72 \times .0254} = 133.5$$

and

$$S_{cr} = \frac{1 \times \pi^2 \times 210 \times 10^9}{3(133.5)^2} = 38.76 \text{ MPa}$$

Along the y-y axis, $L/r = \frac{16}{2}/0.98 \times 0.0254 = 321.4$ for the unsupported length. At this point, the design must stop since L/r exceeds 200, and this beam is not to be used as a compression member. Let us modify the design to allow the beam to be supported at the *third points in the weak direction*. Thus

$$\frac{L}{r} = \frac{\frac{16}{3}}{0.98 \times 0.0254} = 214.3$$

which we shall accept for our purposes. Note that the code would not allow the 214.3 based upon the unsupported length. Proceeding,

$$S_{cr} = \frac{1 \times \pi^2 \times 210 \times 10^9}{3(214.3)^2} = 15.04 \text{ MPa}$$

The "weak" direction is weak, and the y-y axis limits us to 15.04 MPa in this application even with bracing at the third points.

9.4 INTERMEDIATE STEEL COLUMNS—HISTORICAL

Prior to the early 1960's, a large body of empirical data had been accumulated, and as a consequence, design formulas were proposed for the intermediate column range ($L/r < 120$). The problem that arises when this regime of column design is studied is that the strength of the material is a governing parameter. Because of this, one finds that there are in existence differing design formulas, each based upon empirical data, which depend upon the various legal jurisdictions (cities, states, federal, etc.). After 1961, the AISC Code was changed to a form having a variable factor of safety, which will be more fully discussed in the next section of this chapter. Since the earlier formulas are still widely used and are part of the building codes of many municipalities, we shall first study them and their application.

The formulas proposed and used can be broken into the following categories based upon their mathematical form:

1. The *straight line formula*
2. The *parabolic formula*
3. The *Rankine-Gordon formula* (so named for the English engineers who first proposed it).

The straight line formula, first proposed in about 1890, gives the critical stress on a column in the following form

$$\left(\frac{F}{A}\right)_{cr} = S_{cr} = a - b\left(\frac{L}{r}\right) \tag{9.3}$$

where a and b are constants of the material. For most practical purposes the yield point is the value taken for S_{cr}. In design, it is more customary to use an equation of the form of equation (9.3), based upon the allowable stress. One of the most common of these formulas is the one known as the *Chicago formula* since it was part of the Chicago Building Code for many years. In terms of the allowable stress S_{all}

$$\left(\frac{F}{A}\right)_{all} = S_{all} = 16{,}000 - 70\left(\frac{L}{r}\right) \text{ psi} \tag{9.4}$$

$$= 110 - 0.483\left(\frac{L}{r}\right) \text{ MPa}$$

and the formula can be used for $30 < L/r < 120$ for main members and for values of $L/r < 30$, $S_{all} = 14{,}000$ psi (96.5 MPa). This equation has been plotted on Fig. 9.5 where it will be noted that it intersects the Euler curve at $L/r = 126$, when the Euler curve is based upon a factor of safety of 2.5. Below L/r values of 126, the allowable stress increases to an L/r of 30 where it is 14,000 psi (96.5 MPa) and remains constant for all values of $L/r < 30$. Since equation (9.4) contains both the area and radius of gyration, its use (and the use of most column equations) usually entails a trial and error procedure. Note that the conversion factor (multiplier) to convert from psi to newtons is 6895.

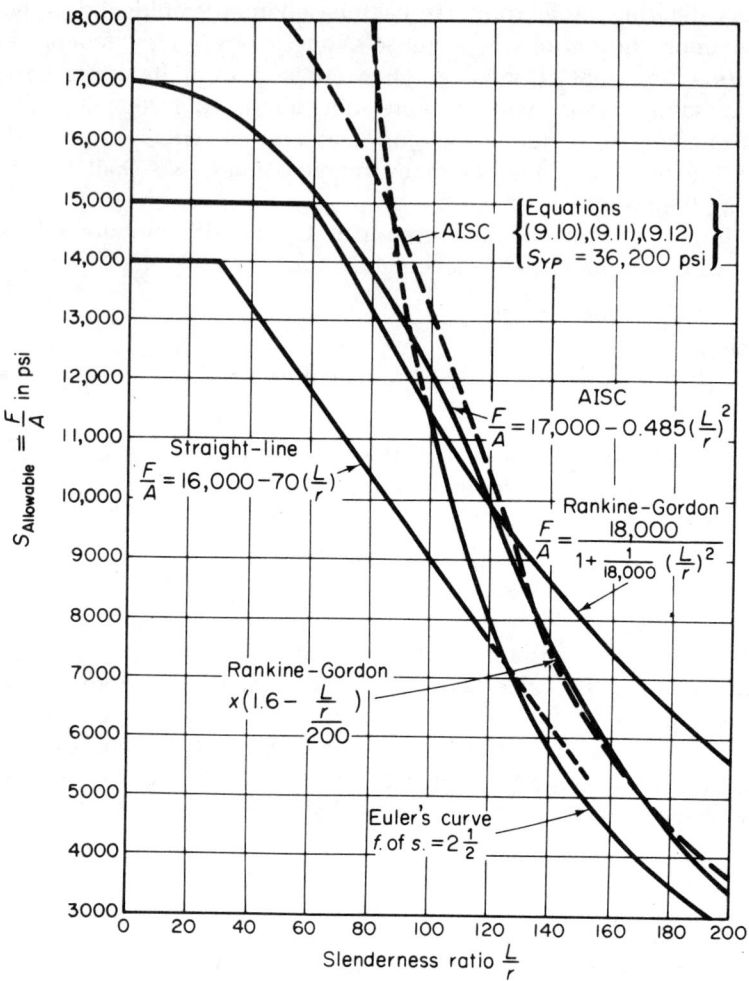

Fig 9.5 Comparison of unit loads of various column formulas

Illustrative problem 9.7

A column 8 ft long is to be used to carry a load of 50,000 lb. An S8 × 23 beam is proposed for this application. Is it suitable? Use Chicago formula.

Solution:

Let us first obtain the value of L/r that governs this design. For this beam

$r_{xx} = 3.10$ and $r_{yy} = 0.798$.

$$\left(\frac{L}{r}\right)_{xx} = \frac{8 \times 12}{3.10} = 31$$

$$\left(\frac{L}{r}\right)_{yy} = \frac{8 \times 12}{0.798} = 120.3$$

Thus the governing value of L/r is

$$\left(\frac{L}{r}\right)_{yy} = 120.3$$

Since

$$A = 6.77 \text{ sq. in.}, \quad \left(\frac{F}{A}\right) = \frac{50{,}000}{6.77} = 7380 \text{ psi (50.9 MPa)}$$

But

$$\left(\frac{F}{A}\right)_{\text{all}} = 16{,}000 - 70\frac{L}{r} = 16{,}000 - 70(120.3) = 7580 \text{ psi. (52.3 MPa)}$$

The allowable stress exceeds the imposed stress, and we therefore conclude that the beam selected is suitable for the application. Also note that L/r is at the 120 limit.

Illustrative problem 9.8

Select a wide flange column to carry a load of 57,500 lb if the column is to be 10 ft long. Use the Chicago formula.

Solution:

It may be necessary to try several sections before a selection is made. As a starting point let us evaluate r for $L/r = 120$. Thus $r = L/120 = (10 \times 12)/120 = 1$; if the straight line formula is to be used, $r \geq 1$. From Appendix C, "Column Tables," we note that a $W8 \times 20$ beam seems suitable. Let us try it: $L/r = (12 \times 10)/1.25 = 96$; $A = 5.89$; $F/A_{\text{applied}} = 57{,}500/5.89 = 9760$ psi; $S_{\text{all}} = 16{,}000 - 70(L/r) = 16{,}000 - 70(96) = 9280$ psi.

The section selected is not suitable since the applied stress exceeds the allowable stress. As a next iteration, let us try a W8 × 24 section: $L/r = (12 \times 10)/1.61 = 74.5$; $A = 7.06$; $F/A_{applied} = 57,500/7.06 = 8140$ psi. $S_{all} = 16,000 - 70(L/r) = 16,000 - 70(74.5) = 10,780$ psi. This selection is satisfactory.

The parabolic formula expresses the critical stress of an intermediate column in the form

$$\left(\frac{F}{A}\right)_{cr} = S_{cr} = a - b\left(\frac{L}{r}\right)^2 \tag{9.5}$$

where a and b are constants depending upon the physical properties of the material. A working equation of the form of equation (9.5) specified in the earlier (pre-1960) editions of the AISC Manual of Steel Construction gives the allowable (working) stress on columns that have $L/r < 120$ as

$$\left(\frac{F}{A}\right)_{all} = S_{all} = 17,000 - 0.485\left(\frac{L}{r}\right)^2 \text{ psi} \tag{9.6}$$

Equation (9.6) is also shown on Fig. 9.5. This curve gives a value of $S_{all} = 17,000$ at $L/r = 0$, and 10,000 at $L/r = 120$. In order to facilitate calculations, we can use Table 9.1 from the AISC Manual (pre-1960), which gives S_{all} for values of L/r from 1 to 120 in L/r increments of 1. This table can be interpolated if necessary.

Illustrative problem 9.9

Repeat Illustrative Problem 9.8 using the AISC formula.

Solution:

If we proceed as before, using $L/r = 120$ as the starting point, we again require $r \geq 1$. Selecting a W8 × 20 beam as in Illustrative Problem 9.8, we apply F/A and again obtain 9760 psi, but from Table 9.1 for $L/r = 96$, $S_{all} = 12,530$ psi. Thus, we can use a W8 × 20 beam in this case, where the Chicago formula would not permit its use. We would reach the same conclusion by using Fig. 9.5.

Table 9.1* Allowable Stresses per Square Inch for Compression Members

Main and Secondary Members, L/r not over 120, $S_{all} = 17000 - 0.485\left(\dfrac{L}{r}\right)^2$						Secondary Members, L/r 121 to 200, $S_{all} = \dfrac{18000}{1 + \dfrac{L^2}{18{,}000\,r^2}}$				Main Members, L/r 121 to 200, $D_0 \times \left(1.6 - \dfrac{L/r}{200}\right)$			
$\dfrac{L}{r}$	Unit Stress 1000 psi	$\dfrac{L}{r}$	Unit Stress 1000 psi	$\dfrac{L}{r}$	Unit Stress 1000 psi	$\dfrac{L}{r}$	Unit Stress 1000 psi	$\dfrac{L}{r}$	Unit Stress 1000 psi	$\dfrac{L}{r}$	Unit Stress 1000 psi	$\dfrac{L}{r}$	Unit Stress 1000 psi
1	17.00	41	16.19	81	13.82	121	9.93	161	7.38	121	9.88	161	5.87
2	17.00	42	16.14	82	13.74	122	9.85	162	7.32	122	9.75	162	5.78
3	17.00	43	16.10	83	13.66	123	9.78	163	7.27	123	9.63	163	5.71
4	16.99	44	16.06	84	13.58	124	9.71	164	7.22	124	9.52	164	5.63
5	16.99	45	16.02	85	13.50	125	9.64	165	7.16	125	9.40	165	5.53
6	16.98	46	15.97	86	13.41	126	9.56	166	7.11	126	9.27	166	5.47
7	16.98	47	15.93	87	13.33	127	9.49	167	7.06	127	9.16	167	5.40
8	16.97	48	15.88	88	13.24	128	9.42	168	7.01	128	9.04	168	5.33
9	16.96	49	15.84	89	13.16	129	9.35	169	6.96	129	8.93	169	5.25
10	16.95	50	15.79	90	13.07	130	9.28	170	6.91	130	8.82	170	5.18
11	16.94	51	15.74	91	12.98	131	9.22	171	6.86	131	8.71	171	5.11
12	16.93	52	15.69	92	12.90	132	9.15	172	6.81	132	8.60	172	5.04
13	16.92	53	15.64	93	12.81	133	9.08	173	6.76	133	8.49	173	4.97
14	16.91	54	15.59	94	12.72	134	9.01	174	6.71	134	8.38	174	4.90
15	16.89	55	15.53	95	12.62	135	8.94	175	6.66	135	8.27	175	4.83
16	16.88	56	15.48	96	12.53	136	8.88	176	6.62	136	8.17	176	4.77
17	16.86	57	15.42	97	12.44	137	8.81	177	6.57	137	8.06	177	4.70
18	16.84	58	15.37	98	12.34	138	8.75	178	6.52	138	7.96	178	4.63
19	16.83	59	15.31	99	12.25	139	8.68	179	6.48	139	7.86	179	4.57
20	16.81	60	15.25	100	12.15	140	8.62	180	6.43	140	7.76	180	4.50
21	16.79	61	15.20	101	12.05	141	8.55	181	6.38	141	7.65	181	4.43
22	16.77	62	15.14	102	11.95	142	8.49	182	6.34	142	7.56	182	4.37
23	16.74	63	15.08	103	11.86	143	8.43	183	6.29	143	7.46	183	4.31
24	16.72	64	15.01	104	11.75	144	8.36	184	6.25	144	7.36	184	4.25
25	16.70	65	14.95	105	11.65	145	8.30	185	6.20	145	7.26	185	4.19
26	16.67	66	14.89	106	11.55	146	8.24	186	6.16	146	7.17	186	4.13
27	16.65	67	14.82	107	11.45	147	8.18	187	6.12	147	7.08	187	4.07
28	16.62	68	14.76	108	11.34	148	8.12	188	6.07	148	6.98	188	4.01
29	16.59	69	14.69	109	11.24	149	8.06	189	6.03	149	6.89	189	3.95
30	16.56	70	14.62	110	11.13	150	8.00	190	5.99	150	6.80	190	3.89
31	16.53	71	14.56	111	11.02	151	7.94	191	5.95	151	6.71	191	3.84
32	16.50	72	14.49	112	10.92	152	7.88	192	5.91	152	6.62	192	3.78
33	16.47	73	14.42	113	10.81	153	7.82	193	5.86	153	6.53	193	3.72
34	16.44	74	14.34	114	10.70	154	7.77	194	5.82	154	6.45	194	3.67
35	16.41	75	14.27	115	10.59	155	7.71	195	5.78	155	6.36	195	3.61

Table 9.1 (continued)

Main and Secondary Members, L/r not over 120, $S_{all} = 17000 - 0.485 \left(\dfrac{L}{r}\right)^2$						Secondary Members, L/r 121 to 200, $S_{all} = \dfrac{18000}{1 + \dfrac{L^2}{18{,}000 r^2}}$				Main Members, L/r 121 to 200, $D_0 \times \left(1.6 - \dfrac{L/r}{200}\right)$			
$\dfrac{L}{r}$	Unit Stress 1000 psi	$\dfrac{L}{r}$	Unit Stress 1000 psi	$\dfrac{L}{r}$	Unit Stress 1000 psi	$\dfrac{L}{r}$	Unit Stress 1000 psi	$\dfrac{L}{r}$	Unit Stress 1000 psi	$\dfrac{L}{r}$	Unit Stress 1000 psi	$\dfrac{L}{r}$	Unit Stress 1000 psi
36	16.37	76	14.20	116	10.47	156	7.65	196	5.74	156	6.27	196	3.56
37	16.34	77	14.12	117	10.36	157	7.60	197	5.70	157	6.19	197	3.51
38	16.30	78	14.05	118	10.25	158	7.54	198	5.66	158	6.11	198	3.45
39	16.26	79	13.97	119	10.13	159	7.49	199	5.62	159	6.03	199	3.40
40	16.22	80	13.90	120	10.02	160	7.43	200	5.59	160	5.94	200	3.35

*Reproduced with permission from the *Steel Construction Manual of the American Institute of Steel Construction*, 5th ed., 1960.
Note: psi × 6895 = Pa.

The third equation, the Rankine-Gordon equation, has the form

$$\left(\frac{F}{A}\right)_{cr} = S_{cr} = \frac{a}{1 + b(L/r)^2} \qquad (9.7)$$

where a and b are (once more) constants depending upon the properties of the material. The AISC Code specified that axial-loaded columns that are used for bracing, or as other secondary members, should have allowable stresses that do not exceed those given by equation (9.8), which is of the Rankine-Gordon form:

$$\left(\frac{F}{A}\right)_{all} = S_{all} = \frac{18000}{1 + \dfrac{1}{18{,}000}\left(\dfrac{L}{r}\right)^2} \qquad (9.8)$$

Equation (9.8) can be used for values of $120 < L/r < 200$ for bracing or secondary members, i.e., those members that are used to lend stiffness and/or stability to a structure but do not carry the permanent load of the structure.

The AISC Code permitted the further use of a Rankine-Gordon type equation [equation (9.7)] for main compression members for values of $120 < L/r < 200$ provided that they would not ordinarily be subject to shock or vibratory loads if a derating factor were used in conjunction with equation (9.8) for this condition. The derating factor is

$$\left[1.6 - \frac{(L/r)}{200} \right] \qquad (9.9)$$

and multiplies the right side of equation (9.8) by a factor effectively less than unity. Numerical values of S_{all} are tabulated in Table 9.1 for equation (9.8) and also for the derating of equation (9.8) by the factor given by equation (9.9). Also, equations (9.8) and (9.9) are shown on Fig. 9.5. The AISC Code prohibited the use of columns in all cases with L/r values greater than 200.

Illustrative problem 9.10

Determine the maximum load that a $W12 \times 65$ column can carry if it is 35 feet long. Assume it to be: (a) a secondary member; (b) a main member.

Solution:

The greatest L/r for this column is $(35 \times 12)/3.02 = 139$. Equation (9.8) applies, and we obtain, from Fig. 9.5 or Table 9.1, $S_{all} = 8,680$ psi for a secondary member and $S_{all} = 7,860$ psi for a main member. Since the area of the column is 19.1 sq. in., the loads are

(a) secondary member $= 19.1 \times 8680 = 165,800$ lb
(b) main member $= 19.1 \times 7860 = 150,100$ lb

The design of columns, other than steel columns, is usually done by empirical equations for the specific material established either by the local building code authority or, in its absence, by a federal or nationally known engineering society or association. Thus, timber columns are usually designed to the requirements of the Forest Products Laboratory or the National Lumber Manufacturer's Association. Aluminum columns are often designed to requirements of the American Society of Civil Engineers or Alcoa Structural Handbook formulas. Due to many such organizations, as well as the large number of materials used as structural columns, it would not be possible to do more than just list formulas without the necessary precautions or detailed discussions of each formula that should be given. For any column design, the student (as well as the practicing engineer) should consult the governing authority.

9.5 INTERMEDIATE STEEL COLUMNS—AISC DESIGN PROCEDURE

In 1961 the AISC adopted a set of empirical design formulas for intermediate steel columns after extensive studies of all available column data. The results of these studies were incorporated into the 1963 edition of the AISC Steel Construction Manual and the AISC design specifications. In addition, the 1963 sixth edition of the Steel Construction Manual provided for the use of structural steel having a specified minimum yield point up to, but not exceeding, 50,000 psi (344.8 MPa). The 1970 seventh edition of this Manual has approved for use high strength steels with yield strengths of up to 100,000 psi (689.5 MPa).

Before considering the present design procedure, a further discussion of the end restraints of columns is in order since the AISC Manual now contains provisions for various end restraints. The basic column considered is one that has pinned or hinged ends and is braced against side (lateral) displacement. This corresponds to Fig. 9.4(a) where $n = 1$ and $L_e = L$. Where the ends of the column are restrained, Figs. 9.4(b) and (c) indicate effective lengths less than the actual length of the column. For these cases, the decrease in effective length should be equivalent to permitting a larger value of allowable loading on the column. Where there is no end restraint but only lateral restraint [such as at the free end of the cantilevered column shown in Fig. 9.4(d)], the effective length is greater than the actual length of the column, and this effect should be equivalent to decreasing the allowable loading on the column. The AISC design procedure allows for this effect by defining a slenderness ratio KL/r, where K is the ratio of effective column length to actual unbraced length. Thus, for the cases indicated in Figs. 9.4(a), (b), and (c), $K \leqq 1$; while for the column shown in Fig. 9.4(d), $K = 1$. Figure 9.6 shows both the theoretical K values for six idealized conditions and the recommended design values for use when ideal conditions are approached in actual design. Since the end fixity of a column is not readily apparent, we shall retain K in the design procedure, but take its value to be unity (i.e., $K = 1$) in all problems.

The AISC design procedure for main and secondary members whose slenderness ratios (KL/r) are less than 120 recommends the following formulas

$$\left(\frac{F}{A}\right)_{all} = S_{all} = \frac{\left[1 - \frac{(KL/r)^2}{2C_c^2}\right]}{f} S_{yp} \qquad (9.10)$$

where f, the factor of safety, is given by

$$f = \frac{5}{3} + \frac{3(KL/r)}{8C_c} - \frac{(KL/r)^3}{8C_c^3} \qquad (9.11)$$

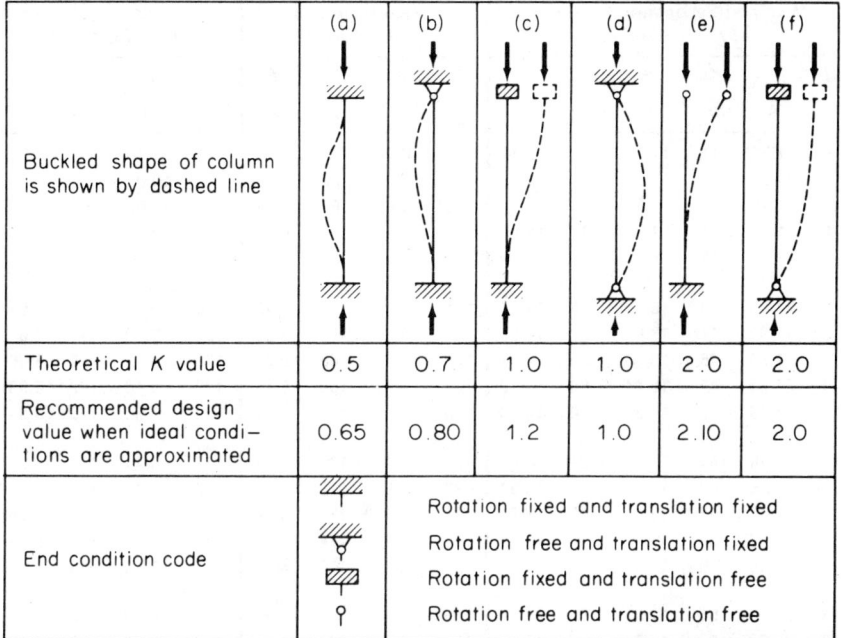

Fig 9.6 Values of K for column design. Reproduced with permission from the *AISC Manual of Steel Construction*, 7th edition, 1970.

and

$$C_c = \sqrt{\frac{2\pi^2 E}{S_{yp}}} \qquad (9.12)$$

The term S_{yp} is the yield point stress of the material while S_{all} is the allowable axial stress. It is interesting to note that the factor of safety f has a minimum value of $\frac{5}{3}$ for a slenderness ratio of zero and increases as the slenderness ratio increases. Thus the maximum allowable stress from equation (9.10) is $\frac{3}{5} S_{yp}$, and it decreases as the slenderness ratio increases. Equations (9.10), (9.11), and (9.12) can be used for all main and secondary members whose slenderness ratios are 120 or less ($KL/r \leqq 120$) and for main members whose slenderness ratios are equal to or less than the value of C_c, i.e., $(KL/r) \leqq C_c$.

If KL/r, the slenderness ratio, exceeds C_c, the recommended equation for the allowable stress is

$$\left(\frac{F}{A}\right)_{all} = S_{all} = \frac{12\pi^2 E}{23(KL/r)^2} \qquad (9.13)$$

Table 9.2 Allowable Stress, 1000 psi, for Compression Members of 36,000 psi Yield Point Steel[*]

Main and Secondary Members KL/r not over 120						Main Members KL/r 121 to 200				Secondary Members L/r 121 to 200			
$\dfrac{KL}{r}$	$\dfrac{F}{A}$	$\dfrac{KL}{r}$	$\dfrac{F}{A}$	$\dfrac{KL}{r}$	$\dfrac{F}{A}$	$\dfrac{KL}{r}$	$\dfrac{F}{A}$	$\dfrac{KL}{r}$	$\dfrac{F}{A}$	$\dfrac{L}{r}$	$\dfrac{F}{A}$	$\dfrac{L}{r}$	$\dfrac{F}{A}$
1	21.56	41	19.11	81	15.24	121	10.14	161	5.76	121	10.19	161	7.25
2	21.52	42	19.03	82	15.13	122	9.99	162	5.69	122	10.09	162	7.20
3	21.48	43	18.95	83	15.02	123	9.85	163	5.62	123	10.00	163	7.16
4	21.44	44	18.86	84	14.90	124	9.70	164	5.55	124	9.90	164	7.12
5	21.39	45	18.78	85	14.79	125	9.55	165	5.49	125	9.80	165	7.08
6	21.35	46	18.70	86	14.67	126	9.41	166	5.42	126	9.70	166	7.04
7	21.30	47	18.61	87	14.56	127	9.26	167	5.35	127	9.59	167	7.00
8	21.25	48	18.53	88	14.44	128	9.11	168	5.29	128	9.49	168	6.96
9	21.21	49	18.44	89	14.32	129	8.97	169	5.23	129	9.40	169	6.93
10	21.16	50	18.35	90	14.20	130	8.84	170	5.17	130	9.30	170	6.89
11	21.10	51	18.26	91	14.09	131	8.70	171	5.11	131	9.21	171	6.85
12	21.05	52	18.17	92	13.97	132	8.57	172	5.05	132	9.12	172	6.82
13	21.00	53	18.08	93	13.84	133	8.44	173	4.99	133	9.03	173	6.79
14	20.95	54	17.99	94	13.72	134	8.32	174	4.93	134	8.94	174	6.76
15	20.89	55	17.90	95	13.60	135	8.19	175	4.88	135	8.86	175	6.73
16	20.83	56	17.81	96	13.48	136	8.07	176	4.82	136	8.78	176	6.70
17	20.78	57	17.71	97	13.35	137	7.96	177	4.77	137	8.70	177	6.67
18	20.72	58	17.62	98	13.23	138	7.84	178	4.71	138	8.62	178	6.64
19	20.66	59	17.53	99	13.10	139	7.73	179	4.66	139	8.54	179	6.61
20	20.60	60	17.43	100	12.98	140	7.62	180	4.61	140	8.47	180	6.58
21	20.54	61	17.33	101	12.85	141	7.51	181	4.56	141	8.39	181	6.56
22	20.48	62	17.24	102	12.72	142	7.41	182	4.51	142	8.32	182	6.53
23	20.41	63	17.14	103	12.59	143	7.30	183	4.46	143	8.25	183	6.51
24	20.35	64	17.04	104	12.47	144	7.20	184	4.41	144	8.18	184	6.49
25	20.28	65	16.94	105	12.33	145	7.10	185	4.36	145	8.12	185	6.46
26	20.22	66	16.84	106	12.20	146	7.01	186	4.32	146	8.05	186	6.44
27	20.15	67	16.74	107	12.07	147	6.91	187	4.27	147	7.99	187	6.42
28	20.08	68	16.64	108	11.94	148	6.82	188	4.23	148	7.93	188	6.40
29	20.01	69	16.53	109	11.81	149	6.73	189	4.18	149	7.87	189	6.38
30	19.94	70	16.43	110	11.67	150	6.64	190	4.14	150	7.81	190	6.36
31	19.87	71	16.33	111	11.54	151	6.55	191	4.09	151	7.75	191	6.35
32	19.80	72	16.22	112	11.40	152	6.46	192	4.05	152	7.69	192	6.33
33	19.73	73	16.12	113	11.26	153	6.38	193	4.01	153	7.64	193	6.31
34	19.65	74	16.01	114	11.13	154	6.30	194	3.97	154	7.59	194	6.30
35	19.58	75	15.90	115	10.99	155	6.22	195	3.93	155	7.53	195	6.28
36	19.50	76	15.79	116	10.85	156	6.14	196	3.89	156	7.48	196	6.27
37	19.42	77	15.69	117	10.71	157	6.06	197	3.85	157	7.43	197	6.26
38	19.35	78	15.58	118	10.57	158	5.98	198	3.81	158	7.39	198	6.24
39	19.27	79	15.47	119	10.43	159	5.91	199	3.77	159	7.34	199	6.23
40	19.19	80	15.36	120	10.28	160	5.83	200	3.73	160	7.29	200	6.22

[*]Reproduced with permission of the American Institute of Steel Construction, New York, N.Y., from the *Steel Construction Manual of the American Institute of Steel Construction*, 7th ed., 1970.

For secondary members or braces subjected to compression loads with slenderness ratios in excess of C_c, i.e., $KL/r > C_c$, equations (9.10) and (9.13) are divided by the derating factor given in equation (9.9), which yields equation (9.14). This procedure, in effect, gives a larger permissible value of S_{all} for those members. Thus

$$\left(\frac{F}{A}\right)_{all} = S_{all}$$

$$= [\text{equation (9.10) or (9.13)}] \left[\frac{1}{1.6 - (L/r)/200}\right] \quad (9.14)$$

It should be noted that K does not appear in the derating term, which is equivalent to assuming $K = 1$ in this term. The more liberal working stress allowed for these members has been justified by the relative unimportance of such members and also by the greater effectiveness of end restraint likely to be present at their ends. Due to this, the full unbraced length of the member should always be used and equation (9.14) should be restricted to those columns whose ends are more or less fixed against rotation and translation at end points. It is interesting to note that equation (9.13), which is used for columns that can fail by elastic buckling, is Euler's Equation with a constant factor of safety of $23/12$.

Table 9.2 gives values of the allowable stress on compression members having a yield point of 36,000 psi and $E = 29 \times 10^6$, which corresponds closely to A36 steel. It should be noted here that the maximum value of L/r for any member is 200. Similar tables for steels having yield points up to 100,000 psi are given in Appendix C. Table 9.3 gives values of

Table 9.3 Values of C_c

S_{yp} (psi)	C_c	S_{yp}	C_c
33,000	131.7	46,000	111.6
35,000	129.7	50,000	107.0
36,000	126.1	55,000	102.0
39,000	121.2	60,000	97.7
42,000	116.7	65,000	93.8
45,000	112.8	90,000	79.8
		100,000	75.7

C_c as a function of the yield stress of the material. For this table, E has been taken to be 29×10^6 psi.

Figure 9.7 shows the allowable stress for columns that have $KL/r \leq 120$ as a function of the yield point of the steel. Reference is again made

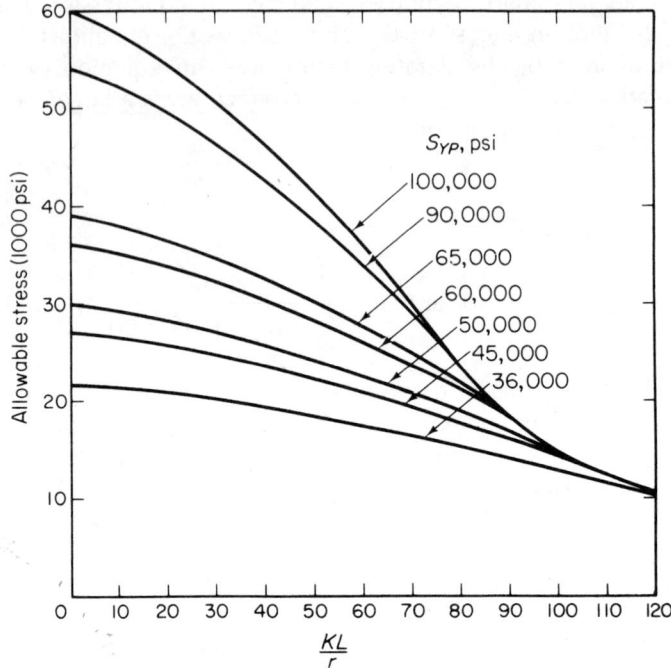

Fig 9.7 Allowable stress for columns

to Fig. 9.5 where equations (9.10), (9.11), and (9.12) have been plotted for a steel having a yield point of 36,000 psi. It is quite apparent that the allowable stress (for slenderness ratios $\leq$ 120) based upon the latest AISC formulas exceeds those previously permitted by earlier codes and formulas. Also, for short struts, the present AISC procedure allows a limiting stress of 21,600 psi against a maximum value of 15,000 psi by earlier AISC manuals and the value of 14,000 psi allowed by the Chicago formula. For $L/r > 126$, the allowable stress by the latest procedure yields a value of allowable stress very close to that obtained by earlier AISC procedures.

Illustrative problem 9.11

Determine the maximum load that a $W12 \times 65$ column can carry if it is 35 feet long. Assume it to be: (a) a secondary member; (b) a main member. Compare the results with those found in Illustrative Problem 9.9. Use S_{yp} = 36,000 psi, (248.2 MPa) and $E = 29 \times 10^6$ psi (200 GPa) (A 36 steel).

Solution:

As in Illustrative Problem 9.9, we find $KL/r = 139$, with $K = 1$. For this steel $C_c = \sqrt{\dfrac{2\pi^2 E}{S_{yp}}} = \sqrt{(2\pi^2 \times 29 \times 10^6)/36{,}000} = 126.1$. Since the slenderness ratio exceeds C_c, we shall have to use equation (9.13) for a main member or equation (9.14) for a secondary member. For this steel we can also use Table 9.2 or Fig. 9.5. Therefore

(a) for a secondary member S_{all} at a KL/r value of 139 is 8,540 psi (58.9 MPa) (Table 9.2), and $F = 19.10 \times 8540 = 163{,}100$ lb. (725.5 N).

(b) for a main member S_{all} at a KL/r value of 139 is 7,730 psi (53.3 MPa) (Table 9.2) and $F = 19.10 \times 7730 = 147{,}600$ lb (656.5 kN).

When these values are compared to those obtained in Illustrative Problem 9.9, it is found that for both (a) and (b) the values are quite close, but that for this case the allowable load is slightly less by the newer procedure.

9.6 ECCENTRICALLY LOADED COLUMNS

In the preceding sections of this chapter, we have considered only the centric loading of columns; that is, the resultant force acting through the centroid of the column section. From a practical standpoint it is nearly impossible to avoid small eccentricities in the application of central loads on a column. These effects have been considered, and the empirical equations discussed in Sections 9.4 and 9.5 incorporate in their safe load values factors of safety against such unavoidable load eccentricities. There are cases, however, where a column is loaded eccentrically due to its usage. Figure 9.8 shows such a situation where the column is eccentrically

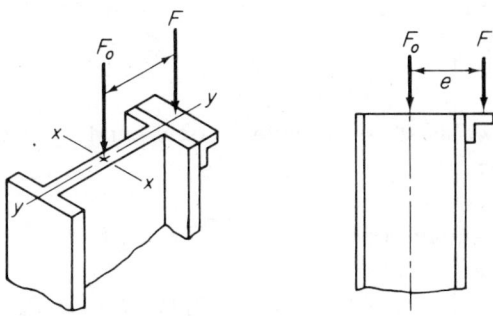

Fig 9.8 Eccentrically loaded columns

loaded about the x-x axis by a nonaxial load on a bracket attached to the side of the column. It is possible to extend the theoretical considerations that lead to the Euler Equation to take into account the eccentricity of the load. The resulting equation is known as the *secant formula,* due to the appearance of the trigonometric secant function in the equation's final form. This equation is cumbersome to use, and direct solutions are not possible. Frequently, design curves are generated to facilitate the numerical calculations.

An alternate design method can be obtained by considering the action of the central load F_o and the eccentric load F separately. The load on the column is taken to be a direct load due to the sum of $F + F_o$ and a moment Fe. The stresses due to the axial loads and the moment are assumed to be additive and their arithmetic sum does not exceed the allowable stress. Thus

$$\frac{F + F_o}{A} + \frac{(Fe)C}{I} = S_{all} \tag{9.15}$$

However, the allowable stress for a column is a function of its slenderness ratio, while the allowable stress in bending is not a function of length. In order to account for this condition, both sides of equation (9.15) are divided by S_{all}, yielding

$$\frac{(F + F_o)/A}{S_{all}} + \frac{(Fe)C/I}{S_{all}} = 1 \tag{9.16}$$

If we now denote

$\dfrac{F + F_o}{A}$ = actual axial stress = $(S_{act})_A$

S_{all} = allowable stress permitted if only axial stress existed = $(S_{all})_A$

$\dfrac{(Fe)C}{I}$ = actual bending unit stress = $(S_{act})_B$

S_{all} = allowable stress permitted if only bending existed = $(S_{all})_B$
$= 0.6 S_{yp}$

We can rewrite equation (9.16) in the following form

$$\frac{(S_{act})_A}{(S_{all})_A} + \left[\frac{(S_{act})_B}{(S_{all})_B}\right]_x + \left[\frac{(S_{act})_B}{(S_{all})_B}\right]_y \leq 1 \tag{9.17}$$

Equation (9.17) is the form given by the AISC for $\dfrac{(S_{act})_A}{(S_{all})_A} \le 0.15$. For mild steel, the allowable permitted stress in bending can be taken as 20,000 psi, and the allowable permitted stress due to the axial load is determined by one of the suitable column formulas for the slenderness ratio of the column. When calculating the bending stress, the moment of inertia you must use is the one that is about the axis on which the eccentric load causes bending, but the slenderness ratio to use should be based upon the least radius of gyration regardless of the axis about which bending may occur. The following illustrative problem will illustrate the foregoing considerations and procedures.

Illustrative problem 9.12

A $W14$ column having an unbraced length of 12 feet carries a concentric load of 300,000 pounds and an eccentric load of 50,000 pounds applied at 18 inches from the x-x axis (see Fig. 9.8). Select a suitable column assuming the material has a yield point of 50,000 psi and an allowable bending stress of 30,000 psi.

Solution:

It is not possible to obtain a direct solution to this problem. Let us, as a first guess, say that the average allowable stress is of the order of 10,000 psi. Then $(300,000 + 50,000)/10,000 = 35$ sq. in. area, giving us $W14 \times 119$ as a first selection. For this column $L/r = (12 \times 12)/3.75 = 38.4$. From Fig. 9.7 for $S_{yp} = 50,000$, $(S_{all})_A = 26,000$ psi; $(S_{act})_A = 350,000/35.0 = 10,000$; for a $W14 \times 119$ column, $I_{xx} = 1370.0$; $C = 7.25$. Therefore: $(S_{act})_B = (50,000 \times 18 \times 7.25)/1370.0 = 4760$ psi. $[(S_{act})_B = FeC/I]$. Using these values, we find

$$\frac{10,000}{26,000} + \frac{4760}{30,000} = 0.543$$

While this column meets the necessary requirements, it is obviously greatly overdesigned for this application. Let us, as a second trial, try a $W14 \times 61$ (almost half the weight per foot). Proceeding as before,

$$(S_{act})_A = \frac{350,000}{17.9} = 19,550 \text{ psi}; \quad (S_{act})_B = \frac{50,000 \times 18 \times 13.91/2}{641}$$

$$= 9765 \text{ psi}$$

$$\frac{L}{r} = \frac{12 \times 12}{2.45} = 58.8; \quad (S_{all})_A \cong 23{,}000 \text{ psi}$$

Therefore

$$\frac{19{,}550}{23{,}000} + \frac{9765}{30{,}000} = 1.175$$

We have selected a column that is not adequate for the loading: a heavier column is required. The final selection should be closer to the $W14 \times 61$ column than to the $W14 \times 119$ column. Selecting a $W14 \times 74$,

$$(S_{act})_A = \frac{350{,}000}{21.8} = 16{,}055 \text{ psi}; \quad (S_{act})_B = \frac{50{,}000 \times 18 \times 14.19/2}{797}$$

$$= 8{,}010 \text{ psi}$$

$$\frac{L}{r} = \frac{12 \times 12}{2.48} = 58; \quad (S_{all})_A = 23{,}000$$

Thus

$$\frac{16{,}055}{23{,}000} + \frac{8010}{30{,}000} = 0.965$$

This result, 0.965 is satisfactory, however, the first term exceeds 0.15. For this case it is necessary to refer to the AISC code (pp. 5-22, 5-23) for further specific restrictions and requirements.

9.7 CLOSURE

Long slender columns can be designed using Euler's Equation as a basis; intermediate columns cannot be designed satisfactorily by any single rational approach. Once again, in the realm of actual design, we find that we are faced with the reality that design equations are mandated by the legal authority that has jurisdiction. However, the student should note that the members of code committees that are charged with the responsibility of formulating the design rules are invariably engineers with the resources of an entire industry at their disposal. Their responsibility is really quite simple—the public must be protected at all costs. For this reason, code rules are invariably conservative and only after adequate tests or operational experience will these code committees alter their rules. It is most unusual to hear of a structural failure in a structure designed to code. Unfortunately, this same conservative approach does tend to limit original designs not contemplated when a specific code is formulated. For this reason (among many others) codes are constantly being revised and updated.

At best, this chapter is designed to give the student a brief introduction to those approaches most widely used in the design of columns. It is not expected that the expertise gained from this chapter will make anyone a column designer; on the contrary, it is hoped that the work of this chapter will put the student on the alert to go to the proper authority (AISC Code, etc.) if and when he must execute a column design.

PROBLEMS

9.1 Determine the maximum dimension of a square column if it is 10 ft long and L/r, the slenderness ratio, is not to exceed 120.

9.2 If a column is a rectangle having one side twice the other, determine the dimensions of the column if it is 8 ft long and L/r is not to exceed 120. Check both the x-x and y-y axes.

9.3 If a solid circular cylinder is to be used as a column, what maximum length should it have if its diameter is 4 in. and L/r is to be limited to 120?

9.4 Solve Problem 9.2 if L/r is permitted to be 200 as for a secondary member.

9.5 Four 2-in. square bars are to be used to support a compressive load. Using a limiting L/r value for the assembly of 120, determine the maximum length for this combination. Is this for the x-x or y-y axis?

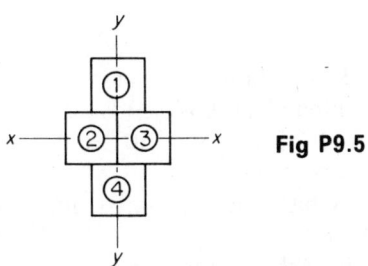

Fig P9.5

9.6 Can a 12-in. Schedule 80 pipe be used as a main member if it is 25 ft long?

9.7 A solid circular cylinder is to be used as a structural member, with L/r not to exceed 120. If it is 7 ft long, determine the required diameter of the cylinder.

9.8 Determine the required diameter of a solid circular cylinder that is used as a main member. (Use Euler's Equation.) Both ends of the cylinder are fixed. It is made of a steel alloy having a yield stress of 45,000 psi and $E = 30 \times 10^6$ psi. A factor of safety of $2\frac{1}{2}$ is taken to be appropriate for the application. The column is 10 ft long.

9.9 Solve Problem 9.8 if the yield stress of the steel is 100,000 psi.

9.10 Determine the required diameter of the column in Problem 9.8 if both ends are hinged.

9.11 If a 4-in. Schedule 40 steel pipe is used as a column and its L/r is to be 150, determine the safe load it will carry. Use Euler's Equation and a factor of safety of $2\frac{1}{2}$ and assume the ends to be hinged.

9.12 If the column in Problem 9.11 has both ends built-in, determine its safe load.

9.13 Solve Problem 9.11 if one end is hinged and the other end is built-in.

9.14 A cantilevered beam is subject to a direct axial load. What maximum value of load can this member support? Use a factor of safety of 3; $E = 10 \times 10^6$. Assume Euler's Equation applies.

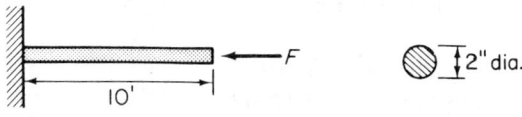

Fig P9.14

9.15 A $W8 \times 31$ beam is to be used as a column. What is the allowable load that it can carry if it is hinged at both ends of its 25 foot length? Use a factor of safety of $2\frac{1}{2}$ and $E = 30 \times 10^6$. Assume Euler's Equation is applicable.

9.16 If the column in Problem 9.15 has one end fixed and one end hinged, what allowable load can it carry?

9.17 Solve Problem 9.16 if both ends of the column are fixed.

9.18 A $W10 \times 33$ column, 50 ft long and hinged at both ends, is braced at its midpoint in its weakest direction and is unbraced in its strongest direction. Using a factor of safety of 2 in Euler's Equation, determine the maximum load that it will carry.

9.19 Using the Chicago formula, determine the maximum load that a 4-in. Schedule 40 pipe can sustain as a main member.

9.20 Use the Chicago formula to determine the maximum load that a solid 4-in.-diameter steel bar can sustain as a main member.

9.21 If two C5 × 9 channels are set up to carry an axial load, determine the maximum load this configuration can carry if it is 10 ft long with the load at the center of gravity of the section.

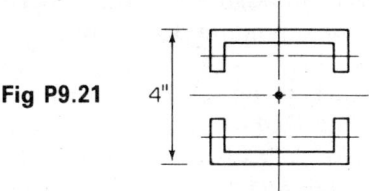

Fig P9.21

9.22 What is the maximum allowable load that a secondary member can carry if it is an S12 × 50 beam? Assume the column to be 15 ft long.

For the following problems, assume that the AISC method of Section 9.4 can be used. Table 9.1 and Fig. 9.5 can be used to solve or check the solutions to these problems.

9.23 A W12 × 58 column is to be 22 ft long. If it is a main member of the structure, what allowable load can it sustain?
9.24 A column is to support a load of 200,000 lb. If it is 24 ft long, determine the most suitable W14 section to use as a main member.
9.25 Is an S12 × 31.8 beam suitable to support a 10,000-lb load if it is 10 ft long and is a main member?
9.26 Is an S5 × 14.7 beam suitable to support a 10,000-lb load if it is 10 ft long and is a secondary member?
9.27 Select a W14 beam to support an axial load of 250,000 lb if the column is 30 ft long and is a main member.
9.28 A C6 × 13 channel is to be used as a column. Determine the maximum axial load it can carry if it is 8 ft long and the load is at the center of gravity of the section (r_{yy} = 0.525 in.).
9.29 A W8 × 31 structural member is 25 ft long and is to be used as a column. Determine the maximum load that it can support as a main member.
9.30 A W12 × 58 structural member is 20 ft long and is to be used as a column. Determine the maximum load that it will support.
9.31 Determine the maximum load that a W10 × 89 structural member can carry if it is 15 ft long and acts as a column.
9.32 A column is to support a centric load of 150,000 lb. If it is 21 ft long, select a suitable S section for this application.
9.33 Select a column for Problem 9.29 if the yield point of the material is 60,000 psi.
9.34 Is an S12 × 35 beam suitable to carry a 100,000-lb centric load as a column if it is 12 ft long?

Problems 9.23 through 9.34 are repeated below and are to be solved using Table 9.2 and the methods of Section 9.5. If desired, any problem assigned from problems 9.23 through 9.34 can be compared to its counterpart in the set from problems 9.35 through 9.46. This procedure corresponds to the use of steel having a yield point of 36,000 psi.

9.35 A W12 × 58 column is to be 22 ft long. If it is a main member of the structure, what allowable load can it sustain?

9.36 A column is to support a load of 200,000 lb. If it is 24 ft long, determine the most suitable W14 section to use as a main member.

9.37 Is an S12 × 31.8 beam suitable to support a 10,000-lb load if it is 10 ft long and is a main member?

9.38 Is an S5 × 14.7 beam suitable to support a 10,000-lb load if it is 10 ft long and is a secondary member?

9.39 Select a W14 beam to support an axial load of 250,000 lb if the column is 30 ft long and is a main member.

9.40 A C6 × 13 channel is to be used as a column. Determine the maximum axial load it can carry if it is 8 ft long and the load is at the center of gravity of the section (r_{yy} = 0.525 in.).

9.41 A W8 × 31 structural member is 25 ft long and is to be used as a column. Determine the maximum load that it can support as a main member.

9.42 A W12 × 58 structural member is 20 ft long and is to be used as a column. Determine the maximum load that it will support.

9.43 Determine the maximum load that a W10 × 89 structural member can carry if it is 15 ft long and acts as a column.

9.44 A column is to support a centric load of 150,000 lb. If it is 21 ft long, select a suitable S section for this application.

9.45 Select a column for Problem 9.29 if the yield point of the material is 60,000 psi.

9.46 Is an S12 × 35 beam suitable to carry a 100,000-lb centric load as a column if it is 12 ft long?

Use the latest AISC procedures and the tables in Appendix C to solve Problems 9.47 through 9.52. The methods of Section 9.5 and Figure 9.7 may also be used to solve these problems.

9.47 A W8 × 31 structural member is 25 ft long and is to be used as a column. Using a material having a yield point stress of 90,000 psi, determine the maximum load that it can support. It is a main member.

9.48 A W12 × 58 structural member is 20 ft long and is to be used as a column. If the yield point stress of the material is 60,000 psi, determine the maximum load that it will support.

Chapter 9, Columns   389

9.49 Determine the maximum load that a $W10 \times 89$ structural member made of 65,000-psi yield point material can carry if it is 15 ft long.

9.50 A column is to support a centric load of 150,000 lb. If it is 21 ft long, select a suitable S section for this application. Assume the material to have a yield point stress of 100,000.

9.51 Select a column for Problem 9.47 if the yield point of the material is 60,000 psi.

9.52 Is an $S12 \times 35$ beam suitable to carry a 100,000-lb centric load if the material has a yield point of 65,000 psi, and the column is 12 ft long?

Use the method of Section 9.5, Fig. 9.7 and/or the tables in Appendix B and C for the following problems. The material can be taken to have a yield stress of 50,000 psi and an allowable bending stress of 30,000 psi. Note that the allowable tensile stress equals $0.6 \times S_{yp}$.

9.53 A $W14 \times 103$ column is to carry a concentric load of 200,000 lb and an eccentric load of 100,000 lb. If the column is 10 ft long, what is the maximum eccentricity from the x-x axis that this column can be designed for?

9.54 Can a $W12 \times 22$ column support a 30,000-lb load eccentrically applied 8 in. from the x-x of the column? The column is to be 13 ft long.

9.55 In addition to a 50,000-lb eccentrically applied load at 6 in. from the x-x axis of an $S15 \times 50$ beam, there is a centric load. If the column is 10 ft long, what maximum centric load can it support?

9.56 A $W12 \times 79$ structural section is to be used to carry a centric load of 500,000 lb and an eccentric load of 100,000 lb applied 4 in. from the x-x axis. Determine whether this section is suitable for use as a column 9 ft long if the material yield point is 50,000 psi, and the allowable bending stress is taken as $0.6S_{yp}$.

9.57 If the column in Problem 9.56 is 16 ft long, and the material has a yield point of 100,000 psi and an allowable bending stress of 60,000 psi, is this column suitable for the application?

REFERENCES

American Institute of Steel Construction. *Manual of Steel Construction*. 7th ed., 1970.

Arges, K. P. and A. E. Palmer. *Mechanics of Materials*. New York: McGraw-Hill Book Co., 1963.

Bassin, M. E. and S. M. Brodsky. *Statics and Strength of Materials*. 2nd ed. New York: McGraw-Hill Book Co., 1969.

Doughty, V. L. and A. Vallance. *Design of Machine Members.* 4th ed. New York: McGraw-Hill Book Co., 1964.

Higdon, A., E. E. Ohlsen, and W. B. Stiles. *Mechanics of Materials.* 3rd ed. New York: John Wiley & Sons, 1976.

Jensen, A. and H. H. Chenoweth. *Applied Strength of Materials.* 3rd ed. New York: McGraw-Hill Book Co., 1971.

Levinson, Irving J. *Mechanics of Materials.* 2nd ed. Englewood Cliffs, NJ: Prentice-Hall, Inc., 1970.

Olsen, G. A. *Elements of Mechanics of Materials.* 3rd ed. Englewood Cliffs, NJ: Prentice-Hall, Inc., 1974.

Singer, F. L. *Strength of Materials.* 2nd ed. New York: Harper and Row, 1962.

Timoshenko, S. and Donovan H. Young. *Elements of Strength of Materials.* 5th ed. New York: D. Van Nostrand Co., 1968.

10
Combined stresses

10.1 INTRODUCTION

In Chapters 3, 5, 7, and 9, we briefly looked at several loading conditions that give rise to a combination of shear, tension, and/or compression. It is rare indeed to find actual structural or machine parts that are loaded by a single, well-defined load that gives rise to a simple, single state of stress. More frequently, the member in question is loaded in such a manner as to induce a combination of stresses such as shear, tension, etc. Under certain types of loading, it is permissible simply to add algebraically the resulting stresses to obtain the maximum stress at a point, but this is generally not possible. Our task in this chapter will be to investigate combinations of stresses that exist at a given section and the resultant effects induced.

The formulas that we have derived elsewhere in this book, for the most part, enable us to determine the stresses on certain planes passing through a point. In this chapter we shall further evaluate the maximum stress conditions as well as determine the planes on which they occur. While it is possible, in principle, to mathematically determine the state of stress at a given section in a member, it is much more difficult to incorporate these results into meaningful theories or criteria of failure for such members under combined loads. Even simple cases of loading cause internal stress distributions sufficiently complex so as to prevent us from using the data from simple tension tests to predict the onset of failure (inelastic action).

10.2 BENDING AND AXIAL LOADS COMBINED

In Section 9.6 we considered the eccentrically loaded column and developed a rational approach to the design of such structural members by considering the columns to be loaded by a central load F_o and an eccentric load F as shown in Fig. 9.8 (Repeated). Basically, we considered

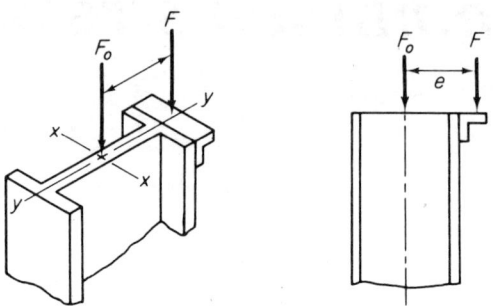

Fig 9.8 (Repeated) Eccentrically loaded columns

each load separately and added the effects with proper consideration of the sense of each effect, i.e., compression or tension. Due to axial loading, the member is in compression and the resulting compression stress is

$$S_{axial} = \frac{F_o + F}{A} \qquad (10.1)$$

The eccentric load F gives rise to a moment about the x-x axis causing flexural stresses to exist, which have maximum values given by the Flexure Formula as

$$S_{flexure} = \frac{MC}{I} = \frac{(Fe)C}{I} \qquad (10.2)$$

At plane B, the element is in compression and at plane A, it is in tension. We now assume that it is possible to algebraically sum the axial and flexural stresses to obtain

$$S = \pm \frac{F_o + F}{A} \pm \frac{MC}{I} = \pm \frac{F_o + F}{A} \pm \frac{(Fe)C}{I} \qquad (10.3)$$

where the $\pm$ notation indicates that compression is to be considered (algebraically) as a negative effect, and tension is to be considered as a positive effect. Equation (10.3) is satisfactory for those cases where the member can be considered to be rigid, such as for posts or short compression members. Where the member is long and slender and its deflection under load is significant, the method of superposition of stresses used above is not applicable.

Illustrative problem 10.1

A clamp is made as shown in Fig. 10.1. If the jaws are tightened so that the clamp is subjected to a load of 2200 N, determine the stresses induced on plane z-z.

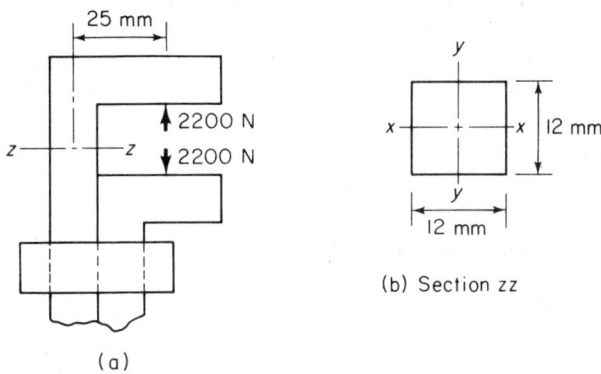

Fig 10.1   Illustrative problem 10.1

Solution:

The clamp is made of 0.012-m × 0.012-m stock; its area A is 1.44×10^{-4} m^2 and its moment of inertia about x-x or y-y is $bh^3/12 = 0.012(0.012)^3/12 = 1.728 \times 10^{-9}$ m^4. Evaluating the direct load and the flexural load, we find

$$S = \frac{F}{A} = \frac{2200}{1.44 \times 10^{-4}} = 15.28 \text{ MPa (tension)}$$

$$S = \frac{MC}{I} = \frac{2200 \times 0.025 \times 0.012/2}{1.728 \times 10^{-9}} = 190.97 \text{ MPa}$$

At A, bending causes compression, therefore

$S = +15.28$ MPa $- 190.97$ MPa $= 175.69$ MPa (compression)

At *B*, bending causes tension, therefore

$S = 15.28$ MPa $+ 190.97$ MPa $= 206.25$ MPa (tension)

Illustrative problem 10.2

A hanger is designed as shown using an S5 × 10 beam. Determine the stresses at sections *A* and *B* if the hung load weighs 20 000 N.

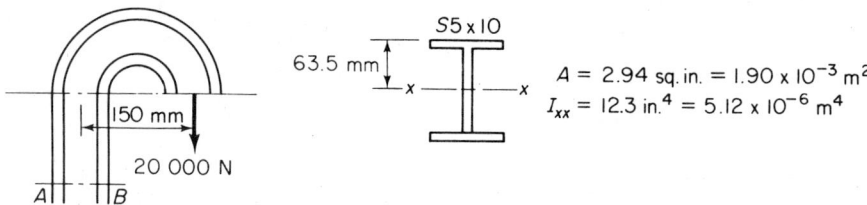

Fig 10.2 Illustrative problem 10.2

Solution:

The direct stress on the section due to the two-ton load is

$$S = \frac{F}{A} = \frac{20\,000}{1.90 \times 10^{-3}} = 10.53 \text{ MPa (compression)}$$

The bending stress is $S = MC/I = 0.150(20\,000)(0.0635)/5.12 \times 10^{-6}$ = 37.2 MPa. At *A*, the section is in tension due to bending, therefore

$S = -10.53$ MPa $+ 37.21$ MPa $= 26.68$ MPa (tension)

At *B*, the section is in compression, therefore

$S = -10.53$ MPa $- 37.21$ MPa $= -47.74$ MPa (compression)

10.3 PURE SHEAR

Let us now re-examine the conclusions we reached in Chapters 5 and 7 concerning the existence of a shear stress on a plane through a given point in a body. We showed earlier that if a shear stress occurs on a plane, a shear stress of equal intensity exists on a plane at right angles to the first plane and passing through the same point as the first plane. When shearing stresses occur only on two perpendicular planes at a point

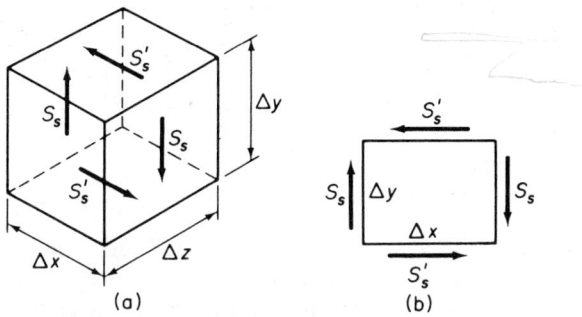

Fig 5.16 (Repeated) Stresses on an element

in the body, that point is said to be "subject to pure shear." Referring to Fig. 5.16 (Repeated), we can write

$$S_s = S'_s \tag{10.4}$$

If any plane is now passed through the point, there will be both normal and shear stresses acting on this oblique plane. In order to determine the stress conditions on any oblique plane, we shall use the following technique: An imaginary plane will be passed through the body at the point in question. We shall then isolate the resulting small free body, and on it all *stresses* will be shown. Since a free-body diagram shows all of the forces acting upon the body, it becomes necessary to multiply the stresses by the areas over which they act and then to apply the conditions for static equilibrium. Regardless of the distribution of the stresses at a point in a body, there exist three mutually perpendicular planes through the point on which the stresses are normal to the planes. These planes, therefore, have no shearing stresses on them, and the normal stresses on these planes are called *principal stresses*. The importance of this concept of principal stress lies in the fact that the maximum normal stress at a point is always a principal stress. We shall return to this point later in this chapter.

Figure 10.3 shows a free body diagram of an element cut by an oblique plane. We shall endeavor to determine the angle φ for the plane of principal stress as well as the value of the principal stress S_N. Since S_N is a principal stress, there will be zero shear on this plane. The free body diagram [Fig.

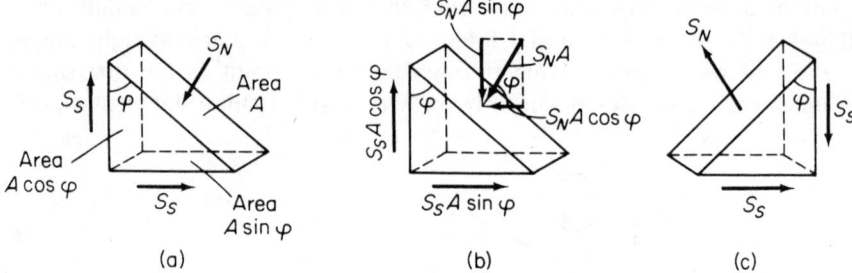

Fig 10.3 Pure shear

10.3(b)] therefore shows only S_N on this oblique plane, and the force due to S_N, $(S_N A)$, has been resolved into its component parts as indicated. Taking a force summation in the horizontal direction yields

$$S_s A \sin \varphi = S_N A \cos \varphi \qquad (10.5)$$

and a force summation in the vertical direction gives us

$$S_s A \cos \varphi = S_N A \sin \varphi \qquad (10.6)$$

From equation (10.5)

$$S_N = S_s \frac{\sin \varphi}{\cos \varphi} \qquad (10.7)$$

and from equation (10.6)

$$S_N = S_s \frac{\cos \varphi}{\sin \varphi} \qquad (10.8)$$

From the simultaneous conditions shown by equations (10.7) and (10.8), we conclude that $\sin \varphi = \cos \varphi$, requiring φ to be 45° and that $S_N = S_s$. Note that S_N is a compressive stress. Had we cut the element as shown in Fig. 10.3(c), we would also have concluded φ to be 45° and $S_N = S_s$, where S_N is a tensile stress.

We can summarize the foregoing by stating that if a body is subjected to a pure shear at a point, there exist planes having only normal stresses (principal stresses) numerically equal to the applied shearing stress. These planes make angles of 45° to both the horizontal and vertical and are at right angles to each other. One important conclusion that we reach is that a material that is weak in tension will fail in tension on a plane that is inclined at 45° to the plane of shear when the body is subject to pure shear. Figure 10.4 shows the conditions and types of failure one obtains when a brittle material is subject to pure shear.

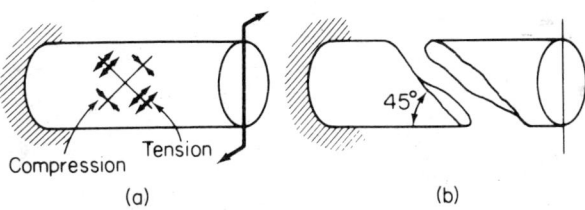

Fig 10.4 Brittle failure in pure shear

The inverse of the situation discussed above occurs when an element of area is subjected to a vertical compressive stress and an equal horizontal tensile stress. This situation is illustrated in Fig. 10.5(a). Due to these applied

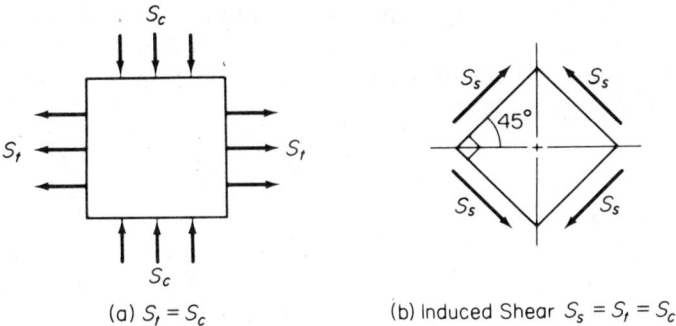

(a) $S_t = S_c$ (b) Induced Shear $S_s = S_t = S_c$

Fig 10.5 Equivalent shearing stresses

stresses, there must exist planes on which pure shear stresses act. These planes are rotated at 45° to the planes of tension and compression as shown in Fig. 10.5(b), and the stresses on these planes are known as *equivalent shearing stresses*.

Illustrative problem 10.3

A bar is subjected to axial tension as shown in Fig. 10.6. Determine the stresses on a plane *a-a* making an angle φ with the horizontal.

Strength of Materials for Engineering Technology

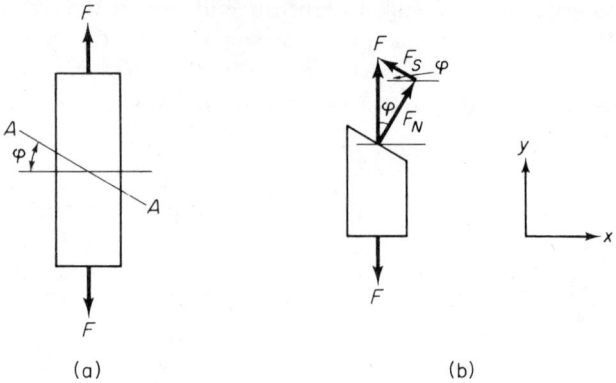

Fig 10.6 Illustrative problem 10.3

Solution:

In order to solve this problem, we shall consider that a hypothetical cutting plane is passed through the bar coinciding with plane a-a. In order for the bar to be in equilibrium, let us postulate that a force F_N exists normal to the plane in question, and a force F_S exists parallel to the plane. In order for the cut bar to be in equilibrium, it is necessary that $\Sigma F_y = 0$ and $\Sigma F_x = 0$. Thus

$$F_S \cos \varphi = F_N \sin \varphi$$

and

$$F = F_S \sin \varphi + F_N \cos \varphi$$

Solving for F_N from the first equation and substituting it into the second equation yields

$$F = F_S \sin \varphi + \left(F_S \frac{\cos \varphi}{\sin \varphi}\right) \cos \varphi = \frac{F_S(\sin^2 \varphi + \cos^2 \varphi)}{\sin \varphi} = \frac{F_S}{\sin \varphi}$$

yielding

$$F_S = F \sin \varphi$$

Solving for F_N

$$F_N = F \cos \varphi$$

The stress that each force causes equals the force divided by the area over which the force acts. If the area normal to F is denoted as A, the area of plane a-a is $A/\cos \varphi$. The shear stress, S_s, is given by

$$S_s = \frac{F_s}{A_A} = \frac{F \sin \varphi}{A/\cos \varphi} = \frac{F}{A} \sin \varphi \cos \varphi$$

Using the trigonometric relation $\sin 2\varphi = 2 \sin \varphi \cos \varphi$

$$S_s = \frac{F \sin 2\varphi}{2A}$$

Similarly, $S_N = F_N/A_N = F \cos \varphi/(A/\cos \varphi) = (F/A) \cos^2 \varphi$. Using the trigonometric identity

$$\cos^2 \varphi = \frac{1 + \cos 2\varphi}{2}, \quad S_N = \frac{F}{A}\left(\frac{1 + \cos 2\varphi}{2}\right)$$

We can check our results in this problem by noting S_N is maximum when $\varphi = 0$ and equals F/A (the principal stress). Also, the maximum value of S_s occurs when $\varphi = 45°$ ($2\varphi = 90°$) and equals $F/2A$ or $S_t/2$. These results agree with our previous general conclusions. Note that for this problem φ is the angle the plane makes with the horizontal and is the angle that the normal to the plane makes with the vertical.

10.4 COMBINED NORMAL STRESSES

We have seen that pure shear on an element causes tensile and compressive stresses within the element. The next condition that we shall consider is the case of a cube of material subjected to normal stresses on *only* two perpendicular sets of its faces as shown in Fig. 10.7. The

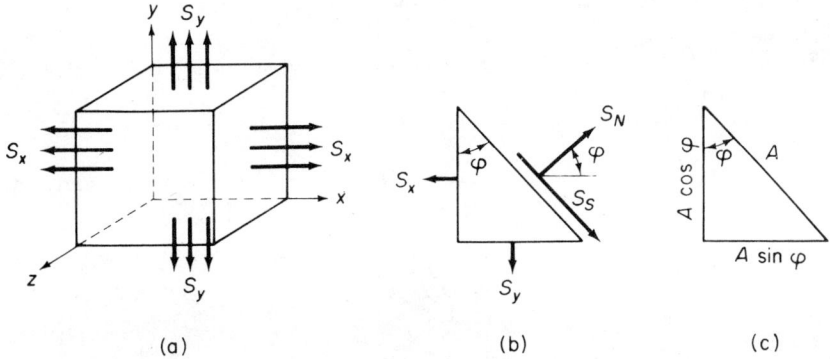

Fig 10.7 Biaxial normal plane stresses

third set of faces is assumed to have no stresses. This condition of stress is called *biaxial plane stress* since the stress vectors can be portrayed on one plane as shown in Fig. 10.7(b). In Fig. 10.7(b) the stresses are shown for the case where the material has been cut by an imaginary plane making an angle φ with the vertical. The stress vectors are shown on this figure with S_S indicating the shear stress on the cutting plane, and S_N indicating the normal stress on this plane. Note that the angle φ is also the angle between the normal to the plane and the x axis. Since S_x and S_y are normal stresses that act on planes of zero shear, they are also the principal stresses. The first step in determining S_S and S_N is to note that Fig. 10.7(b) is not a free-body diagram; it is a *stress diagram*. In order to express the forces acting on each face of the element it is necessary to multiply each stress by the area over which it acts. Choosing A to be the area of the inclined plane gives us $A_x = A \sin \varphi$, and $A_y = A \cos \varphi$ as shown in Fig. 10.7(c). Figure 10.8 shows the

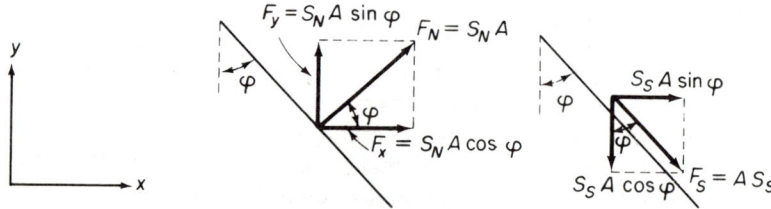

Fig 10.8 Resolution of forces on cutting plane

forces on the inclined plane due to S_S and S_N and the resolution of these forces into components in the x and y directions. If we now apply the equilibrium requirement that $\Sigma F_x = 0$ and $\Sigma F_y = 0$, we obtain

for the x direction: $\quad S_N A \cos \varphi + S_S A \sin \varphi = S_x A \cos \varphi \quad$ (10.9)

for the y direction: $\quad S_N A \sin \varphi - S_S A \cos \varphi = S_y A \sin \varphi \quad$ (10.10)

Solving for S_S from equation (10.10) and substituting the results into equation (10.9)

$$S_S = \frac{\sin \varphi}{\cos \varphi} (S_N - S_y) \quad (10.11a)$$

and

$$S_N = S_x \cos^2 \varphi + S_y \sin^2 \varphi \quad (10.11b)$$

Using the trigonometric identities: $\cos^2 \varphi = (\cos 2\varphi + 1)/2$ and $\sin^2 \varphi = (1 - \cos 2\varphi)/2$, and equation (10.11b)

$$S_N = \frac{S_x + S_y}{2} + \frac{(S_x - S_y)}{2}(\cos 2\varphi) \qquad (10.12)$$

If we now solve equation (10.9) for S_N and substitute the result into equation (10.10), we obtain an expression for S_S

$$S_S = S_x \sin \varphi \cos \varphi - S_y \sin \varphi \cos \varphi \qquad (10.13)$$

But $\sin 2\varphi = 2 \sin \varphi \cos \varphi$. Therefore

$$S_S = \frac{S_x - S_y}{2}(\sin 2\varphi) \qquad (10.14)$$

Equations (10.12) and (10.14) give us the components of the stress vector on any plane in a body subject to biaxial normal stresses. From equation (10.12), we can see that the maximum normal stress S_N occurs on the planes of φ at 0° and 90°, and the maximum value that stress can have is the greatest value of S_x or S_y. From equation (10.14), we can conclude that the maximum value of S_S is half the algebraic difference of S_x and S_y, the principal stresses. The student should note that a compressive stress is to be considered as a negative tensile stress. Also, the maximum value of S_S occurs on each of two planes that bisect the angles between the planes on which the maximum and minimum principal stresses occur. In terms of our notation this occurs when $\varphi = 45°$ and when $\varphi = 135°$. We can further conclude that if $S_x = S_y$, there will be no shear on any planes within a body and the normal stress on all such planes will be either S_x or S_y. If the member is subjected to equal but opposite normal stresses ($S_x = -S_y$), as is the case where one stress is tension and the other is compression, the resultant shear stress will be greater than the case where the stresses have like senses. In other words, compression-tension or tension-compression produces greater shear stresses than tension-tension or compression-compression.

Illustrative problem 10.4

An element is shown in Fig. 10.9 subjected to equal perpendicular loads. Determine the equivalent shearing stress within the element. Also determine the stresses on an interior element rotated 30° as shown in Fig. 10.9.

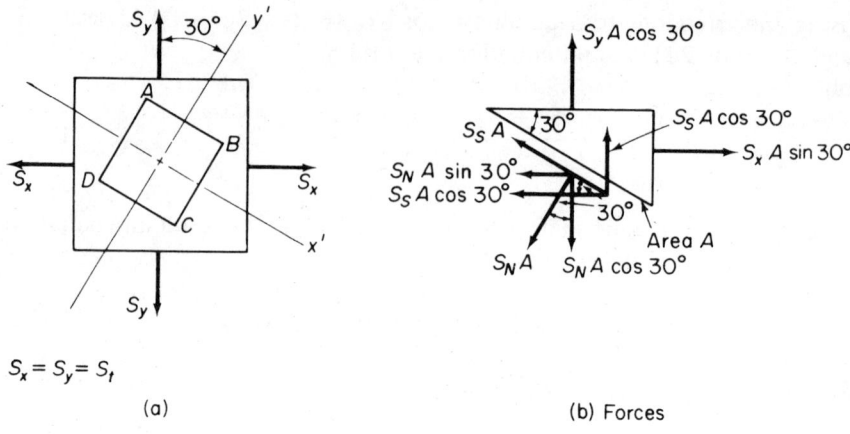

$S_x = S_y = S_t$

(a) (b) Forces

Fig 10.9 Illustrative problem 10.4

Solution:

Since the perpendicular tensile stresses are shown to occur on planes having no shear stresses, they must be principal stresses. Thus, the equivalent shearing stresses must equal S_t and must occur on perpendicular planes making angles of 45° with the horizontal and vertical, respectively. In order to determine the state of stress on the rotated element, we shall consider the element to be cut by a plane at 30° and to coincide with face AB. While it is possible to solve this problem by simply substituting the proper values into equations (10.12) and (10.14), we will analyze it from basic considerations and then show the agreement with the equations and conclusions already reached. Thus, if we consider the element shown in Fig. 10.9(b) and write the force equilibrium conditions in the x and y directions:

In the x direction:

$$AS_S \cos 30° + AS_N \sin 30° = S_x(A \sin 30°)$$

In the y direction:

$$AS_S \sin 30° - AS_N \cos 30° = S_y(A \cos 30°)$$

and

$$S_x = S_y = S_t$$

Substituting S_t for S_x and S_y and solving the two force equations for S_N in terms of S_t yield $S_N = S_t = S_x = S_y$. Placing this result into the equation for the x directed forces gives us $S_s \cos 30° + S_t \sin 30° = S_t \sin 30°$ and $S_s = 0$. Had we gone directly to equation (10.12), we would have obtained $S_N = S_t$, and equation (10.14) would have given us $S_s = 0$. We also reached these conclusions when we discussed equations (10.12) and (10.14) earlier. The student should carefully note which angle corresponds to φ for this problem.

At this time we shall introduce a graphical construction, known as *Mohr's Circle*, which greatly simplifies the calculation of combined stresses once the construction has been fully understood. Basically, the construction is a graphical portrayal of equations (10.12) and (10.14) for the case of biaxial normal stresses. Later, we shall extend it to the solution of more generalized cases of plane stresses. Let us choose our coordinate axes: S_S as the ordinate and the principal (normal) stress as the abscissa. We now construct a circle with its center O on the abscissa, a distance $(S_x + S_y)/2$ from the origin and having a radius equal to $(S_x - S_y)/2$ as shown in Fig. 10.10. If a radius is now drawn from the center of the

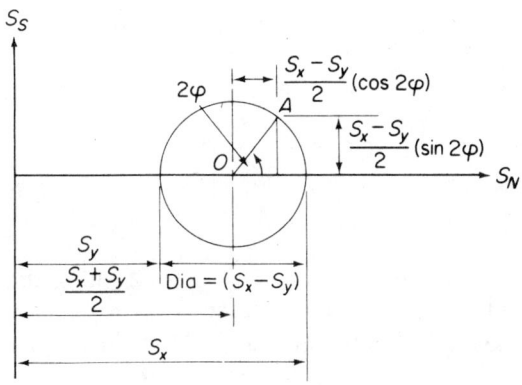

Fig 10.10 Mohr's circle

circle at an angle 2φ to the S_N axis, we intersect the circle at point A. As we see in Fig. 10.10 the x and y coordinates of point A are

in the S_N (horizontal) direction:

$$S_N = \frac{S_x + S_y}{2} + \frac{S_x - S_y}{2} (\cos 2\varphi) \qquad (10.12a)$$

in the S_S (vertical) direction:

$$S_S = \frac{S_x - S_y}{2} (\sin 2\varphi) \qquad (10.14a)$$

It is evident that equations (10.12a) and (10.14a) are identical to equations (10.12) and (10.14).

Illustrative problem 10.5

Solve Illustrative Problem 10.3 using Mohr's Circle.

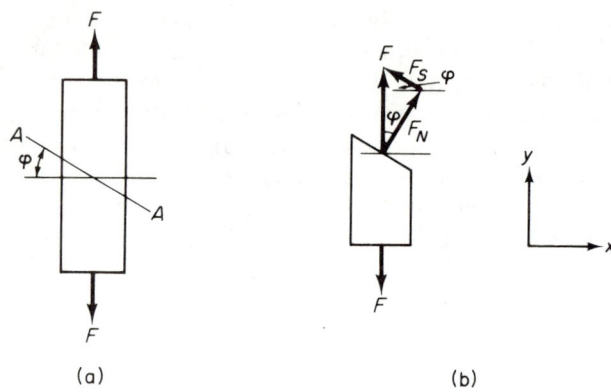

Fig 10.6 (Repeated) Illustrative problem 10.5

Solution:

Since $S_x = 0$, the center of the circle O is located at $S_y/2$ along the S_N axis, and the radius of the circle has a magnitude equal to $S_y/2$. The Mohr Cricle for these conditions is shown in Fig. 10.11. From this figure

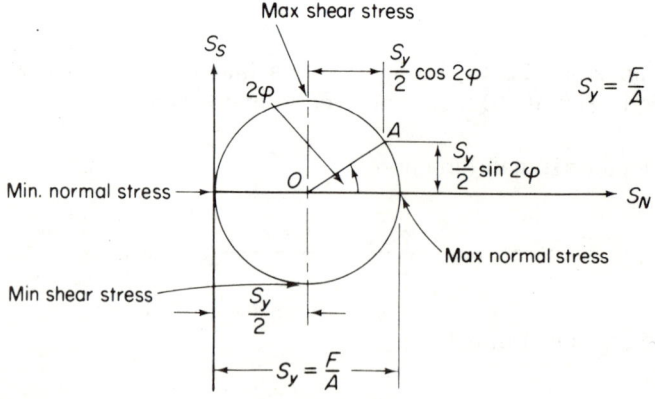

Fig 10.11 Illustrative problem 10.5

we can write the following relations for S_N and S_S directly for any element located at an angle from the vertical

$$S_N = \frac{S_y}{2} + \frac{S_y}{2}(\cos 2\varphi) = \frac{S_y}{2}(1 + \cos 2\varphi)$$

$$S_S = \frac{S_y}{2}\sin 2\varphi$$

These results are the same that we obtained in Illustrative Problem 10.3 when F/A was substituted for S_y.

Illustrative problem 10.6

Use Mohr's Circle to solve Illustrative Problem 10.3 if the applied load is compressive.

Solution:

The solution to this problem differs from that of Illustrative Problem 10.5 in that here the center of the circle is located $-S_y/2$ from the origin since the applied load is compressive. The angle 2φ is measured counter-clockwise from the x axis in the third quadrant. The Mohr Circle for this

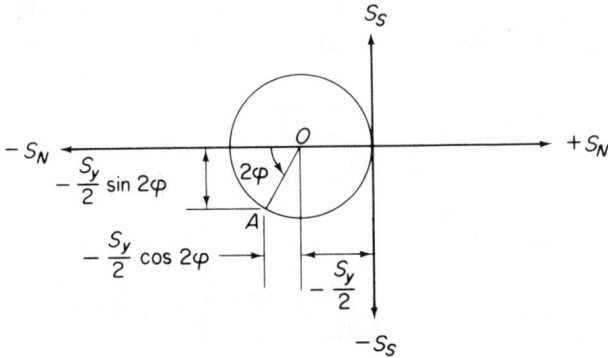

Fig 10.12 Illustrative problem 10.6

problem is shown in Fig. 10.12, and we have as the normal and shearing stresses

$$S_N = \frac{S_y}{2} - \frac{S_y}{2}(\cos 2\varphi) = -\frac{S_y}{2}(1 + \cos 2\varphi)$$

$$S_S = -\frac{S_y}{2}\sin 2\varphi$$

This result is the same as the previous cases with $-S_y$ substituted where we had S_y.

Illustrative problem 10.7

Construct Mohr's Circle for the case of biaxial normal stresses S_x and S_y where one of these stresses (say S_x) is negative due to the fact that it is a compressive load.

Solution:

The construction can proceed as for the general case illustrated in Fig. 10.10 with S_x negative. However, the construction can be somewhat simplified by noting that in general the diameter of the Mohr Circle $= (S_x - S_y)$, and the center O is located a distance $S_y \pm (D/2)$ from the origin where D = diameter of Mohr's Circle $= (S_x - S_y)$. Thus, we proceed by first laying out the S_S, S_N coordinate system. Along the S_N axis place S_y to the right along the S_N axis from the origin C since S_y is positive. Now lay out S_x to the left from C along the S_N axis since S_x is negative. Since $(S_x - S_y)$ is the diameter D, bisecting $(S_x - S_y)$ yields the radius of the Mohr Circle and also locates its center as shown in Fig. 10.13.

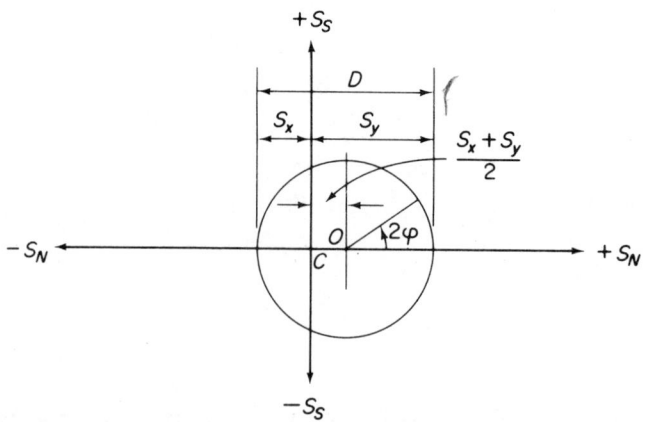

Fig 10.13   Illustrative problem 10.7

Notice also that this procedure correctly yields the results shown in Fig. 10.10.

Illustrative problem 10.8

A member is subjected to a tensile stress of 70 MPa in the y direction and a compressive stress of 35 MPa in the x direction. Determine the normal and shear stresses on a plane whose normal makes an angle of +15° to the x axis as shown in Fig. 10.14.

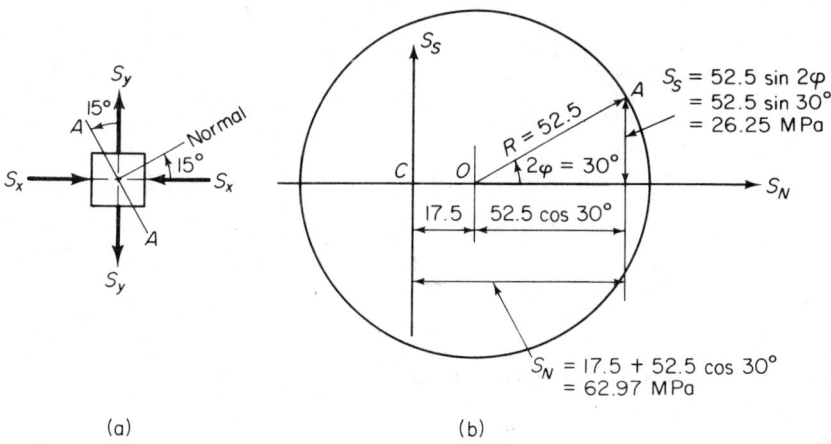

(a)   (b)

Fig 10.14 Illustrative problem 10.8

Solution:

As noted previously, we first lay out our S_s, S_N coordinate system. Then the S_y value of 70 MPa is scaled off to the right of the origin since it is positive by our convention, and S_x is scaled off as 35 MPa to the left of the origin since it is negative by convention. Bisecting the distance between the ends of these points yields a radius of 52.5 located at 17.5 along the x axis. The value of S_N for an angle of +15° is found by laying out 2φ or 30° from the origin O counter-clockwise to give us $S_N = 17.5 + 52.5 \cos 30° = 62.97$ MPa. Similarly for S_s, $S_s = 52.5 \sin 30° = 26.25$ MPa.

At this point the student should be aware that we have adopted the convention that the angle in Mohr's Circle has the same sense as the angle from the normal to the plane in question and the x axis, measured from the x axis.

10.5 GENERAL CASE OF PLANE STRESS

The cases studied in the previous paragraphs of this chapter all can be considered special cases of the general case of *plane stress*. If we consider an element subjected to both shear and normal stresses on its faces, as shown in Fig. 10.15, we have the general conditions of plane

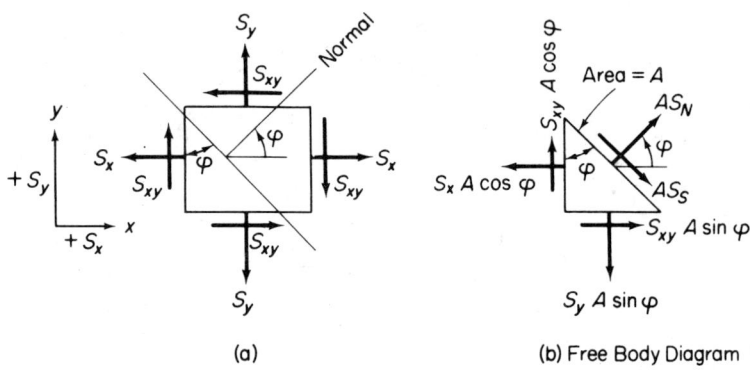

(a) (b) Free Body Diagram

Fig 10.15 General case of plane stress

stress. For this case we shall adopt the conventions that S_S is the shear stress on any inclined plane, and that S_N is the normal stress on this plane. S_{xy} will be used to denote the shear stress acting on the x plane in a direction parallel to the y plane. Thus, S_{yx} denotes the shearing stress on the y plane parallel to the x plane. Since $S_{xy} = S_{yx}$, we will use S_{xy} for both stresses. Figure 10.15(b) is a free-body diagram of a wedge cut out of the element shown in Fig. 10.15(a). The area of the inclined plane has been denoted as A, making $A_x = A \sin \varphi$ and $A_y = A \cos \varphi$. Note that on the free-body diagram forces, not stresses, are shown. Resolving forces into components in the x and y direction and applying the condition for equilibrium yield

in the x direction:

$$S_x A \cos \varphi - S_{xy} A \sin \varphi - AS_N \cos \varphi - AS_S \sin \varphi = 0 \quad (10.15)$$

in the y direction:

$$S_{xy}A \cos \varphi - S_y A \sin \varphi + AS_N \sin \varphi - AS_S \cos \varphi = 0 \quad (10.16)$$

The relations involving the double angle are $\cos^2 \varphi = (1 + \cos 2\varphi)/2$, $\sin^2 \varphi = (1 - \cos 2\varphi)/2$. Therefore we can obtain the relations for the normal and shear stresses on the cutting plane of Fig. 10.15 as

$$S_N = \frac{S_x + S_y}{2} + \frac{S_x - S_y}{2} (\cos 2\varphi) - S_{xy} \sin 2\varphi \quad (10.17)$$

and

$$S_S = \frac{S_x - S_y}{2} (\sin 2\varphi) + S_{xy} \cos 2\varphi \quad (10.18)$$

The planes of maximum and minimum normal stresses can be obtained by the methods of the calculus giving

$$\tan 2\varphi_N = -\frac{2S_{xy}}{S_x - S_y} \quad (10.19)$$

By similar techniques we can obtain the planes of maximum shear as

$$\tan 2\varphi_S = \frac{S_x - S_y}{2S_{xy}} \quad (10.20)$$

From these equations we can conclude that

1. Planes of maximum and minimum normal stresses occur 90° apart.
2. Planes of maximum shear occur 90° apart.
3. Maximum and minimum normal stresses occur on planes of zero shear and are therefore principal stresses.
4. Planes of maximum shearing stress form angles of 45° with the planes of principal stress.

By substitution of the angles defined by equations (10.19) and (10.20) into equations (10.17) and (10.18) respectively, we can obtain the maximum stresses as

$$S_N \text{ (max or min)} = \frac{S_x + S_y}{2} \pm \sqrt{\left(\frac{S_x - S_y}{2}\right)^2 + (S_{xy})^2} \quad (10.21)$$

$$S_S \text{ (max)} = \pm \sqrt{\left(\frac{S_x - S_y}{2}\right)^2 + (S_{xy})^2} \quad (10.22)$$

The foregoing analysis can be used for obtaining solutions to any general case of plane stress. However, it is usually much easier to use Mohr's Circle for this situation than to attempt to commit these formulas to memory. In order to apply Mohr's Circle, we shall adopt one further convention at this time: that shearing stress is positive when it causes a clockwise movement about the center of an element. We can now proceed to construct Mohr's Circle for the case of generalized plane stresses by the following steps:

1. Locate the points on the plot of S_N versus S_S of points having the coordinates S_x, S_{xy}, and S_y, S_{yx} as shown in Fig. 10.16.

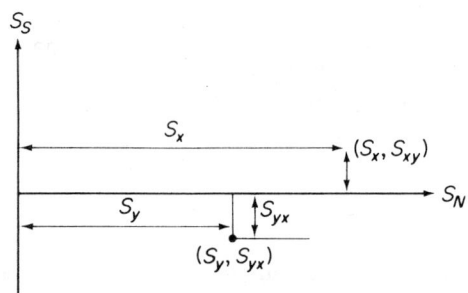

Fig 10.16 First step in the construction of generalized Mohr's circle

2. The diameter of the Mohr Circle is the line joining (S_x, S_{xy}) and (S_y, S_{yx}).
3. The principal maximum and minimum stresses occur on planes of zero shear. Thus, points A and B on the S_N axis of Fig. 10.17 correspond to the maximum and minimum principal stresses.

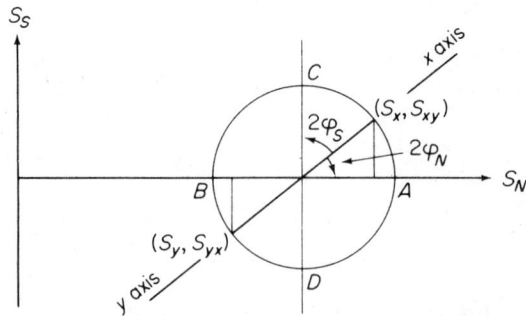

Fig 10.17 Second step in the construction of generalized Mohr's circle

4. $\text{Tan } 2\varphi_N = -\dfrac{S_{xy}}{(S_x - S_y)/2} = -\dfrac{2S_{xy}}{S_x - S_y}$
5. The maximum shear stress occurs at points C and D.
6. $\text{Tan } 2\varphi_s = \dfrac{S_x - S_y}{2S_{xy}}$

Illustrative problem 10.9

A plane element is subjected to the stresses shown in Fig. 10.18. Determine the maximum and minimum normal stresses, the maximum shear stress, and locate the planes of maximum shear stress and the principal planes.

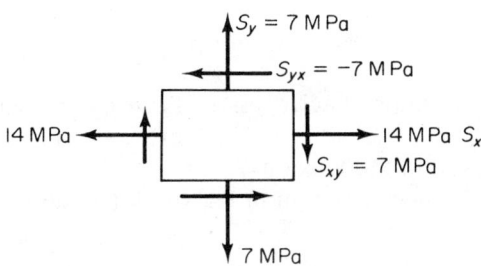

Fig 10.18 Illustrative problem 10.9

Solution:

We construct Mohr's Circle using the rules developed previously.

The radius of the circle = $\sqrt{(7)^2 + (3.5)^2}$ = 7.826 MPa
Point A = Max S_N = 10.5 + 7.826 = 18.326 MPa
Point B = Min S_N = 10.5 − 7.826 = 2.674 MPa
Point C = Max S_s = +7.826 MPa on a plane having S_N = 10.5 MPa
Point D = Min S_s = −7.826 MPa on a plane having S_N = 10.5 MPa
Tan $2\varphi_N$ = 7/3.5 = 2; $2\varphi_N$ = 63.5°; φ_N = 31.75° and (31.75 + 90)
  = 121.75°

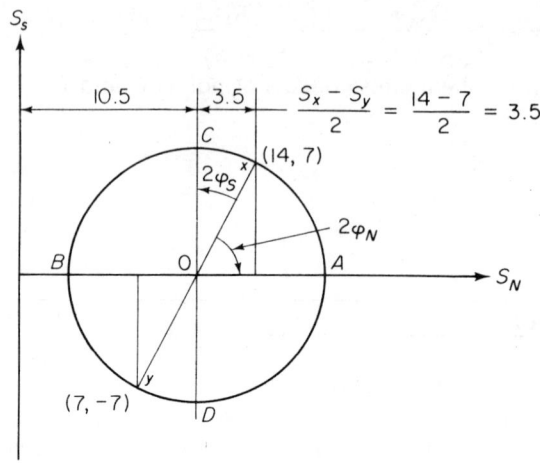

Fig 10.19 Mohr's circle for illustrative problem 10.9

The angle φ_N is measured clockwise from $O\text{-}x$, the positive x axis to the S_N axis.

The plane of maximum shear stress is 45° from the plane of maximum normal stress, i.e., $\varphi_N + \varphi_S = 45°$. Therefore $\varphi_S = 45 - 31.75 = 13.25°$ counter-clockwise from the x axis.

The complete solution is shown in Fig. 10.20(a) and (b).

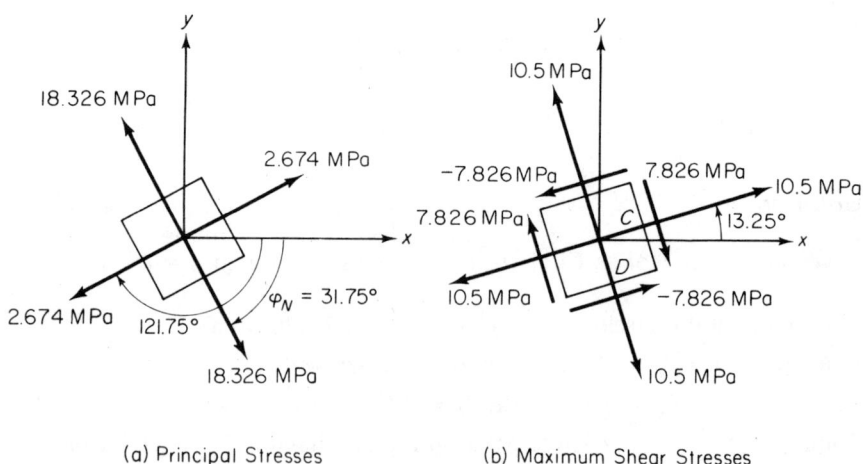

(a) Principal Stresses (b) Maximum Shear Stresses

Fig 10.20 Solution to illustrative problem 10.9

10.6 COMBINED STRESSES IN CIRCULAR SHAFTS

In our discussion of pure shear we noted that a body subjected to pure shear has principal stresses in tension and compression equal numerically to the applied shear stress acting on planes inclined at 45° to the shaft axis. If in addition to being subjected to the pure shear, the body is also acted upon by a bending moment, there will be a longitudinal stress induced in the body. Consider that the body is a circular shaft subject to a torque T in./lb and a bending moment of M in./lb. For this case the principal stresses will occur at the top and bottom of the shaft. Due to the applied torque T, $S_s = TC/J$; and due to the bending moment M, $S_b = MC/I$. The bending action causes either tension or compression at the top or bottom of the shaft, respectively. The principal stress due to this action,

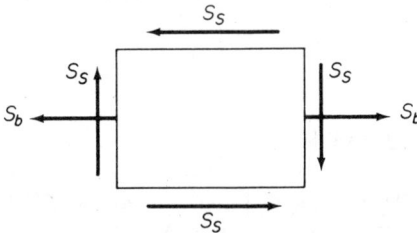

Fig 10.21 Shaft subject to bending and twisting

which is shown in Fig. 10.21, can be obtained from equation (10.21) as

$$S_N \text{ (max or min)} = \frac{S_b}{2} \pm \frac{1}{2}\sqrt{S_b^2 + 4S_s^2} \qquad (10.23)$$

Let us now recall that the polar moment of inertia J equals the sum of the rectangular moments of inertia $I_x + I_y$. For a symmetrical shape $J = 2I$ where I is the rectangular moment of inertia about either axis of symmetry. We can therefore write

$$S_b = \frac{MC}{I}, \quad S_s = \frac{TC}{J}, \quad J = 2I$$

and

$$\frac{S_b}{S_s} = \frac{MC/I}{TC/2I} = \frac{2M}{T} \quad \text{or} \quad S_s = S_b\left(\frac{T}{2M}\right) \qquad (10.24)$$

If the results of equation (10.24) are combined with equation (10.23)

$$S_N = \frac{S_b}{2} \pm \frac{1}{2}\sqrt{S_b^2 + 4S_b^2\left(\frac{T}{2M}\right)^2} = \frac{S_b}{2} \pm \frac{1}{2}\sqrt{\frac{(M^2+T^2)S_b^2}{M^2}}$$

and

$$S_N = \frac{S_b}{2}\left[1 \pm \sqrt{\frac{M^2+T^2}{M^2}}\right] \tag{10.25}$$

At this point we will define an *equivalent bending moment* M_e as that moment which, if it had acted alone on the shaft, would produce the same maximum principal stress in the shaft as that produced when the shaft is subjected to bending and twisting. Using this definition

$$S_N = \frac{M_e C}{I} \tag{10.26}$$

Substitution into (10.25) yields

$$\frac{M_e C}{I} = \frac{MC}{2I}\left[1 + \sqrt{\frac{M^2+T^2}{M^2}}\right] \tag{10.27}$$

From which M_e is obtained as

$$M_e = \frac{1}{2}(M + \sqrt{M^2+T^2}) \tag{10.28}$$

The maximum shearing stress on the element can be directly obtained by use of equation (10.22)

$$S_s(\text{max}) = \pm\frac{1}{2}\sqrt{S_b^2 + 4S_s^2} = \pm\frac{1}{2}S_b\sqrt{\frac{M^2+T^2}{M^2}} \tag{10.29}$$

Defining T_e as the equivalent torque that would produce the same maximum shearing stress as is given by equation (10.29) yields

$$\frac{T_e C}{J} = \frac{S_b T_e}{2M} = \frac{S_b}{2M}\sqrt{M^2+T^2} \tag{10.30}$$

where equation (10.24) gave us the relation between S_b and S_s. Simplifying equation (10.30)

Chapter 10, Combined Stresses   415

$$T_e = \sqrt{M^2 + T^2} \qquad (10.31)$$

Equation (10.31) can now be combined with equation (10.28) to yield

$$M_e = \frac{1}{2}(M + T_e) \qquad (10.32)$$

Equations (10.31) and (10.32) can be applied in the usual manner as if the equivalent moments and equivalent torques had acted alone.

Illustrative problem 10.10

A shaft having a 50 mm diameter is subjected to a torque of 900 N·m and a bending moment of 1200 N·m. Determine the maximum shear and bending stresses in the shaft.

Solution:

The equivalent torque and equivalent bending moment are determined from equations (10.31) and (10.32), respectively

$$T_e = \sqrt{M^2 + T^2} = \sqrt{(1200)^2 + (900)^2} = 1500 \text{ N·m}$$

$$M_e = \frac{1}{2}(M + T_e) = \frac{1}{2}(1200 + 1500) = 1350 \text{ N·m}$$

The maximum shear stress is given by

$$S_s = \frac{T_e C}{J} = \frac{1500 \times 0.050/2}{\pi(0.050)^4/32} = 61.12 \times 10^6 \text{ Pa} = 61.12 \text{ MPa}$$

The maximum principal stress is

$$S = \frac{M_e C}{I} = \frac{1350 \times 0.050/2}{\pi(0.050)^4/64} = 110.0 \times 10^6 \text{ Pa} = 110.0 \text{ MPa}$$

Illustrative problem 10.11

Design a circular shaft to carry a torque of 1000 N·m and a bending moment of 1500 N·m if the allowable bending stress shall not exceed 140 MPa and the allowable shear stress shall not exceed 84 MPa.

Strength of Materials for Engineering Technology

Solution:

In order to solve this problem, we shall have to size the shaft for bending and shear, and use the condition that gives us the greater shaft size. The equivalent torque is

$$T_e = \sqrt{M^2 + T^2} = \sqrt{(1500)^2 + (1000)^2} = 1802.8 \text{ N} \cdot \text{m}$$

$$M_e = \frac{1}{2}(M + T_e) = \frac{1}{2}(1500 + 1802.8) = 1651.4 \text{ N} \cdot \text{m}$$

For shear

$$S_s = 84 \times 10^6 = \frac{T_e C}{J} = \frac{1802.8 \times d/2}{\pi d^4/32} = \frac{918.6}{d^3}$$

$$d^3 = 9181.6/84 \times 10^6 = 1.093 \times 10^{-4}$$

$$d = 4.78 \times 10^{-2} \text{ m} = 47.8 \text{ mm, say 48 mm}$$

For bending

$$S = 140 \times 10^6 = \frac{M_e C}{I} = \frac{1651.4 \times d/2}{\pi d^4/64} = \frac{16\,821}{d^3}$$

$$d^3 = \frac{16\,821}{140 \times 10^6} = 1.2 \times 10^{-4}$$

$$d = 4.93 \times 10^{-2} \text{ m} = 49.3 \text{ mm}$$

The shaft size to use is approximately 50 mm.

There are many cases in which shafts are subjected to twisting and axial loads. Propeller shafts are just one example of this type of loading. The state of stress in the element of such a shaft will appear to be the same as that shown in Fig. 10.21, and we can use our foregoing analysis for the calculation of the principal stresses and the maximum shear stresses. Thus

$$S_N = \frac{F}{2A} \pm \frac{1}{2}\sqrt{\left(\frac{F}{A}\right)^2 + 4\left(\frac{TC}{J}\right)^2} \qquad (10.33)$$

$$S_{S(\max)} = \pm \frac{1}{2}\sqrt{\left(\frac{F}{A}\right)^2 + 4\left(\frac{TC}{J}\right)^2} \qquad (10.34)$$

where F/A is the direct axial stress. In the case of axial compression, it may be necessary to consider the shaft as a long column. However, we shall not consider it as such in our present discussion. Mohr's Circle is instructive in visualizing the state of stress for the situation we have when a shaft is subject to an axial load as well as to twisting. The same diagram can also be used for bending and twisting. Figure 10.22 shows

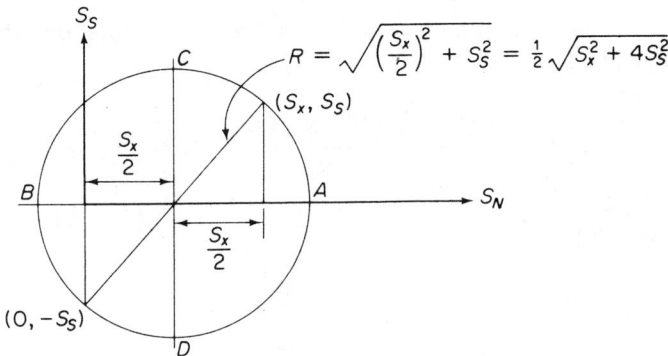

Fig 10.22 Mohr's circle for combined axial loading and twisting

the construction of Mohr's Circle for this case. Point A locates $S_{N(max)}$ as

$$S_{N(max)} = \frac{S_x}{2} + \frac{1}{2}\sqrt{S_x^2 + 4S_S^2} \qquad (10.21a)$$

Points C and D locate the maximum shear stresses as

$$S_{S(max)} = R = \pm\frac{1}{2}\sqrt{S_x^2 + 4S_S^2} \qquad (10.22a)$$

Equations (10.21a) and (10.22a) are obviously in agreement with equations (10.21) and (10.22).

Illustrative problem 10.12

A solid propeller shaft must transmit 800 H.P. at 300 R.P.M. at the same time it is subjected to a 45 kN axial load. If the maximum shear stress is considered to be the critical stress and is limited to 84 MPa, compute the necessary shaft size.

Solution:

The direct axial load is

$$\frac{F}{A} = \frac{45\,000}{\pi d^2/4} = \frac{57\,296}{d^2}$$

The shear due to twisting is found from the torque equation. The torque to be transmitted is

$$T = \frac{7124 \text{ H.P.}}{N} = \frac{7124 \times 800}{300} = 18\,997 \text{ N·m}$$

The stress, S_s, is

$$S_s = \frac{TC}{J} = \frac{18\,997 \times d/2}{\pi d^4/32} = \frac{96\,751}{d^3}$$

$$S_{s(max)} = 84 \times 10^6 = \frac{1}{2}\sqrt{\left(\frac{57\,296}{d^2}\right)^2 + 4\left(\frac{96\,751}{d^3}\right)^2}$$

This equation cannot be solved directly, but must be solved by a trial and error procedure. A close approximation to the solution can be obtained directly by neglecting the d^2 term. This gives us an answer of 0.105 m, or 105 mm.

When several forces acting on a shaft cause bending in different planes, it is necessary to determine the component bending moments in the horizontal and vertical directions. The resultant moment at any section is given by $\sqrt{(M_{vertical})^2 + (M_{horizontal})^2}$. We proceed with the combination of this resultant moment with the shear stress as we did earlier.

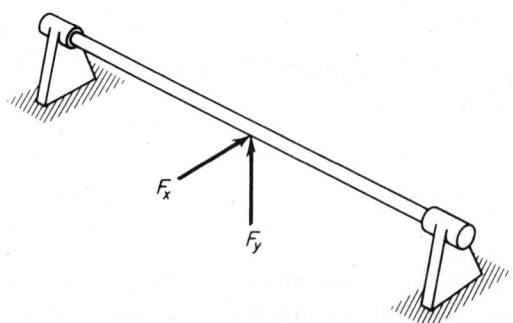

Fig 10.23 Combined horizontal and vertical bending

Illustrative problem 10.13

At a given section of a shaft there is a horizontal bending moment of 1400 N·m and a vertical bending moment of 800 N·m. If the shaft is 50 mm in diameter and must transmit a torque of 2500 N·m, determine the principal stress and shear stress in the shaft.

Solution:

The resultant bending moment is $\sqrt{(1400)^2 + (800)^2} = 1612.5$ N·m lb.

$$T_e = \sqrt{M^2 + T^2} = \sqrt{(1612.5)^2 + (2500)^2} = 2974.9 \text{ N·m}$$

$$S_{S(max)} = \frac{T_e C}{J} = \frac{2974.9 \times 0.050/2}{\pi(0.050)^4/32} = 121.2 \text{ MPa}$$

$$M_e = \frac{1}{2}(M + T_e) = \frac{1}{2}(1612.5 + 2974.9) = 2293.7 \text{ N·m}$$

$$S_{N(max)} = \frac{M_e C}{I} = \frac{2293.7 \times 0.050/2}{\pi(0.050)^4/64} = 186.9 \text{ MPa}$$

10.7 CLOSURE

This chapter has been divided into two basic parts: the mathematical derivation of principal and shear stresses, and the use of Mohr's Circle to accomplish the same end. It is always tempting to the student to solve problems by substituting numbers into equations. Unfortunately, this procedure does not lend itself to the learning process and very often is self-defeating. The construction of Mohr's Circle is more time consuming, but if it is used frequently, the student will find that he will have greater and greater facility in the use of the construction and will have a better grasp of the principles studied in this chapter.

One further caution must be stated at this time: The use of the free-body diagram requires that all of the *forces* acting on a body be properly shown. It is an all too common error to show the stresses acting on a body and to apply the equations of equilibrium to the stresses rather than the forces. By systematically showing the stresses and the areas that these stresses act upon, we can avoid this error. The drawing and sketching of loads and areas should be done with care, and angles should be carefully labelled to avoid error in problem solution.

PROBLEMS

10.1 A square bar is subjected to an axial load and a transverse load as shown in Fig. P10.1. Determine the maximum and minimum stress that occurs at the built-in end.

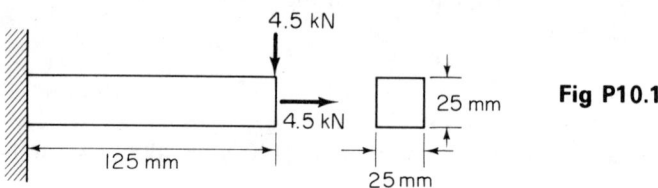

Fig P10.1

10.2 For the bar shown, determine the maximum and minimum stresses at the built-in end.

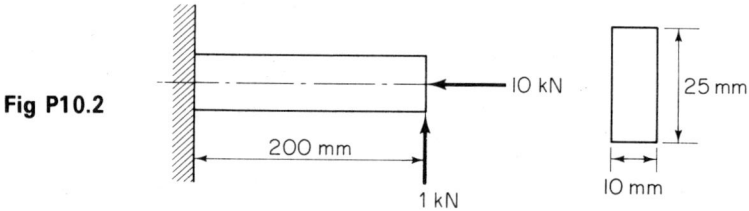

Fig P10.2

10.3 A short 125-mm-diameter solid steel bar is used to carry a 90-kN centric compression load and a 45-kN eccentric load located 150 mm from its center. Determine the maximum stress in the bar.

10.4 A 4-in. Schedule 40 pipe is used as a hanger. What is the maximum stress at the built-in end?

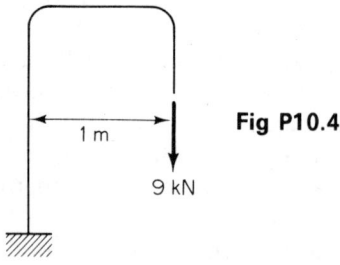

Fig P10.4

10.5 A short strut is subjected to the load shown in Fig. P10.5 through its centroid. Determine the maximum and minimum stresses at its base.

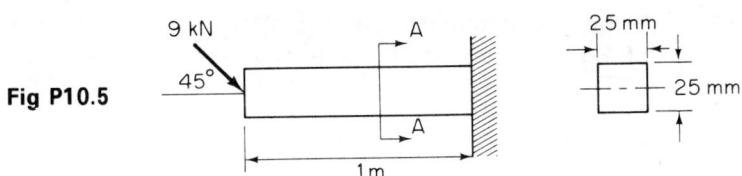

Fig P10.5

10.6 If the angle that the 9-kN force makes with the horizontal in Problem 10.5 is 30°, determine the maximum and minimum stresses at the base of the bar.

10.7 A W14 × 103 short strut is to carry a concentric load of 900 kN and an eccentric load of 450 kN, having an eccentricity of 200 mm, giving a net moment about the x-x axis as shown in Fig. 9.8 (repeated). Determine the maximum and minimum combined stress in the member.

10.8 A W12 × 50 short column carries a centric load of 90 kN and an eccentric load of 135 kN, 125 mm from the x-x axis. Determine the maximum stress in the column.

10.9 An S5 × 10 short strut is loaded as shown in Fig. P10.9. Determine the maximum stress at the base of the strut.

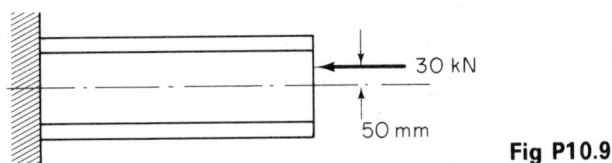

Fig P10.9

10.10 If the 30-kN force in Problem 10.9 causes bending about the y-y axis, solve for the maximum stress at the base of the strut.

10.11 If an S5 × 10 strut is loaded as shown in Fig. P10.11, determine the maximum stress at the base of the strut.

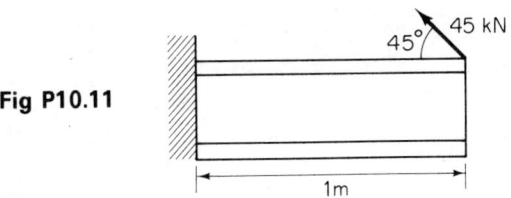

Fig P10.11

10.12 Solve Problem 10.11 for an angle of 30°.

10.13 If the C-clamp shown in Fig. P10.13 is loaded with a force F of 4.5 kN, determine the maximum stress across Section A-A.

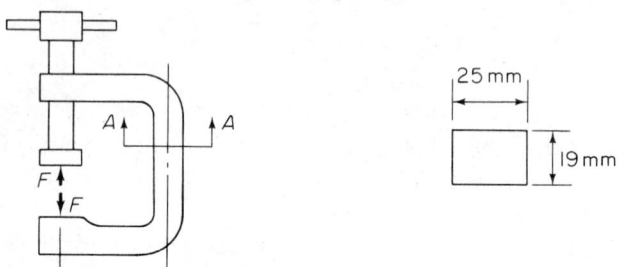

Fig P10.13

10.14 If section A-A of Problem 10.13 is a "T" section, determine the maximum stress across section A-A. $F = 4.5$ kN.

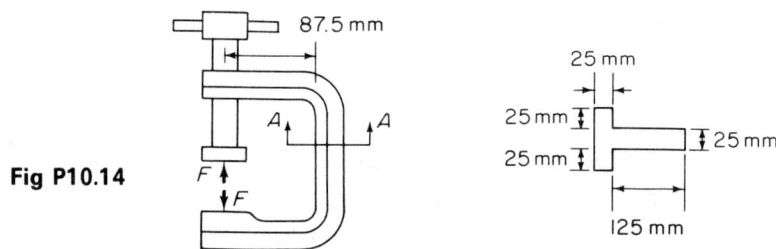

Fig P10.14

10.15 A centric load of 22 kN acts as shown in Fig. P10.15. Determine the normal and shear stresses on planes making 0°, 30°, 45°, 60°, and 90° to the horizontal as shown.

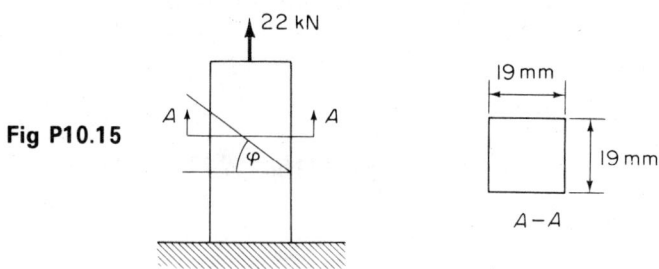

Fig P10.15

10.16 Solve Problem 10.15 using Mohr's Circle.

10.17 A short square bar with a side of 25 mm has a maximum allowable compressive stress of 140 MPa and a maximum allowable shear stress of 70 MPa. What axial load can this strut support?

10.18 A short strut having a rectangular cross section of 25 mm × 50 mm has a maximum allowable compressive stress of 200 MPa and a maximum allowable shear stress of 100 MPa. What axial load can this strut support?

10.19 If an element is loaded so that $S_x = 100$ MPa and $S_y = 50$ MPa (both tension and both principal stresses), determine the normal and shear stresses on a plane whose normal makes an angle of $+30°$ with the x-plane (angle φ in Fig. 10.7). Use equations (10.12) and (10.14) to obtain the solution.

10.20 Solve Problem 10.19 if S_y is compressive.

10.21 Solve Problem 10.19 using Mohr's Circle.

10.22 Solve Problem 10.20 using Mohr's Circle.

10.23 A centric load of 45 kN acts on a 12.5-mm × 12.5-mm bar as shown in Fig. P10.23. Determine the normal and shearing stresses acting on a plane making an angle of 35° with the x plane (φ in Fig. 10.7). Use equations (10.12) and (10.14).

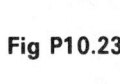
Fig P10.23
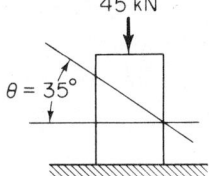
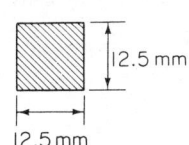

10.24 Solve Problem 10.23 using an angle of 25°.

10.25 Solve Problem 10.24 using Mohr's Circle.

10.26 An element is loaded so that $S_x = 75$ MPa and $S_y = 50$ MPa. Determine the normal and shearing stresses on a plane making an angle of 15° with the x plane (φ in Fig. 10.7). Use equations (10.12) and (10.14).

10.27 Solve Problem 10.26 using Mohr's Circle.

10.28 Determine the maximum principal stress and the maximum shear stress on the element of Problem 10.26 using equations (10.17), (10.18), and (10.19).

10.29 Solve Problem 10.28 using Mohr's Circle.

10.30 An element is subjected to the stresses shown in Fig. P10.30. Determine the stress components (S_N and S_s) along the diagonal 1-1. Use equations (10.12) and (10.14).

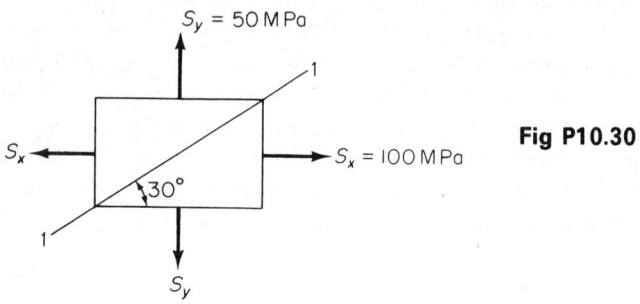

Fig P10.30

10.31 Solve Problem 10.30 if S_y is compressive.
10.32 Solve Problem 10.30 using Mohr's Circle.
10.33 Solve Problem 10.31 using Mohr's Circle.
10.34 For the element shown in Fig. P10.34, determine the principal stresses, the maximum shear stresses, and the orientation of the planes of maximum normal stress. Use equations (10.17), (10.18), and (10.19) to obtain the solution.

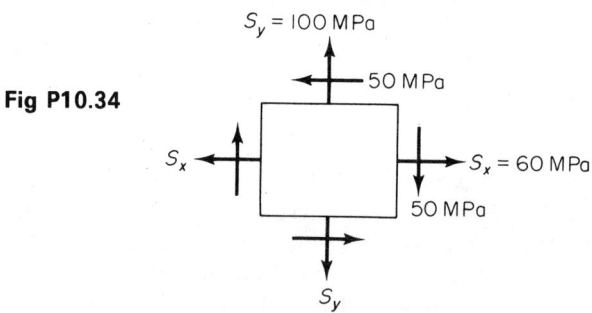

Fig P10.34

10.35 Solve Problem 10.34 using Mohr's Circle.
10.36 For the element shown in Fig. P10.22, determine the principal stresses, the maximum shear stress, and the orientation of the plane of maximum normal stress. Use equations (10.17), (10.18) and (10.19) to obtain the solution.

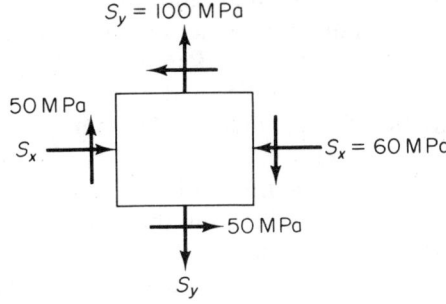

Fig P10.36

10.37 Solve Problem 10.36 using Mohr's Circle.
10.38 For the element shown in Fig. P10.38, determine the principal stresses, the maximum shear stress, and the orientation of the plane of maximum normal stress. Use equations (10.17), (10.18), and (10.19).

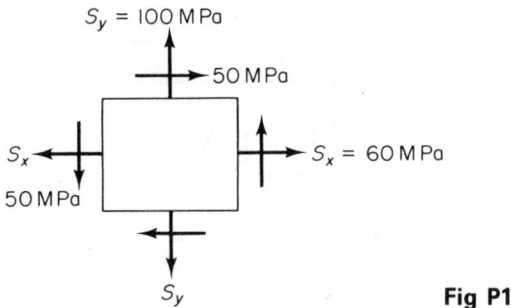

Fig P10.38

10.39 Solve Problem 10.38 using Mohr's Circle.
10.40 A 25 mm diameter shaft is subjected to the simultaneous action of a 40 N·m torque and an 80 N·m bending moment. Determine the maximum shear and bending stresses in the shaft.
10.41 A 50 mm diameter shaft is subject to a torque of 100 N·m and a bending moment of 150 N·m. What is the maximum shear and bending stresses in the shaft?
10.42 A 40 mm diameter shaft must transmit 400 H.P. at 400 R.P.M. If the shaft is also subjected to an axial load of 500 kN, determine the maximum shear stress in the shaft.
10.43 A 75 mm diameter shaft must transmit 500 H.P. at 500 R.P.M. while it is concurrently subjected to an axial load of 450 kN. Determine the maximum shear stress in the shaft.

10.44 A shaft must transmit 50 H.P. at 500 R.P.M. If the shaft is 25 mm in diameter and must simultaneously resist a bending moment of 100 N·m determine the maximum shear and maximum normal stress in the shaft.

10.45 A shaft that is 100 mm in diameter must transmit 500 H.P. at 500 R.P.M. If the shaft is also subjected to a 500 N·m bending moment, determine the maximum shear stress and the maximum normal stress in the shaft.

10.46 The shaft shown in Fig. 10.23 is 3 m long between supports. If $F_x = F_y = 4.5$ kN and if the beam is considered to be simply supported, determine the principal stress and shear stress in the shaft. The shaft must simultaneously transmit a torque of 1500 N·m and 37.5 mm in diameter. Assume the loads act at the center of the shaft.

10.47 A shaft is loaded as shown in Fig. P10.47. Determine the maximum shear and principal stress in the shaft. The bearings provide simple support. Neglect the weight of the pulleys and shaft.

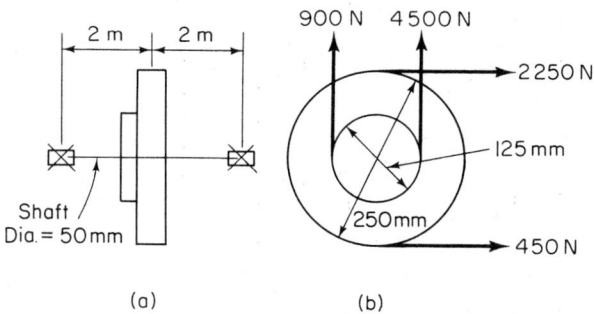

Fig P10.47

10.48 A shaft transmits 250 H.P. at 120 R.P.M. If the shaft is 62.5 mm in diameter and is also subjected to an axial load of 315 kN, determine the maximum shear and maximum normal stress in the shaft.

10.49 Solve Problem 10.48 using Mohr's Circle.

10.50 Solve Problem 10.42 using Mohr's Circle.

10.51 At a given section of a shaft, there is a horizontal bending moment of 1200 N·m and a vertical bending moment of 600 N·m. If the shaft is 37.5 mm in diameter and transmits a torque of 1500 N·m determine the maximum bending stress and the maximum shear stress.

10.52 Two pulleys, each 250 mm in diameter are placed at the third points of a shaft, one pulley being the driver and the other the driven. If the belts are taken off so that the belt tensions are in the horizontal direction, determine the normal and shear stresses in the shaft. The shaft is to transmit 100 H.P. at 500 R.P.M. and is 50 mm in diameter and 4 m long. Since one pulley is the driver and the other the driven, each will give an equal, but opposite net tensile force acting in the horizontal plane due to the unbalanced belt tensions.

REFERENCES

Arges, K. P. and A. E. Palmer. *Mechanics of Materials.* New York: McGraw-Hill Book Co., 1963.

Bassin, M. E. and S. M. Brodsky. *Statics and Strength of Materials.* 2nd ed. New York: McGraw-Hill Book Co., 1969.

Conway, H. D. *Mechanics of Materials.* Englewood Cliffs, NJ: Prentice-Hall, Inc.

Jensen, A. and H. H. Chenoweth. *Applied Strength of Materials.* 3rd ed. New York: McGraw-Hill Book Co., 1971.

Levinson, Irving J. *Mechanics of Materials.* 2nd ed. Englewood Cliffs, NJ: Prentice-Hall, Inc., 1970.

Olsen, G. A. *Strength of Materials.* Englewood Cliffs, NJ: Prentice-Hall, Inc.

Seely, F. B. and J. O. Smith. *Advanced Mechanics of Materials.* 2nd ed. New York: John Wiley & Sons, Inc., 1952.

Singer, F. L. *Strength of Materials.* 2nd ed. New York: Harper and Row, 1962.

Timoshenko, S. and Donovan H. Young. *Elements of Strength of Materials.* 5th ed. New York: D. Van Nostrand Co., 1968.

Answers to even-numbered problems

CHAPTER 1

1.2 61.8 mm
1.4 19.05 kN
1.6 A-A   57.3 MPa
     B-B   229.2 MPa
     C-C   101.9 MPa
1.8 yes   0.28%
1.10 2.42 MPa
1.12 1.3 MPa
1.14 5.29×10^{-3} m
1.16 134.96 kN
1.18 1.344 MN
1.20 118.42 MPa (column to plate)
     5 MPa (plate to footing)
     1.25 MPa (footing to earth)
1.22 $S_{steel} = 9.4$ MPa
     $S_{concrete} = 3.13$ MPa
1.24 1.719×10^{-3} m
1.26 $\dfrac{EA}{100}$
1.28 210 MPa
1.30 $\mu = 0.364$
1.32 3.333×10^{-8} m^3
1.34 2.015×10^{-4} radians
1.36 1.526×10^{-5} m

1.38 4.796×10^{-4}
1.40 34.97×10^{9} Pa
1.42 0.000136
1.44 110.6 MPa
1.46 8.00018 m
1.48 93.86 MPa
1.50 0.00524 m
1.52 $P_{al} = -2342$ N
     $P_{steel} = 51.17$ kN
1.54 $P_{al} = -14\ 686$ N
     $P_{steel} = 57\ 343$ N
1.56 $S_{steel} = 728.9$ MPa
     $S_{al} = 242.97$ MPa
1.58 $S_{steel} = 24.57$ MPa
     $S_{al} = 38.48$ MPa
1.60 $S_{al} = 80.38$ MPa
     $S_{copper} = 80.38$ MPa
1.62 5.56 mm
1.64 15 mm
1.66 59.73 kPa
1.68 $S_{mean} = 31.5$ MPa
     $S_{inside} = 28$ MPa
     $S_{actual} = 39.2$ MPa

CHAPTER 2

2.2   509.3 MPa
2.4 $\mu = 0.32$
2.6 $\mu = 0.40$
2.8   40%
2.10 1.386 GPa (failure)
2.12 135,870 psi, 64%
2.14 % elongation = 37.5%
      % reduction in area = 59.04%
2.16 % elongation = 39.5%
      % reduction in area = 61.8%
2.18 $\varepsilon = 0.0003125$ mm/mm
      $F = 8.053$ kN
2.20 $E = 169.7$ GPa
      $\mu = 0.267$
2.22 $E = 50$ GPa
2.24 Prop. limit = 37,783 psi
      yield stress = 45,045 psi
      breaking stress = 90,090 psi
      true breaking stress = 216,200 psi
      % change in area = 58.3%
2.26 Prop. limit = 30,000 psi
      $E = 30.3 \times 10^6$ psi
      % elongation = 39.5%
      % reduction of area = 64.2%
      ultimate stress = 60,000 psi (on original area)
      breaking stress = 46,000 psi (on original area)

CHAPTER 3

3.2   69.3%
3.4   213.8 MPa
3.6   58.69 kN
3.8   83.16 kN
3.10 29 mm (closely)
3.12 34.2%
3.14 110.53 kN
3.16 none
3.18 351.8 kN
3.20 47.43 kN
3.22 27.04 kN
3.24 61.52 kN
3.26 (a)
3.28 639.9 kN
3.30 1.08 Mn
3.32 $L = 41$ mm
3.34 $L = 0.162$ m
3.36 same
3.38 11 mm
3.40 discussion
3.42 $A = 200$ mm
      $B = 100$ mm
3.44 discussion
3.46 2.4 × direct stress
3.48 71.53 MPa

CHAPTER 4

- 4.2 88.69 mm (up from base)
- 4.4 $\bar{x} = 0$, $\bar{y} = 68.75$ mm (up from base)
- 4.6 44.53 mm
- 4.8 45.29 mm
- 4.10 $\bar{x} = 60.71$ mm
  $\bar{y} = 41.07$ mm
- 4.12 $\bar{y} = 50$ mm
  $\bar{x} = 51.44$ mm
- 4.14 $\bar{x} = 0$
  $\bar{y} = 27.47$ mm (from bottom)
- 4.16 $\bar{x} = 50$ mm
  $\bar{y} = 48.86$ mm
- 4.18 $\bar{x} = 44.67$ mm
  $\bar{y} = 155.32$ mm
- 4.20 $R = 118.35$ mm
- 4.22 6.328×10^8 mm^4
- 4.24 $\bar{k} = 96.82$ mm
  $k_{base} = 178.54$ mm
- 4.26 $I = 23.31 \times 10^6$ mm^4
- 4.28 19.74×10^8 mm^4
- 4.30 513.7×10^6 mm^4
- 4.32 59.96×10^6 mm^4
- 4.34 7.097×10^6 mm^4
- 4.36 45.58×10^6 mm^4
- 4.38 39.41×10^6 mm^4
- 4.40 112 mm
- 4.42 108.32 mm
- 4.44 27.8×10^6 mm^4
- 4.46 46.14 mm
- 4.48 28.98×10^6 mm^4
- 4.50 34.49×10^6 mm^4
- 4.52 $I = 19.96 \times 10^8$ mm^4
  $k = 158.86$ mm

CHAPTER 5

- 5.2 derivation
- 5.4 1425 RPM
- 5.6 57.93 mm
- 5.8 77.81 mm
- 5.10 308.8 HP
- 5.12 13.13 MPa
- 5.14 0.00417 radians/m
- 5.16 0.02037 radians/m
- 5.18 48.38 GPa
- 5.20 0.39 radians
- 5.22 34.148×10^{-3} radians
- 5.24 derivation
- 5.26 271.3 HP
- 5.28 432 mm
- 5.30 8819 N·m
- 5.32 6187 N·m
- 5.34 $3582 \dfrac{N \cdot m}{radian}$
- 5.36 2.03 N·m
- 5.38 $1036 \dfrac{N \cdot m}{radian}$
- 5.40 62.7 mm
- 5.42 734 N/m
- 5.44 820.3 N/m
- 5.46 694.7 MPa
- 5.48 23.1 mm
- 5.50 $d \cong 5.3$ mm
- 5.52 6 mm, 36 mm
- 5.54 derivation $k_{eq} = k_1 + k_2$
- 5.56 5.54
- 5.58 2188 N·m
- 5.60 34.13 MPa
- 5.62 30.6 mm

CHAPTER 6

6.2 $Fa\left(1 - \dfrac{x}{a+b}\right)$
6.4 1000 N, 4400 N·m
6.6 5000 N, 26.0 × 10^3 N·m
6.8 1750 N, 11.875 × 10^3 N·m
6.10 5.14 m
6.12 3500 N, 16.250 × 10^3 N·m
6.14 $R_1 = R_2 = F$
6.16 $R_1 = R_2 = 3000$
6.18 $R_1 = 7400, R_2 = 7600$
6.20 $R_1 = 1150, R_2 = 350$
6.22 $R_1 = 333, R_2 = 767, M = 1470$ N·m @ 3.84 m from right support
6.24 $R_1 = 1600, R_2 = 2800, M = 2560$ N·m @ 3.2 m from left support
6.26 $R_1 = 1345, R_2 = 955$
6.28 $R_1 = 1995, R_2 = 805$
6.30 Figures, $R_1 = -75, R_2 = 375$
6.32 Figures, $R_1 = 1031, R_2 = 1669$
6.34 Figures, $R_1 = R_2 = 300$
6.36 Figures, $R_1 = R_2 = 800$
6.38 Figures, $R_1 = 375, R_2 = 825$
6.40 Figures, $R_1 = 1600, R_2 = 1000$
6.42 Figures, $R_1 = 1907, R_2 = 3293$
6.44 214.7 N·m
6.46 3400 N·m
6.48 $R_1 = 955, R_2 = 1845, M = 3404$ N·m
6.50 Figures
6.52 Figures
6.54 $M = 16.74 \times 10^3$ N·m
6.56 $M = 44.181 \times 10^3$ N·m; 9818 N
6.58 $M = 137.4 \times 10^3$ N·m; 12.84 kN

CHAPTER 7

7.2 216.0 MPa
7.4 163.0 MPa
7.6 11 037 N/m
7.8 1622.5 N
7.10 148.4 MPa
7.12 183.2 MPa
7.14 283.2 MPa
7.16 W8 × 13
7.18 35.44 MPa
7.20 15.6 MPa
7.22 36.28 kN·m
7.24 60 mm
7.26 8633 N·m
7.28 110.1 kN·m
7.30 Increase 196%
7.32 3222 N/m

7.34 Increase 277%
7.36 Steel = 197.2 MPa
   Wood = 219.3 kPa
7.38 Aluminum = 56.83 MPa
   Wood = 4.78 MPa
7.40 162 kN·m
7.42 $A_s = 1.409 \times 10^{-3}$ m^2
7.44 concrete limits 29.89 kN·m
7.46 $A_s = 1.11 \times 10^3$ mm^2
7.48 Concrete limits 104.4 kN·m

7.50 1.8 MPa
7.52 1.528 MPa
7.54 Derivation
7.56 53.65 kN
7.58 1071 N/m
7.60 33 kN
7.62 At break in section
7.64 4.23 MPa
7.66 168.7 kN
7.68 11.165 kN

CHAPTER 8

8.2   356.1 m
8.4   400.6 m
8.6   175 MPa
8.8 1.06×10^{-3} m
8.10 0.136 m
8.12 0.227 m, 1.89×10^{-2} radians
8.14 0.653 m
8.16 Derivation
8.18 1.05×10^{-3} m
8.20 $\dfrac{CE}{30 L}$
8.22 1.916×10^{-3} m
8.24 0.0677 m
8.26 7.208×10^{-3} m
8.28 0.00885 m
8.30 3.571×10^{-3} m
8.32 4.677×10^{-3} m
8.34 0.0275 m
8.36 2.68×10^{-3} m
8.38 $(5/16) W L^2$
8.40 0.00229 m
8.42 60.2×10^{-5} m (up)
8.44 yes

8.46 1.429×10^{-3} m
8.48 2.98×10^{-2} m
8.50 6.435×10^{-6} m^4
8.52 2.58×10^{-4} m
8.54 0.0278 m
8.56 2.78×10^{-4} m
8.58 1.04×10^{-4} m
8.60 3.82×10^{-4} m
8.62 derivation
8.64 $R_1 = R_3 = 1250$ N
      $R_{center} = 5500$ N
      $M = -3000$ N·m
8.66 Same as 8.64
8.68 -1266 N·m
8.70 -2189 N·m
8.72 $M_1 = 0$
      $M_2 = -4800$ N·m
      $M_3 = -4800$ N·m
      $M_4 = 0$
8.74 $M_1 = 0$
      $M_2 = -5600$
      $M_3 = -7600$
      $M_4 = 0$

CHAPTER 9

9.2 $b = 2.77''$
9.4 $b = 4.62''$
9.6 yes
9.8 $4.68''$
9.10 $9.36''$
9.12 66,808 lbs
9.14 1346 lbs
9.16 96,914 lbs
9.18 70,400 lbs
9.20 23,876 lbs
9.22 97,902 lbs
9.24 207,200 lbs
9.26 25,350 lbs
9.28 16,507 lbs
9.30 214,263 lbs
9.32 $S18 \times 54.7$ braced against y-y
9.34 no
9.36 214,200 lbs
9.38 27,342 lbs
9.40 17,082 lbs
9.42 230,058 lbs
9.44 Brace y-y − $S18 \times 54.7$
9.46 no
9.48 278,900 lbs
9.50 use bracing $S12 \times 50$
9.52 no
9.54 no
9.56 not satisfactory

CHAPTER 10

10.2 $S_{max} = 232$ MPa
$S_{min} = 152$ MPa
10.4 $S_{max} = 172.8$ MPa
10.6 $S_{max} = 1740.5$ MPa
$S_{min} = 1715.5$ MPa
10.8 39.6 MPa
10.10 $S_{max} = 128.5$ MPa
10.12 $S_{max} = 330.4$ MPa
10.14 $S_{max} = 4.21$ MPa
$S_{min} = 1.27$ MPa
10.16 at 30°, $S_N = 6660$ psi, $S_s = 3840$ psi
10.18 250 kN
10.20 $S_N = 62.5$ MPa, $S_s = 65.0$ MPa
10.22 see problem 10.20
10.24 $S_N = 236.6$ MPa, $S_s = 110.3$ MPa
10.26 $S_N = 73.3$ MPa, $S_s = 6.25$ MPa
10.28 $S_{s\,max} = 12.5$ MPa
10.30 $S_N = 62.5$ MPa, $S_s = -21.65$ MPa
10.32 see problem 10.30
10.34 $S_{N\,max} = 133.85$ MPa, $S_{N\,min} = 26.15$ MPa
$S_{s\,max} = 53.85$ MPa
10.36 $S_{N\,max} = 114.3$ MPa; $S_{N\,min} = -74.3$ MPa
$S_{s\,max} = 94.3$ MPa
10.38 $S_{N\,max} = 133.9$ MPa, $S_{N\,min} = 26.15$ MPa
$S_{s\,max} = 53.9$ MPa
10.40 $S_b = 55.2$ MPa
$S_s = 29.14$ MPa
10.42 $S_{s\,max} = 600.8$ MPa
10.44 $S_s = 234.5$ MPa
$S_b = 267.1$ MPa
10.46 $S_s = 483.2$ MPa
$S_b = 944.1$ MPa
10.48 $S_N = 365.2$ MPa
$S_s = 313.8$ MPa
10.50 see Problem 10.42
10.52 $S_s = 314.3$ MPa
$S_b = 623.1$ MPa

Appendix A

Appendix A   435

AVERAGE MECHANICAL PROPERTIES OF SELECTED MATERIALS

Material	U.S. Customary Units			SI Units				Coefficient of thermal expansion 10^{-6} m/m °C	μ
	Yield stress, psi	Modulus of Elasticity, psi		Yield stress, MPa	Modulus of Elasticity, GPa				
		Tension, E	Shear, G		Tension, E	Shear, G			
Steel (SAE 1020)	36 000	30 000 000	12 000 000	250	210	85		11.7	0.27
Steel (SAE 1090)	65 000	30 000 000	12 000 000	450	210	85		11.7	0.27
Cast Iron	80 000	15 000 000	6 000 000	560	105	40		12.1	0.20
Aluminum, 6061 alloy	35 000	10 000 000	4 000 000	240	70	30		23.6	0.33
Brass	15 000	15 000 000	6 000 000	105	105	40		18.9	0.35
Bronze	20 000	15 000 000	6 500 000	140	105	45			
Copper, hard-drawn	35 000	17 000 000	6 000 000	240	115	40		32.4	
Douglas fir timber (Air dry-parallel to grain)		1 700 000			12			5.4	
Lead	2 000	2 000 000	700 000	14	14	4.9		29.5	0.43
Stainless Steel	80 000	28 000 000	9 500 000	560	193	66.5		17.3	0.30
Titanium alloy	120 000	15 900 000	6 200 000	840	110	43.4		8.3	0.34

Appendix B
Properties of structural shapes

HOT-ROLLED STRUCTURAL STEEL SHAPE DESIGNATIONS

New Designation	Type of Shape	Old Designation
W 24 × 76 W 14 × 26	W shape	24 WF 76 14 B 26
S 24 × 100	S shape	24 I 100
M 8 × 18.5 M 10 × 9 M 8 × 34.3	M shape	8 M 18.5 10 JR 9.0 8 × 8 M 34.3
C 12 × 20.7	American Standard Channel	12 [20.7
MC 12 × 45 MC 12 × 10.6	Miscellaneous Channel	12 × 4 [45.0 12 JR [10.6
HP 14 × 73	HP shape	14 BP 73
L 6 × 6 × ¾ L 6 × 4 × ⅝	Equal Leg Angle Unequal Leg Angle	∠ 6 × 6 × ¾ ∠ 6 × 4 × ⅝
WT 12 × 38 WT 7 × 13	Structural Tee cut from W shape	ST 12 WF 38 ST 7 B 13
ST 12 × 50	Structural Tee cut from S shape	ST 12 I 50
MT 4 × 9.25 MT 5 × 4.5 MT 4 × 17.15	Structural Tee cut from M shape	ST 4 M 9.25 ST 5 JR 4.5 ST 4 M 17.15
PL ½ × 18	Plate	PL 18 × ½
Bar 1 ▢ Bar 1¼ φ Bar 2½ × ½	Square Bar Round Bar Flat Bar	Bar 1 ▢ Bar 1¼ φ Bar 2½ × ½
Pipe 4 Std. Pipe 4 X - Strong Pipe 4 XX - Strong	Pipe	Pipe 4 Std. Pipe 4 X-Strong Pipe 4 XX-Strong
TS 4 × 4 × .375 TS 5 × 3 × .375 TS 3 OD × .250	Structural Tubing: Square Structural Tubing: Rectangular Structural Tubing: Circular	Tube 4 × 4 × .375 Tube 5 × 3 × .375 Tube 3 OD × .250

*Reprinted with the permission of the American Institute of Steel Construction, New York, N.Y., *Manual of Steel Construction*, Seventh edition, 1970.

W SHAPES
Properties for designing

Designation	Area A	Depth d	Flange Width b_f	Flange Thickness t_f	Web Thickness t_w	Axis X-X I	Axis X-X Z	Axis X-X k	Axis Y-Y I	Axis Y-Y Z	Axis Y-Y k
	In.²	In.	In.	In.	In.	In.⁴	In.³	In.	In.⁴	In.³	In.
W 36×300	88.3	36.72	16.655	1.680	0.945	20300	1110	15.2	1300	156	3.83
×280	82.4	36.50	16.595	1.570	0.885	18900	1030	15.1	1200	144	3.81
×260	76.5	36.24	16.551	1.440	0.841	17300	952	15.0	1090	132	3.77
×245	72.1	36.06	16.512	1.350	0.802	16100	894	15.0	1010	123	3.75
×230	67.7	35.88	16.471	1.260	0.761	15000	837	14.9	940	114	3.73
W 36×194	57.2	36.48	12.117	1.260	0.770	12100	665	14.6	375	61.9	2.56
×182	53.6	36.32	12.072	1.180	0.725	11300	622	14.5	347	57.5	2.55
×170	50.0	36.16	12.027	1.100	0.680	10500	580	14.5	320	53.2	2.53
×160	47.1	36.00	12.000	1.020	0.653	9760	542	14.4	295	49.1	2.50
×150	44.2	35.84	11.972	0.940	0.625	9030	504	14.3	270	45.0	2.47
×135	39.8	35.55	11.945	0.794	0.598	7820	440	14.0	226	37.9	2.39
W 33×240	70.6	33.50	15.865	1.400	0.830	13600	813	13.9	933	118	3.64
×220	64.8	33.25	15.810	1.275	0.775	12300	742	13.8	841	106	3.60
×200	58.9	33.00	15.750	1.150	0.715	11100	671	13.7	750	95.2	3.57
W 33×152	44.8	33.50	11.565	1.055	0.635	8160	487	13.5	273	47.2	2.47
×141	41.6	33.31	11.535	0.960	0.605	7460	448	13.4	246	42.7	2.43
×130	38.3	33.10	11.510	0.855	0.580	6710	406	13.2	218	37.9	2.38
×118	34.8	32.86	11.484	0.738	0.554	5900	359	13.0	187	32.5	2.32
W 30×210	61.9	30.38	15.105	1.315	0.775	9890	651	12.6	757	100	3.50
×190	56.0	30.12	15.040	1.185	0.710	8850	587	12.6	673	89.5	3.47
×172	50.7	29.88	14.985	1.065	0.655	7910	530	12.5	598	79.8	3.43
W 30×132	38.9	30.30	10.551	1.000	0.615	5760	380	12.2	196	37.2	2.25
×124	36.5	30.16	10.521	0.930	0.585	5360	355	12.1	181	34.4	2.23
×116	34.2	30.00	10.500	0.850	0.564	4930	329	12.0	164	31.3	2.19
×108	31.8	29.82	10.484	0.760	0.548	4470	300	11.9	146	27.9	2.15
× 99	29.1	29.64	10.458	0.670	0.522	4000	270	11.7	128	24.5	2.10

W SHAPES
Properties for designing

Designation	Area A	Depth d	Flange Width b_f	Flange Thickness t_f	Web Thickness t_w	Axis X-X I	Axis X-X Z	Axis X-X k	Axis Y-Y I	Axis Y-Y Z	Axis Y-Y k
	In.²	In.	In.	In.	In.	In.⁴	In.³	In.	In.⁴	In.³	In.
W 27×177	52.2	27.31	14.090	1.190	0.725	6740	494	11.4	556	78.9	3.26
×160	47.1	27.08	14.023	1.075	0.658	6030	446	11.3	495	70.6	3.24
×145	42.7	26.88	13.965	0.975	0.600	5430	404	11.3	443	63.5	3.22
W 27×114	33.6	27.28	10.070	0.932	0.570	4090	300	11.0	159	31.6	2.18
×102	30.0	27.07	10.018	0.827	0.518	3610	267	11.0	139	27.7	2.15
× 94	27.7	26.91	9.990	0.747	0.490	3270	243	10.9	124	24.9	2.12
× 84	24.8	26.69	9.963	0.636	0.463	2830	212	10.7	105	21.1	2.06
W 24×160	47.1	24.72	14.091	1.135	0.656	5120	414	10.4	530	75.2	3.35
×145	42.7	24.49	14.043	1.020	0.608	4570	373	10.3	471	67.1	3.32
×130	38.3	24.25	14.000	0.900	0.565	4020	332	10.2	412	58.9	3.28
W 24×120	35.4	24.31	12.088	0.930	0.556	3650	300	10.2	274	45.4	2.78
×110	32.5	24.16	12.042	0.855	0.510	3330	276	10.1	249	41.4	2.77
×100	29.5	24.00	12.000	0.775	0.468	3000	250	10.1	223	37.2	2.75
W 24× 94	27.7	24.29	9.061	0.872	0.516	2690	221	9.86	108	23.9	1.98
× 84	24.7	24.09	9.015	0.772	0.470	2370	197	9.79	94.5	21.0	1.95
× 76	22.4	23.91	8.985	0.682	0.440	2100	176	9.69	82.6	18.4	1.92
× 68	20.0	23.71	8.961	0.582	0.416	1820	153	9.53	70.0	15.6	1.87
W 24× 61	18.0	23.72	7.023	0.591	0.419	1540	130	9.25	34.3	9.76	1.38
× 55	16.2	23.55	7.000	0.503	0.396	1340	114	9.10	28.9	8.25	1.34
W 21×142	41.8	21.46	13.132	1.095	0.659	3410	317	9.03	414	63.0	3.15
×127	37.4	21.24	13.061	0.985	0.588	3020	284	8.99	366	56.1	3.13
×112	33.0	21.00	13.000	0.865	0.527	2620	250	8.92	317	48.8	3.10
W 21× 96	28.3	21.14	9.038	0.935	0.575	2100	198	8.61	115	25.5	2.02
× 82	24.2	20.86	8.962	0.795	0.499	1760	169	8.53	95.6	21.3	1.99
W 21× 73	21.5	21.24	8.295	0.740	0.455	1600	151	8.64	70.6	17.0	1.81
× 68	20.0	21.13	8.270	0.685	0.430	1480	140	8.60	64.7	15.7	1.80
× 62	18.3	20.99	8.240	0.615	0.400	1330	127	8.54	57.5	13.9	1.77
× 55	16.2	20.80	8.215	0.522	0.375	1140	110	8.40	48.3	11.8	1.73
W 21× 49	14.4	20.82	6.520	0.532	0.368	971	93.3	8.21	24.7	7.57	1.31
× 44	13.0	20.66	6.500	0.451	0.348	843	81.6	8.07	20.7	6.38	1.27

W SHAPES
Properties for designing

Designation	Area A	Depth d	Flange Width b_f	Flange Thickness t_f	Web Thickness t_w	Elastic Properties Axis X-X			Elastic Properties Axis Y-Y		
						I	Z	k	I	Z	k
	In.²	In.	In.	In.	In.	In.⁴	In.³	In.	In.⁴	In.³	In.
W 18×114	33.5	18.48	11.833	0.991	0.595	2040	220	7.79	274	46.3	2.86
×105	30.9	18.32	11.792	0.911	0.554	1850	202	7.75	249	42.3	2.84
× 96	28.2	18.16	11.750	0.831	0.512	1680	185	7.70	225	38.3	2.82
W 18× 85	25.0	18.32	8.838	0.911	0.526	1440	157	7.57	105	23.8	2.05
× 77	22.7	18.16	8.787	0.831	0.475	1290	142	7.54	94.1	21.4	2.04
× 70	20.6	18.00	8.750	0.751	0.438	1160	129	7.50	84.0	19.2	2.02
× 64	18.9	17.87	8.715	0.686	0.403	1050	118	7.46	75.8	17.4	2.00
W 18× 60	17.7	18.25	7.558	0.695	0.416	986	108	7.47	50.1	13.3	1.68
× 55	16.2	18.12	7.532	0.630	0.390	891	98.4	7.42	45.0	11.9	1.67
× 50	14.7	18.00	7.500	0.570	0.358	802	89.1	7.38	40.2	10.7	1.65
× 45	13.2	17.86	7.477	0.499	0.335	706	79.0	7.30	34.8	9.32	1.62
W 18× 40	11.8	17.90	6.018	0.524	0.316	612	68.4	7.21	19.1	6.34	1.27
× 35	10.3	17.71	6.000	0.429	0.298	513	57.9	7.05	15.5	5.16	1.23
W 16× 96	28.2	16.32	11.533	0.875	0.535	1360	166	6.93	224	38.8	2.82
× 88	25.9	16.16	11.502	0.795	0.504	1220	151	6.87	202	35.1	2.79
W 16× 78	23.0	16.32	8.586	0.875	0.529	1050	128	6.75	92.5	21.6	2.01
× 71	20.9	16.16	8.543	0.795	0.486	941	116	6.71	82.8	19.4	1.99
× 64	18.8	16.00	8.500	0.715	0.443	836	104	6.66	73.3	17.3	1.97
× 58	17.1	15.86	8.464	0.645	0.407	748	94.4	6.62	65.3	15.4	1.96
W 16× 50	14.7	16.25	7.073	0.628	0.380	657	80.8	6.68	37.1	10.5	1.59
× 45	13.3	16.12	7.039	0.563	0.346	584	72.5	6.64	32.8	9.32	1.57
× 40	11.8	16.00	7.000	0.503	0.307	517	64.6	6.62	28.8	8.23	1.56
× 36	10.6	15.85	6.992	0.428	0.299	447	56.5	6.50	24.4	6.99	1.52
W 16× 31	9.13	15.84	5.525	0.442	0.275	374	47.2	6.40	12.5	4.51	1.17
× 26	7.67	15.65	5.500	0.345	0.250	300	38.3	6.25	9.59	3.49	1.12

W SHAPES
Properties for designing

Designation	Area A In.²	Depth d In.	Flange		Web Thickness t_w In.	Elastic Properties					
			Width b_f In.	Thickness t_f In.		Axis X-X			Axis Y-Y		
						I In.⁴	Z In.³	k In.	I In.⁴	Z In.³	k In.
W 14×730	215	22.44	17.889	4.910	3.069	14400	1280	8.18	4720	527	4.69
×665	196	21.67	17.646	4.522	2.826	12500	1150	7.99	4170	472	4.62
×605	178	20.94	17.418	4.157	2.598	10900	1040	7.81	3680	423	4.55
×550	162	20.26	17.206	3.818	2.386	9450	933	7.64	3260	378	4.49
×500	147	19.63	17.008	3.501	2.188	8250	840	7.49	2880	339	4.43
×455	134	19.05	16.828	3.213	2.008	7220	758	7.35	2560	304	4.37
W 14×426	125	18.69	16.695	3.033	1.875	6610	707	7.26	2360	283	4.34
×398	117	18.31	16.590	2.843	1.770	6010	657	7.17	2170	262	4.31
×370	109	17.94	16.475	2.658	1.655	5450	608	7.08	1990	241	4.27
×342	101	17.56	16.365	2.468	1.545	4910	559	6.99	1810	221	4.24
×314	92.3	17.19	16.235	2.283	1.415	4400	512	6.90	1630	201	4.20
×287	84.4	16.81	16.130	2.093	1.310	3910	465	6.81	1470	182	4.17
×264	77.6	16.50	16.025	1.938	1.205	3530	427	6.74	1330	166	4.14
×246	72.3	16.25	15.945	1.813	1.125	3230	397	6.68	1230	154	4.12
W 14×237	69.7	16.12	15.910	1.748	1.090	3080	382	6.65	1170	148	4.11
×228	67.1	16.00	15.865	1.688	1.045	2940	368	6.62	1120	142	4.10
×219	64.4	15.87	15.825	1.623	1.005	2800	353	6.59	1070	136	4.08
×211	62.1	15.75	15.800	1.563	0.980	2670	339	6.56	1030	130	4.07
×202	59.4	15.63	15.750	1.503	0.930	2540	325	6.54	980	124	4.06
×193	56.7	15.50	15.710	1.438	0.890	2400	310	6.51	930	118	4.05
×184	54.1	15.38	15.660	1.378	0.840	2270	296	6.49	883	113	4.04
×176	51.7	15.25	15.640	1.313	0.820	2150	282	6.45	838	107	4.02
×167	49.1	15.12	15.600	1.248	0.780	2020	267	6.42	790	101	4.01
×158	46.5	15.00	15.550	1.188	0.730	1900	253	6.40	745	95.8	4.00
×150	44.1	14.88	15.515	1.128	0.695	1790	240	6.37	703	90.6	3.99
×142	41.8	14.75	15.500	1.063	0.680	1670	227	6.32	660	85.2	3.97
W 14×320	94.1	16.81	16.710	2.093	1.890	4140	493	6.63	1640	196	4.17

W SHAPES
Properties for designing

Designation	Area A	Depth d	Flange Width b_f	Flange Thickness t_f	Web Thickness t_w	Elastic Properties Axis X-X I	Axis X-X Z	Axis X-X k	Axis Y-Y I	Axis Y-Y Z	Axis Y-Y k
	In.²	In.	In.	In.	In.	In.⁴	In.³	In.	In.⁴	In.³	In.
W 14×136	40.0	14.75	14.740	1.063	0.660	1590	216	6.31	568	77.0	3.77
×127	37.3	14.62	14.690	0.998	0.610	1480	202	6.29	528	71.8	3.76
×119	35.0	14.50	14.650	0.938	0.570	1370	189	6.26	492	67.1	3.75
×111	32.7	14.37	14.620	0.873	0.540	1270	176	6.23	455	62.2	3.73
×103	30.3	14.25	14.575	0.813	0.495	1170	164	6.21	420	57.6	3.72
× 95	27.9	14.12	14.545	0.748	0.465	1060	151	6.17	384	52.8	3.71
× 87	25.6	14.00	14.500	0.688	0.420	967	138	6.15	350	48.2	3.70
W 14× 84	24.7	14.18	12.023	0.778	0.451	928	131	6.13	225	37.5	3.02
× 78	22.9	14.06	12.000	0.718	0.428	851	121	6.09	207	34.5	3.00
W 14× 74	21.8	14.19	10.072	0.783	0.450	797	112	6.05	133	26.5	2.48
× 68	20.0	14.06	10.040	0.718	0.418	724	103	6.02	121	24.1	2.46
× 61	17.9	13.91	10.000	0.643	0.378	641	92.2	5.98	107	21.5	2.45
W 14× 53	15.6	13.94	8.062	0.658	0.370	542	77.8	5.90	57.5	14.3	1.92
× 48	14.1	13.81	8.031	0.593	0.339	485	70.2	5.86	51.3	12.8	1.91
× 43	12.6	13.68	8.000	0.528	0.308	429	62.7	5.82	45.1	11.3	1.89
W 14× 38	11.2	14.12	6.776	0.513	0.313	386	54.7	5.88	26.6	7.86	1.54
× 34	10.0	14.00	6.750	0.453	0.287	340	48.6	5.83	23.3	6.89	1.52
× 30	8.83	13.86	6.733	0.383	0.270	290	41.9	5.74	19.5	5.80	1.49
W 14× 26	7.67	13.89	5.025	0.418	0.255	244	35.1	5.64	8.86	3.53	1.08
× 22	6.49	13.72	5.000	0.335	0.230	198	28.9	5.53	7.00	2.80	1.04

W SHAPES
Properties for designing

Designation	Area A	Depth d	Flange Width b_f	Flange Thickness t_f	Web Thickness t_w	Axis X-X I	Axis X-X Z	Axis X-X k	Axis Y-Y I	Axis Y-Y Z	Axis Y-Y k
	In.^{2}	In.	In.	In.	In.	In.^{4}	In.^{3}	In.	In.^{4}	In.^{3}	In.
W 12×190	55.9	14.38	12.670	1.736	1.060	1890	263	5.82	590	93.1	3.25
×161	47.4	13.88	12.515	1.486	0.905	1540	222	5.70	486	77.7	3.20
×133	39.1	13.38	12.365	1.236	0.755	1220	183	5.59	390	63.1	3.16
×120	35.3	13.12	12.320	1.106	0.710	1070	163	5.51	345	56.0	3.13
×106	31.2	12.88	12.230	0.986	0.620	931	145	5.46	301	49.2	3.11
× 99	29.1	12.75	12.192	0.921	0.582	859	135	5.43	278	45.7	3.09
× 92	27.1	12.62	12.155	0.856	0.545	789	125	5.40	256	42.2	3.08
× 85	25.0	12.50	12.105	0.796	0.495	723	116	5.38	235	38.9	3.07
× 79	23.2	12.38	12.080	0.736	0.470	663	107	5.34	216	35.8	3.05
× 72	21.2	12.25	12.040	0.671	0.430	597	97.5	5.31	195	32.4	3.04
× 65	19.1	12.12	12.000	0.606	0.390	533	88.0	5.28	175	29.1	3.02
W 12× 58	17.1	12.19	10.014	0.641	0.359	476	78.1	5.28	107	21.4	2.51
× 53	15.6	12.06	10.000	0.576	0.345	426	70.7	5.23	96.1	19.2	2.48
W 12× 50	14.7	12.19	8.077	0.641	0.371	395	64.7	5.18	56.4	14.0	1.96
× 45	13.2	12.06	8.042	0.576	0.336	351	58.2	5.15	50.0	12.4	1.94
× 40	11.8	11.94	8.000	0.516	0.294	310	51.9	5.13	44.1	11.0	1.94
W 12× 36	10.6	12.24	6.565	0.540	0.305	281	46.0	5.15	25.5	7.77	1.55
× 31	9.13	12.09	6.525	0.465	0.265	239	39.5	5.12	21.6	6.61	1.54
× 27	7.95	11.96	6.497	0.400	0.237	204	34.2	5.07	18.3	5.63	1.52
W 12× 22	6.47	12.31	4.030	0.424	0.260	156	25.3	4.91	4.64	2.31	0.847
× 19	5.59	12.16	4.007	0.349	0.237	130	21.3	4.82	3.76	1.88	0.820
× 16.5	4.87	12.00	4.000	0.269	0.230	105	17.6	4.65	2.88	1.44	0.770
× 14	4.12	11.91	3.968	0.224	0.198	88.0	14.8	4.62	2.34	1.18	0.754

W SHAPES
Properties for designing

Designation	Area A	Depth d	Flange		Web Thickness t_w	Elastic Properties					
			Width b_f	Thickness t_f		Axis X-X			Axis Y-Y		
						I	Z	k	I	Z	k
	In.²	In.	In.	In.	In.	In.⁴	In.³	In.	In.⁴	In.³	In.
W 10×112	32.9	11.38	10.415	1.248	0.755	719	126	4.67	235	45.2	2.67
×100	29.4	11.12	10.345	1.118	0.685	625	112	4.61	207	39.9	2.65
× 89	26.2	10.88	10.275	0.998	0.615	542	99.7	4.55	181	35.2	2.63
× 77	22.7	10.62	10.195	0.868	0.535	457	86.1	4.49	153	30.1	2.60
× 72	21.2	10.50	10.170	0.808	0.510	421	80.1	4.46	142	27.9	2.59
× 66	19.4	10.38	10.117	0.748	0.457	382	73.7	4.44	129	25.5	2.58
× 60	17.7	10.25	10.075	0.683	0.415	344	67.1	4.41	116	23.1	2.57
× 54	15.9	10.12	10.028	0.618	0.368	306	60.4	4.39	104	20.7	2.56
× 49	14.4	10.00	10.000	0.558	0.340	273	54.6	4.35	93.0	18.6	2.54
W 10× 45	13.2	10.12	8.022	0.618	0.350	249	49.1	4.33	53.2	13.3	2.00
× 39	11.5	9.94	7.990	0.528	0.318	210	42.2	4.27	44.9	11.2	1.98
× 33	9.71	9.75	7.964	0.433	0.292	171	35.0	4.20	36.5	9.16	1.94
W 10× 29	8.54	10.22	5.799	0.500	0.289	158	30.8	4.30	16.3	5.61	1.38
× 25	7.36	10.08	5.762	0.430	0.252	133	26.5	4.26	13.7	4.76	1.37
× 21	6.20	9.90	5.750	0.340	0.240	107	21.5	4.15	10.8	3.75	1.32
W 10× 19	5.61	10.25	4.020	0.394	0.250	96.3	18.8	4.14	4.28	2.13	0.874
× 17	4.99	10.12	4.010	0.329	0.240	81.9	16.2	4.05	3.55	1.77	0.844
× 15	4.41	10.00	4.000	0.269	0.230	68.9	13.8	3.95	2.88	1.44	0.809
× 11.5	3.39	9.87	3.950	0.204	0.180	52.0	10.5	3.92	2.10	1.06	0.787

W SHAPES
Properties for designing

Designation	Area A	Depth d	Flange Width b_f	Flange Thickness t_f	Web Thickness t_w	Elastic Properties Axis X-X I	Axis X-X Z	Axis X-X k	Axis Y-Y I	Axis Y-Y Z	Axis Y-Y k
	In.²	In.	In.	In.	In.	In.⁴	In.³	In.	In.⁴	In.³	In.
W 8×67	19.7	9.00	8.287	0.933	0.575	272	60.4	3.71	88.6	21.4	2.12
×58	17.1	8.75	8.222	0.808	0.510	227	52.0	3.65	74.9	18.2	2.10
×48	14.1	8.50	8.117	0.683	0.405	184	43.2	3.61	60.9	15.0	2.08
×40	11.8	8.25	8.077	0.558	0.365	146	35.5	3.53	49.0	12.1	2.04
×35	10.3	8.12	8.027	0.493	0.315	126	31.1	3.50	42.5	10.6	2.03
×31	9.12	8.00	8.000	0.433	0.288	110	27.4	3.47	37.0	9.24	2.01
W 8×28	8.23	8.06	6.540	0.463	0.285	97.8	24.3	3.45	21.6	6.61	1.62
×24	7.06	7.93	6.500	0.398	0.245	82.5	20.8	3.42	18.2	5.61	1.61
W 8×20	5.89	8.14	5.268	0.378	0.248	69.4	17.0	3.43	9.22	3.50	1.25
×17	5.01	8.00	5.250	0.308	0.230	56.6	14.1	3.36	7.44	2.83	1.22
W 8×15	4.43	8.12	4.015	0.314	0.245	48.1	11.8	3.29	3.40	1.69	0.876
×13	3.83	8.00	4.000	0.254	0.230	39.6	9.90	3.21	2.72	1.36	0.842
×10	2.96	7.90	3.940	0.204	0.170	30.8	7.80	3.23	2.08	1.06	0.839
W 6×25	7.35	6.37	6.080	0.456	0.320	53.3	16.7	2.69	17.1	5.62	1.53
×20	5.88	6.20	6.018	0.367	0.258	41.5	13.4	2.66	13.3	4.43	1.51
×15.5	4.56	6.00	5.995	0.269	0.235	30.1	10.0	2.57	9.67	3.23	1.46
W 6×16	4.72	6.25	4.030	0.404	0.260	31.7	10.2	2.59	4.42	2.19	0.967
×12	3.54	6.00	4.000	0.279	0.230	21.7	7.25	2.48	2.98	1.49	0.918
× 8.5	2.51	5.83	3.940	0.194	0.170	14.8	5.08	2.43	1.98	1.01	0.889
W 5×18.5	5.43	5.12	5.025	0.420	0.265	25.4	9.94	2.16	8.89	3.54	1.28
×16	4.70	5.00	5.000	0.360	0.240	21.3	8.53	2.13	7.51	3.00	1.26
W 4×13	3.82	4.16	4.060	0.345	0.280	11.3	5.45	1.72	3.76	1.85	0.991

M SHAPES
Properties for designing

Designation	Area A	Depth d	Flange Width b_f	Flange Thickness t_f	Web Thickness t_w	Elastic Properties Axis X-X I	Axis X-X Z	Axis X-X k	Axis Y-Y I	Axis Y-Y Z	Axis Y-Y k
	In.²	In.	In.	In.	In.	In.⁴	In.³	In.	In.⁴	In.³	In.
M 14× 17.2	5.05	14.00	4.000	0.272	0.210	147	21.1	5.40	2.65	1.33	0.725
M 12× 11.8	3.47	12.00	3.065	0.225	0.177	71.9	12.0	4.55	0.980	0.639	0.532
M 10× 29.1	8.56	9.88	5.937	0.389	0.427	131	26.6	3.92	11.2	3.76	1.14
× 22.9	6.73	9.88	5.752	0.389	0.242	117	23.6	4.16	10.0	3.48	1.22
M 10× 9	2.65	10.00	2.690	0.206	0.157	38.8	7.76	3.83	0.609	0.453	0.480
M 8× 37.7	11.1	8.12	8.002	0.521	0.377	132	32.6	3.46	40.4	10.1	1.91
× 34.3	10.1	8.00	8.003	0.459	0.378	116	29.1	3.40	34.9	8.73	1.86
× 32.6	9.58	8.00	7.940	0.459	0.315	114	28.4	3.44	34.1	8.58	1.89
M 8× 22.5	6.60	8.00	5.395	0.353	0.375	68.2	17.1	3.22	7.48	2.77	1.06
× 18.5	5.44	8.00	5.250	0.353	0.230	62.0	15.5	3.38	6.82	2.60	1.12
M 8× 6.5	1.92	8.00	2.281	0.189	0.135	18.5	4.62	3.10	0.343	0.301	0.423
M 7× 5.5	1.62	7.00	2.080	0.180	0.128	12.0	3.44	2.73	0.249	0.239	0.392
M 6× 33.75	9.93	6.25	6.114	0.605	0.488	64.7	20.7	2.55	21.4	6.99	1.47
× 22.5	6.62	6.00	6.060	0.379	0.372	41.2	13.7	2.49	12.4	4.08	1.37
× 20	5.89	6.00	5.938	0.379	0.250	39.0	13.0	2.57	11.6	3.90	1.40
M 6× 4.4	1.29	6.00	1.844	0.171	0.114	7.20	2.40	2.36	0.165	0.179	0.358
M 5× 18.9	5.55	5.00	5.003	0.416	0.316	24.1	9.63	2.08	7.86	3.14	1.19
M 4× 16.3	4.80	4.20	3.938	0.472	0.312	14.0	6.67	1.71	4.44	2.25	0.962
× 13.8	4.06	4.00	4.000	0.371	0.313	10.8	5.42	1.63	3.58	1.79	0.939
× 13	3.81	4.00	3.940	0.371	0.254	10.5	5.24	1.66	3.36	1.71	0.939

I S SHAPES
Properties for designing

Designation	Area A	Depth d	Flange Width b_f	Flange Thickness t_f	Web Thickness t_w	Axis X-X I	Axis X-X Z	Axis X-X k	Axis Y-Y I	Axis Y-Y Z	Axis Y-Y k
	In.²	In.	In.	In.	In.	In.⁴	In.³	In.	In.⁴	In.³	In.
S 24×120	35.3	24.00	8.048	1.102	0.798	3030	252	9.26	84.2	20.9	1.54
×105.9	31.1	24.00	7.875	1.102	0.625	2830	236	9.53	78.2	19.8	1.58
S 24×100	29.4	24.00	7.247	0.871	0.747	2390	199	9.01	47.8	13.2	1.27
× 90	26.5	24.00	7.124	0.871	0.624	2250	187	9.22	44.9	12.6	1.30
× 79.9	23.5	24.00	7.001	0.871	0.501	2110	175	9.47	42.3	12.1	1.34
S 20× 95	27.9	20.00	7.200	0.916	0.800	1610	161	7.60	49.7	13.8	1.33
× 85	25.0	20.00	7.053	0.916	0.653	1520	152	7.79	46.2	13.1	1.36
S 20× 75	22.1	20.00	6.391	0.789	0.641	1280	128	7.60	29.6	9.28	1.16
× 65.4	19.2	20.00	6.250	0.789	0.500	1180	118	7.84	27.4	8.77	1.19
S 18× 70	20.6	18.00	6.251	0.691	0.711	926	103	6.71	24.1	7.72	1.08
× 54.7	16.1	18.00	6.001	0.691	0.461	804	89.4	7.07	20.8	6.94	1.14
S 15× 50	14.7	15.00	5.640	0.622	0.550	486	64.8	5.75	15.7	5.57	1.03
× 42.9	12.6	15.00	5.501	0.622	0.411	447	59.6	5.95	14.4	5.23	1.07
S 12× 50	14.7	12.00	5.477	0.659	0.687	305	50.8	4.55	15.7	5.74	1.03
× 40.8	12.0	12.00	5.252	0.659	0.472	272	45.4	4.77	13.6	5.16	1.06
S 12× 35	10.3	12.00	5.078	0.544	0.428	229	38.2	4.72	9.87	3.89	0.980
× 31.8	9.35	12.00	5.000	0.544	0.350	218	36.4	4.83	9.36	3.74	1.00
S 10× 35	10.3	10.00	4.944	0.491	0.594	147	29.4	3.78	8.36	3.38	0.901
× 25.4	7.46	10.00	4.661	0.491	0.311	124	24.7	4.07	6.79	2.91	0.954
S 8× 23	6.77	8.00	4.171	0.425	0.441	64.9	16.2	3.10	4.31	2.07	0.798
× 18.4	5.41	8.00	4.001	0.425	0.271	57.6	14.4	3.26	3.73	1.86	0.831
S 7× 20	5.88	7.00	3.860	0.392	0.450	42.4	12.1	2.69	3.17	1.64	0.734
× 15.3	4.50	7.00	3.662	0.392	0.252	36.7	10.5	2.86	2.64	1.44	0.766
S 6× 17.25	5.07	6.00	3.565	0.359	0.465	26.3	8.77	2.28	2.31	1.30	0.675
× 12.5	3.67	6.00	3.332	0.359	0.232	22.1	7.37	2.45	1.82	1.09	0.705
S 5× 14.75	4.34	5.00	3.284	0.326	0.494	15.2	6.09	1.87	1.67	1.01	0.620
× 10	2.94	5.00	3.004	0.326	0.214	12.3	4.92	2.05	1.22	0.809	0.643
S 4× 9.5	2.79	4.00	2.796	0.293	0.326	6.79	3.39	1.56	0.903	0.646	0.569
× 7.7	2.26	4.00	2.663	0.293	0.193	6.08	3.04	1.64	0.764	0.574	0.581
S 3× 7.5	2.21	3.00	2.509	0.260	0.349	2.93	1.95	1.15	0.586	0.468	0.516
× 5.7	1.67	3.00	2.330	0.260	0.170	2.52	1.68	1.23	0.455	0.390	0.522

CHANNELS
AMERICAN STANDARD
Properties for designing

Designation	Area A	Depth d	Flange		Web thickness t_w	$\dfrac{d}{A_f}$	Axis X-X		
			Width b_f	Average thickness t_f			I	Z	k
	In.²	In.	In.	In.	In.		In.⁴	In.³	In.
C 15×50	14.7	15.00	3.716	0.650	0.716	6.21	404	53.8	5.24
×40	11.8	15.00	3.520	0.650	0.520	6.56	349	46.5	5.44
×33.9	9.96	15.00	3.400	0.650	0.400	6.79	315	42.0	5.62
C 12×30	8.82	12.00	3.170	0.501	0.510	7.55	162	27.0	4.29
×25	7.35	12.00	3.047	0.501	0.387	7.85	144	24.1	4.43
×20.7	6.09	12.00	2.942	0.501	0.282	8.13	129	21.5	4.61
C 10×30	8.82	10.00	3.033	0.436	0.673	7.55	103	20.7	3.42
×25	7.35	10.00	2.886	0.436	0.526	7.94	91.2	18.2	3.52
×20	5.88	10.00	2.739	0.436	0.379	8.36	78.9	15.8	3.66
×15.3	4.49	10.00	2.600	0.436	0.240	8.81	67.4	13.5	3.87
C 9×20	5.88	9.00	2.648	0.413	0.448	8.22	60.9	13.5	3.22
×15	4.41	9.00	2.485	0.413	0.285	8.76	51.0	11.3	3.40
×13.4	3.94	9.00	2.433	0.413	0.233	8.95	47.9	10.6	3.48
C 8×18.75	5.51	8.00	2.527	0.390	0.487	8.12	44.0	11.0	2.82
×13.75	4.04	8.00	2.343	0.390	0.303	8.75	36.1	9.03	2.99
×11.5	3.38	8.00	2.260	0.390	0.220	9.08	32.6	8.14	3.11
C 7×14.75	4.33	7.00	2.299	0.366	0.419	8.31	27.2	7.78	2.51
×12.25	3.60	7.00	2.194	0.366	0.314	8.71	24.2	6.93	2.60
× 9.8	2.87	7.00	2.090	0.366	0.210	9.14	21.3	6.08	2.72
C 6×13	3.83	6.00	2.157	0.343	0.437	8.10	17.4	5.80	2.13
×10.5	3.09	6.00	2.034	0.343	0.314	8.59	15.2	5.06	2.22
× 8.2	2.40	6.00	1.920	0.343	0.200	9.10	13.1	4.38	2.34
C 5× 9	2.64	5.00	1.885	0.320	0.325	8.29	8.90	3.56	1.83
× 6.7	1.97	5.00	1.750	0.320	0.190	8.93	7.49	3.00	1.95
C 4× 7.25	2.13	4.00	1.721	0.296	0.321	7.84	4.59	2.29	1.47
× 5.4	1.59	4.00	1.584	0.296	0.184	8.52	3.85	1.93	1.56
C 3× 6	1.76	3.00	1.596	0.273	0.356	6.87	2.07	1.38	1.08
× 5	1.47	3.00	1.498	0.273	0.258	7.32	1.85	1.24	1.12
× 4.1	1.21	3.00	1.410	0.273	0.170	7.78	1.66	1.10	1.17

CHANNELS MISCELLANEOUS
Properties for designing

Designation	Area A	Depth d	Flange Width b_f	Flange Average thickness t_f	Web thickness t_w	$\dfrac{d'}{A_f}$	Axis X-X I	Axis X-X Z	Axis X-X k
	In.²	In.	In.	In.	In.		In.⁴	In.³	In.
MC 18×58	17.1	18.00	4.200	0.625	0.700	6.86	676	75.1	6.29
×51.9	15.3	18.00	4.100	0.625	0.600	7.02	627	69.7	6.41
×45.8	13.5	18.00	4.000	0.625	0.500	7.20	578	64.3	6.56
×42.7	12.6	18.00	3.950	0.625	0.450	7.29	554	61.6	6.64
MC 13×50	14.7	13.00	4.412	0.610	0.787	4.83	314	48.4	4.62
×40	11.8	13.00	4.185	0.610	0.560	5.09	273	42.0	4.82
×35	10.3	13.00	4.072	0.610	0.447	5.23	252	38.8	4.95
×31.8	9.35	13.00	4.000	0.610	0.375	5.33	239	36.8	5.06
MC 12×50	14.7	12.00	4.135	0.700	0.835	4.15	269	44.9	4.28
×45	13.2	12.00	4.012	0.700	0.712	4.27	252	42.0	4.36
×40	11.8	12.00	3.890	0.700	0.590	4.41	234	39.0	4.46
×35	10.3	12.00	3.767	0.700	0.467	4.55	216	36.1	4.59
MC 12×37	10.9	12.00	3.600	0.600	0.600	5.56	205	34.2	4.34
×32.9	9.67	12.00	3.500	0.600	0.500	5.71	191	31.8	4.44
×30.9	9.07	12.00	3.450	0.600	0.450	5.80	183	30.6	4.50
MC 12×10.6	3.10	12.00	1.500	0.309	0.190	25.9	55.4	9.23	4.22
MC 10×41.1	12.1	10.00	4.321	0.575	0.796	4.02	158	31.5	3.61
×33.6	9.87	10.00	4.100	0.575	0.575	4.24	139	27.8	3.75
×28.5	8.37	10.00	3.950	0.575	0.425	4.40	127	25.3	3.89
MC 10×28.3	8.32	10.00	3.502	0.575	0.477	4.97	118	23.6	3.77
×25.3	7.43	10.00	3.550	0.500	0.425	5.63	107	21.4	3.79
×24.9	7.32	10.00	3.402	0.575	0.377	5.11	110	22.0	3.87
×21.9	6.43	10.00	3.450	0.500	0.325	5.80	98.5	19.7	3.91
MC 10× 8.4	2.46	10.00	1.500	0.280	0.170	23.8	32.0	6.40	3.61
MC 10× 6.5	1.91	10.00	1.127	0.202	0.152	43.8	22.1	4.42	3.40

CHANNELS MISCELLANEOUS
Properties for designing

Designation	Area A	Depth d	Flange Width b_f	Flange Average thickness t_f	Web thickness t_w	$\dfrac{d}{A_f}$	Axis X-X I	Axis X-X Z	Axis X-X k
	In.²	In.	In.	In.	In.		In.⁴	In.³	In.
MC 9×25.4	7.47	9.00	3.500	0.550	0.450	4.68	88.0	19.6	3.43
×23.9	7.02	9.00	3.450	0.550	0.400	4.74	85.0	18.9	3.48
MC 8×22.8	6.70	8.00	3.502	0.525	0.427	4.35	63.8	16.0	3.09
×21.4	6.28	8.00	3.450	0.525	0.375	4.42	61.6	15.4	3.13
MC 8×20	5.88	8.00	3.025	0.500	0.400	5.29	54.5	13.6	3.05
×18.7	5.50	8.00	2.978	0.500	0.353	5.37	52.5	13.1	3.09
MC 8×8.5	2.50	8.00	1.874	0.311	0.179	13.7	23.3	5.83	3.05
MC 7×22.7	6.67	7.00	3.603	0.500	0.503	3.89	47.5	13.6	2.67
×19.1	5.61	7.00	3.452	0.500	0.352	4.06	43.2	12.3	2.77
MC 7×17.6	5.17	7.00	3.000	0.475	0.375	4.91	37.6	10.8	2.70
MC 6×18	5.29	6.00	3.504	0.475	0.379	3.60	29.7	9.91	2.37
×15.3	4.50	6.00	3.500	0.385	0.340	4.45	25.4	8.47	2.38
MC 6×16.3	4.79	6.00	3.000	0.475	0.375	4.21	26.0	8.68	2.33
×15.1	4.44	6.00	2.941	0.475	0.316	4.29	25.0	8.32	2.37
MC 6×12	3.53	6.00	2.497	0.375	0.310	6.41	18.7	6.24	2.30
MC 3×9	2.65	3.00	2.122	0.351	0.497	4.02	3.15	2.10	1.09
×7.1	2.09	3.00	1.938	0.351	0.312	4.40	2.73	1.82	1.14

ANGLES
Equal legs
Properties for designing

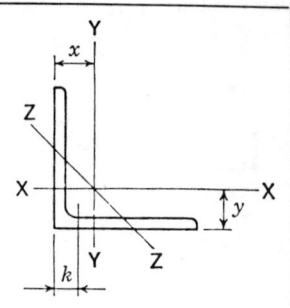

Size and Thickness	k	Weight per Foot	Area	AXIS X-X AND AXIS Y-Y				AXIS Z-Z
				I	Z	r	x or y	r
In.	In.	Lb.	In.²	In.⁴	In.³	In.	In.	In.
L 8 × 8 × 1⅛	1¾	56.9	16.7	98.0	17.5	2.42	2.41	1.56
1	1⅝	51.0	15.0	89.0	15.8	2.44	2.37	1.56
⅞	1½	45.0	13.2	79.6	14.0	2.45	2.32	1.57
¾	1⅜	38.9	11.4	69.7	12.2	2.47	2.28	1.58
⅝	1¼	32.7	9.61	59.4	10.3	2.49	2.23	1.58
9/16	1 3/16	29.6	8.63	54.1	9.34	2.50	2.21	1.59
½	1⅛	26.4	7.75	48.6	8.36	2.50	2.19	1.59
L 6 × 6 × 1	1½	37.4	11.0	35.5	8.57	1.80	1.86	1.17
⅞	1⅜	33.1	9.73	31.9	7.63	1.81	1.82	1.17
¾	1¼	28.7	8.44	28.2	6.66	1.83	1.78	1.17
⅝	1⅛	24.2	7.11	24.2	5.66	1.84	1.73	1.18
9/16	1 1/16	21.9	6.43	22.1	5.14	1.85	1.71	1.18
½	1	19.6	5.75	19.9	4.61	1.86	1.68	1.18
7/16	15/16	17.2	5.06	17.7	4.08	1.87	1.66	1.19
⅜	⅞	14.9	4.36	15.4	3.53	1.88	1.64	1.19
5/16	13/16	12.4	3.65	13.0	2.97	1.89	1.62	1.20
L 5 × 5 × ⅞	1⅜	27.2	7.98	17.8	5.17	1.49	1.57	.973
¾	1¼	23.6	6.94	15.7	4.53	1.51	1.52	.975
⅝	1⅛	20.0	5.86	13.6	3.86	1.52	1.48	.978
½	1	16.2	4.75	11.3	3.16	1.54	1.43	.983
7/16	15/16	14.3	4.18	10.0	2.79	1.55	1.41	.986
⅜	⅞	12.3	3.61	8.74	2.42	1.56	1.39	.990
5/16	13/16	10.3	3.03	7.42	2.04	1.57	1.37	.994
L 4 × 4 × ¾	1⅛	18.5	5.44	7.67	2.81	1.19	1.27	.778
⅝	1	15.7	4.61	6.66	2.40	1.20	1.23	.779
½	⅞	12.8	3.75	5.56	1.97	1.22	1.18	.782
7/16	13/16	11.3	3.31	4.97	1.75	1.23	1.16	.785
⅜	¾	9.8	2.86	4.36	1.52	1.23	1.14	.788
5/16	11/16	8.2	2.40	3.71	1.29	1.24	1.12	.791
¼	⅝	6.6	1.94	3.04	1.05	1.25	1.09	.795

Appendix B 451

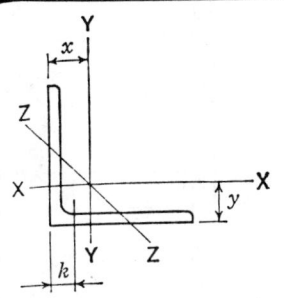

ANGLES
Equal legs
Properties for designing

Size and Thickness	k	Weight per Foot	Area	AXIS X-X AND AXIS Y-Y				AXIS Z-Z
				I	Z	k	x or y	k
In.	In.	Lb.	In.²	In.⁴	In.³	In.	In.	In.
L 3½ × 3½ × ½	⅞	11.1	3.25	3.64	1.49	1.06	1.06	.683
⁷⁄₁₆	¹³⁄₁₆	9.8	2.87	3.26	1.32	1.07	1.04	.684
⅜	¾	8.5	2.48	2.87	1.15	1.07	1.01	.687
⁵⁄₁₆	¹¹⁄₁₆	7.2	2.09	2.45	.976	1.08	.990	.690
¼	⅝	5.8	1.69	2.01	.794	1.09	.968	.694
L 3 × 3 × ½	¹³⁄₁₆	9.4	2.75	2.22	1.07	.898	.932	.584
⁷⁄₁₆	¾	8.3	2.43	1.99	.954	.905	.910	.585
⅜	¹¹⁄₁₆	7.2	2.11	1.76	.833	.913	.888	.587
⁵⁄₁₆	⅝	6.1	1.78	1.51	.707	.922	.869	.589
¼	⁹⁄₁₆	4.9	1.44	1.24	.577	.930	.842	.592
³⁄₁₆	½	3.71	1.09	.962	.441	.939	.820	.596
L 2½ × 2½ × ½	¹³⁄₁₆	7.7	2.25	1.23	.724	.739	.806	.487
⅜	¹¹⁄₁₆	5.9	1.73	.984	.566	.753	.762	.487
⁵⁄₁₆	⅝	5.0	1.46	.849	.482	.761	.740	.489
¼	⁹⁄₁₆	4.1	1.19	.703	.394	.769	.717	.491
³⁄₁₆	½	3.07	0.92	.547	.303	.778	.694	.495
L 2 × 2 × ⅜	⅞	4.7	1.36	.479	.351	.594	.636	.389
⁵⁄₁₆	¹³⁄₁₆	3.92	1.15	.416	.300	.601	.614	.390
¼	¾	3.19	.938	.348	.247	.609	.592	.391
³⁄₁₆	¹¹⁄₁₆	2.44	.715	.272	.190	.617	.569	.394
⅛	⅝	1.65	.484	.190	.131	.626	.546	.398
L 1¾ × 1¾ × ¼	¹¹⁄₁₆	2.77	.813	.227	.186	.529	.529	.341
³⁄₁₆	⅝	2.12	.621	.179	.144	.537	.506	.343
⅛	⁹⁄₁₆	1.44	.422	.126	.099	.546	.484	.347
L 1½ × 1½ × ¼	⅝	2.34	.688	.139	.134	.449	.466	.292
³⁄₁₆	⁹⁄₁₆	1.80	.527	.110	.104	.457	.444	.293
⁵⁄₃₂	⁹⁄₁₆	1.52	.444	.094	.088	.461	.433	.295
⅛	½	1.23	.359	.078	.072	.465	.421	.296
L 1¼ × 1¼ × ¼	⅝	1.92	.563	.077	.091	.369	.403	.243
³⁄₁₆	⁹⁄₁₆	1.48	.434	.061	.071	.377	.381	.244
⅛	½	1.01	.297	.044	.049	.385	.359	.246
L 1 × 1 × ¼	⅝	1.49	.438	.037	.056	.290	.339	.196
³⁄₁₆	⁹⁄₁₆	1.16	.340	.030	.044	.297	.318	.195
⅛	½	.80	.234	.022	.031	.304	.296	.196

ANGLES
Unequal legs
Properties for designing

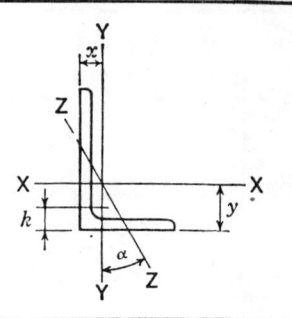

Size and Thickness	k	Weight per Foot	Area	AXIS X-X				AXIS Y-Y				AXIS Z-Z	
				I	Z	k	y	I	Z	k	x	k	Tan α
In.	In.	Lb.	In.²	In.⁴	In.³	In.	In.	In.⁴	In.³	In.	In.	In.	
L 9 × 4 × 1	1½	40.8	12.0	97.0	17.6	2.84	3.50	12.0	4.00	1.00	1.00	.834	.203
⅞	1⅜	36.1	10.6	86.8	15.7	2.86	3.45	10.8	3.56	1.01	.953	.836	.208
¾	1¼	31.3	9.19	76.1	13.6	2.88	3.41	9.63	3.11	1.02	.906	.841	.212
⅝	1⅛	26.3	7.73	64.9	11.5	2.90	3.36	8.32	2.65	1.04	.858	.847	.216
9/16	1 1/16	23.8	7.00	59.1	10.4	2.91	3.33	7.63	2.41	1.04	.834	.850	.218
½	1	21.3	6.25	53.2	9.34	2.92	3.31	6.92	2.17	1.05	.810	.854	.220
L 8 × 6 × 1	1½	44.2	13.0	80.8	15.1	2.49	2.65	38.8	8.92	1.73	1.65	1.28	.543
⅞	1⅜	39.1	11.5	72.3	13.4	2.51	2.61	34.9	7.94	1.74	1.61	1.28	.547
¾	1¼	33.8	9.94	63.4	11.7	2.53	2.56	30.7	6.92	1.76	1.56	1.29	.551
⅝	1⅛	28.5	8.36	54.1	9.87	2.54	2.52	26.3	5.88	1.77	1.52	1.29	.554
9/16	1 1/16	25.7	7.56	49.3	8.95	2.55	2.50	24.0	5.34	1.78	1.50	1.30	.556
½	1	23.0	6.75	44.3	8.02	2.56	2.47	21.7	4.79	1.79	1.47	1.30	.558
7/16	15/16	20.2	5.93	39.2	7.07	2.57	2.45	19.3	4.23	1.80	1.45	1.31	.560
L 8 × 4 × 1	1½	37.4	11.0	69.6	14.1	2.52	3.05	11.6	3.94	1.03	1.05	.846	.247
⅞	1⅜	33.1	9.73	62.5	12.5	2.53	3.00	10.5	3.51	1.04	.999	.848	.253
¾	1¼	28.7	8.44	54.9	10.9	2.55	2.95	9.36	3.07	1.05	.953	.852	.258
⅝	1⅛	24.2	7.11	46.9	9.21	2.57	2.91	8.10	2.62	1.07	.906	.857	.262
9/16	1 1/16	21.9	6.43	42.8	8.35	2.58	2.88	7.43	2.38	1.07	.882	.861	.265
½	1	19.6	5.75	38.5	7.49	2.59	2.86	6.74	2.15	1.08	.859	.865	.267
7/16	15/16	17.2	5.06	34.1	6.60	2.60	2.83	6.02	1.90	1.09	.835	.869	.269
L 7 × 4 × ⅞	1⅜	30.2	8.86	42.9	9.65	2.20	2.55	10.2	3.46	1.07	1.05	.856	.318
¾	1¼	26.2	7.69	37.8	8.42	2.22	2.51	9.05	3.03	1.09	1.01	.860	.324
⅝	1⅛	22.1	6.48	32.4	7.14	2.24	2.46	7.84	2.58	1.10	.963	.865	.329
9/16	1 1/16	20.0	5.87	29.6	6.48	2.24	2.44	7.19	2.35	1.11	.940	.868	.332
½	1	17.9	5.25	26.7	5.81	2.25	2.42	6.53	2.12	1.11	.917	.872	.335
7/16	15/16	15.8	4.62	23.7	5.13	2.26	2.39	5.83	1.88	1.12	.893	.876	.337
⅜	⅞	13.6	3.98	20.6	4.44	2.27	2.37	5.10	1.63	1.13	.870	.880	.340

Appendix B 453

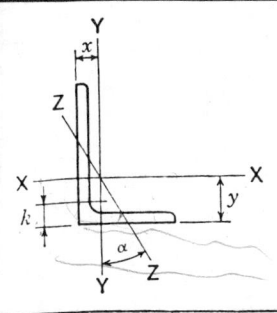

ANGLES
Unequal legs
Properties for designing

Size and Thickness	k	Weight per Foot	Area	AXIS X-X				AXIS Y-Y				AXIS Z-Z	
				I	Z	k	y	I	Z	k	x	k	Tan α
In.	In.	Lb.	In.²	In.⁴	In.³	In.	In.	In.⁴	In.³	In.	In.	In.	
L 6 × 4 × 7/8	1 3/8	27.2	7.98	27.7	7.15	1.86	2.12	9.75	3.39	1.11	1.12	.857	.421
3/4	1 1/4	23.6	6.94	24.5	6.25	1.88	2.08	8.68	2.97	1.12	1.08	.860	.428
5/8	1 1/8	20.0	5.86	21.1	5.31	1.90	2.03	7.52	2.54	1.13	1.03	.864	.435
9/16	1 1/16	18.1	5.31	19.3	4.83	1.90	2.01	6.91	2.31	1.14	1.01	.866	.438
1/2	1	16.2	4.75	17.4	4.33	1.91	1.99	6.27	2.08	1.15	.987	.870	.440
7/16	15/16	14.3	4.18	15.5	3.83	1.92	1.96	5.60	1.85	1.16	.964	.873	.443
3/8	7/8	12.3	3.61	13.5	3.32	1.93	1.94	4.90	1.60	1.17	.941	.877	.446
5/16	13/16	10.3	3.03	11.4	2.79	1.94	1.92	4.18	1.35	1.17	.918	.882	.448
1/4	3/4	8.3	2.44	9.27	2.26	1.95	1.89	3.41	1.10	1.18	.894	.887	.451
L 6 × 3 1/2 × 1/2	1	15.3	4.50	16.6	4.24	1.92	2.08	4.25	1.59	.972	.833	.759	.344
3/8	7/8	11.7	3.42	12.9	3.24	1.94	2.04	3.34	1.23	.988	.787	.767	.350
5/16	13/16	9.8	2.87	10.9	2.73	1.95	2.01	2.85	1.04	.996	.763	.772	.352
1/4	3/4	7.9	2.31	8.86	2.21	1.96	1.99	2.34	0.847	1.01	.740	.777	.355
L 5 × 3 1/2 × 3/4	1 1/4	19.8	5.81	13.9	4.28	1.55	1.75	5.55	2.22	.977	.996	.748	.464
5/8	1 1/8	16.8	4.92	12.0	3.65	1.56	1.70	4.83	1.90	.991	.951	.751	.472
1/2	1	13.6	4.00	9.99	2.99	1.58	1.66	4.05	1.56	1.01	.906	.755	.479
7/16	15/16	12.0	3.53	8.90	2.64	1.59	1.63	3.63	1.39	1.01	.883	.758	.482
3/8	7/8	10.4	3.05	7.78	2.29	1.60	1.61	3.18	1.21	1.02	.861	.762	.486
5/16	13/16	8.7	2.56	6.60	1.94	1.61	1.59	2.72	1.02	1.03	.838	.766	.489
1/4	3/4	7.0	2.06	5.39	1.57	1.62	1.56	2.23	.830	1.04	.814	.770	.492
L 5 × 3 × 1/2	1	12.8	3.75	9.45	2.91	1.59	1.75	2.58	1.15	.829	.750	.648	.357
7/16	15/16	11.3	3.31	8.43	2.58	1.60	1.73	2.32	1.02	.837	.727	.651	.361
3/8	7/8	9.8	2.86	7.37	2.24	1.61	1.70	2.04	.888	.845	.704	.654	.364
5/16	13/16	8.2	2.40	6.26	1.89	1.61	1.68	1.75	.753	.853	.681	.658	.368
1/4	3/4	6.6	1.94	5.11	1.53	1.62	1.66	1.44	.614	.861	.657	.663	.371

ANGLES
Unequal legs
Properties for designing

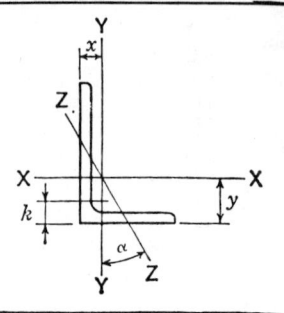

Size and Thickness	k	Weight per Foot	Area	AXIS X-X				AXIS Y-Y				AXIS Z-Z	
				I	Z	k	y	I	Z	k	x	k	Tan α
In.	In.	Lb.	In.²	In.⁴	In.³	In.	In.	In.⁴	In.³	In.	In.	In.	
L 4 × 3½ × ⅝	1 1/16	14.7	4.30	6.37	2.35	1.22	1.29	4.52	1.84	1.03	1.04	.719	.745
½	15/16	11.9	3.50	5.32	1.94	1.23	1.25	3.79	1.52	1.04	1.00	.722	.750
7/16	⅞	10.6	3.09	4.76	1.72	1.24	1.23	3.40	1.35	1.05	.978	.724	.753
⅜	13/16	9.1	2.67	4.18	1.49	1.25	1.21	2.95	1.17	1.06	.955	.727	.755
5/16	¾	7.7	2.25	3.56	1.26	1.26	1.18	2.55	.994	1.07	.932	.730	.757
¼	11/16	6.2	1.81	2.91	1.03	1.27	1.16	2.09	.808	1.07	.909	.734	.759
L 4 × 3 × ⅝	1 1/16	13.6	3.98	6.03	2.30	1.23	1.37	2.87	1.35	.849	.871	.637	.534
½	15/16	11.1	3.25	5.05	1.89	1.25	1.33	2.42	1.12	.864	.827	.639	.543
7/16	⅞	9.8	2.87	4.52	1.68	1.25	1.30	2.18	.992	.871	.804	.641	.547
⅜	13/16	8.5	2.48	3.96	1.46	1.26	1.28	1.92	.866	.879	.782	.644	.551
5/16	¾	7.2	2.09	3.38	1.23	1.27	1.26	1.65	.734	.887	.759	.647	.554
¼	11/16	5.8	1.69	2.77	1.00	1.28	1.24	1.36	.599	.896	.736	.651	.558
L 3½ × 3 × ½	15/16	10.2	3.00	3.45	1.45	1.07	1.13	2.33	1.10	.881	.875	.621	.714
7/16	⅞	9.1	2.65	3.10	1.29	1.08	1.10	2.09	.975	.889	.853	.622	.718
⅜	13/16	7.9	2.30	2.72	1.13	1.09	1.08	1.85	.851	.897	.830	.625	.721
5/16	¾	6.6	1.93	2.33	.954	1.10	1.06	1.58	.722	.905	.808	.627	.724
¼	11/16	5.4	1.56	1.91	.776	1.11	1.04	1.30	.589	.914	.785	.631	.727
L 3½ × 2½ × ½	15/16	9.4	2.75	3.24	1.41	1.09	1.20	1.36	.760	.704	.705	.534	.486
7/16	⅞	8.3	2.43	2.91	1.26	1.09	1.18	1.23	.677	.711	.682	.535	.491
⅜	13/16	7.2	2.11	2.56	1.09	1.10	1.16	1.09	.592	.719	.660	.537	.496
5/16	¾	6.1	1.78	2.19	.927	1.11	1.14	.939	.504	.727	.637	.540	.501
¼	11/16	4.9	1.44	1.80	.755	1.12	1.11	.777	.412	.735	.614	.544	.506
L 3 × 2½ × ½	⅞	8.5	2.50	2.08	1.04	.913	1.00	1.30	.744	.722	.750	.520	.667
7/16	13/16	7.6	2.21	1.88	.928	.920	.978	1.18	.664	.729	.728	.521	.672
⅜	¾	6.6	1.92	1.66	.810	.928	.956	1.04	.581	.736	.706	.522	.676
5/16	11/16	5.6	1.62	1.42	.688	.937	.933	.898	.494	.744	.683	.525	.680
¼	⅝	4.5	1.31	1.17	.561	.945	.911	.743	.404	.753	.661	.528	.684
3/16	9/16	3.39	.996	.907	.430	.954	.888	.577	.310	.761	.638	.533	.688

Appendix B 455

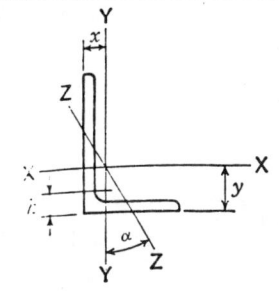

ANGLES
Unequal legs
Properties for designing

L

Size and Thickness	k	Weight per Foot	Area	AXIS X-X				AXIS Y-Y				AXIS Z-Z	
				I	Z	r	y	I	Z	r	x	r	Tan α
In.	In.	Lb.	In.²	In.⁴	In.³	In.	In.	In.⁴	In.³	In.	In.	In.	
L 3 × 2 × ½	13/16	7.7	2.25	1.92	1.00	.924	1.08	.672	.474	.546	.583	.428	.414
7/16	¾	6.8	2.00	1.73	.894	.932	1.06	.609	.424	.553	.561	.429	.421
⅜	11/16	5.9	1.73	1.53	.781	.940	1.04	.543	.371	.559	.539	.430	.428
5/16	⅝	5.0	1.46	1.32	.664	.948	1.02	.470	.317	.567	.516	.432	.435
¼	9/16	4.1	1.19	1.09	.542	.957	.993	.392	.260	.574	.493	.435	.440
3/16	½	3.07	.902	.842	.415	.966	.970	.307	.200	.583	.470	.439	.446
L 2½ × 2 × ⅜	11/16	5.3	1.55	.912	.547	.768	.831	.514	.363	.577	.581	.420	.614
5/16	⅝	4.5	1.31	.788	.466	.776	.809	.446	.310	.584	.559	.422	.620
¼	9/16	3.62	1.06	.654	.381	.784	.787	.372	.254	.592	.537	.424	.626
3/16	½	2.75	.809	.509	.293	.793	.764	.291	.196	.600	.514	.427	.631
L 2½ × 1½ × 5/16	⅝	3.92	1.15	.711	.444	.785	.898	.191	.174	.408	.398	.322	.349
¼	9/16	3.19	.938	.591	.364	.794	.875	.161	.143	.415	.375	.324	.357
3/16	½	2.44	.715	.461	.279	.803	.852	.127	.111	.422	.352	.327	.364
L 2 × 1½ × ¼	½	2.77	.813	.316	.236	.623	.663	.151	.139	.432	.413	.320	.543
3/16	7/16	2.12	.621	.248	.182	.632	.641	.120	.108	.440	.391	.322	.551
⅛	⅜	1.44	.422	.173	.125	.641	.618	.085	.075	.448	.368	.326	.558
L 2 × 1¼ × ¼	½	2.55	.750	.296	.229	.628	.708	.089	.097	.344	.333	.269	.378
3/16	7/16	1.96	.574	.232	.177	.636	.686	.071	.075	.351	.311	.271	.387
⅛	⅜	1.33	.391	.163	.122	.645	.663	.050	.052	.359	.287	.274	.396
L 1¾ × 1¼ × ¼	7/16	2.34	.688	.202	.176	.543	.602	.085	.095	.352	.352	.267	.486
3/16	⅜	1.80	.527	.160	.137	.551	.580	.068	.074	.359	.330	.269	.496
⅛	5/16	1.23	.359	.113	.094	.560	.557	.049	.051	.368	.307	.272	.506

Appendix C

Column tables

*Reprinted with the permission of the American Institute of Steel Construction, New York, N.Y., *Manual of Steel Construction*, Seventh edition, 1970.

TABLE C.1 ALLOWABLE STRESS 1000 psi FOR COMPRESSION MEMBERS OF 36,000 SPECIFIED YIELD STRESS STEEL

$S_{yp} = 36,000$ psi

Main and Secondary Members KL/r not over 120								Main Members KL/r 121 to 200				Secondary Members* L/r 121 to 200			
KL/r	F/A 1000 psi	KL/r	F/A 1000 psi	KL/r	F/A 1000 psi	KL/r	F/A 1000 psi	KL/r	F/A 1000 psi	KL/r	F/A 1000 psi	L/r	F/A 1000 psi	L/r	F/A 1000 psi
1	21.56	41	19.11	81	15.24	121	10.14	161	5.76	121	10.19	161	7.25
2	21.52	42	19.03	82	15.13	122	9.99	162	5.69	122	10.09	162	7.20
3	21.48	43	18.95	83	15.02	123	9.85	163	5.62	123	10.00	163	7.16
4	21.44	44	18.86	84	14.90	124	9.70	164	5.55	124	9.90	164	7.12
5	21.39	45	18.78	85	14.79	125	9.55	165	5.49	125	9.80	165	7.08
6	21.35	46	18.70	86	14.67	126	9.41	166	5.42	126	9.70	166	7.04
7	21.30	47	18.61	87	14.56	127	9.26	167	5.35	127	9.59	167	7.00
8	21.25	48	18.53	88	14.44	128	9.11	168	5.29	128	9.49	168	6.96
9	21.21	49	18.44	89	14.32	129	8.97	169	5.23	129	9.40	169	6.93
10	21.16	50	18.35	90	14.20	130	8.84	170	5.17	130	9.30	170	6.89
11	21.10	51	18.26	91	14.09	131	8.70	171	5.11	131	9.21	171	6.85
12	21.05	52	18.17	92	13.97	132	8.57	172	5.05	132	9.12	172	6.82
13	21.00	53	18.08	93	13.84	133	8.44	173	4.99	133	9.03	173	6.79
14	20.95	54	17.99	94	13.72	134	8.32	174	4.93	134	8.94	174	6.76
15	20.89	55	17.90	95	13.60	135	8.19	175	4.88	135	8.86	175	6.73
16	20.83	56	17.81	96	13.48	136	8.07	176	4.82	136	8.78	176	6.70
17	20.78	57	17.71	97	13.35	137	7.96	177	4.77	137	8.70	177	6.67
18	20.72	58	17.62	98	13.23	138	7.84	178	4.71	138	8.62	178	6.64
19	20.66	59	17.53	99	13.10	139	7.73	179	4.66	139	8.54	179	6.61
20	20.60	60	17.43	100	12.98	140	7.62	180	4.61	140	8.47	180	6.58
21	20.54	61	17.33	101	12.85	141	7.51	181	4.56	141	8.39	181	6.56
22	20.48	62	17.24	102	12.72	142	7.41	182	4.51	142	8.32	182	6.53
23	20.41	63	17.14	103	12.59	143	7.30	183	4.46	143	8.25	183	6.51
24	20.35	64	17.04	104	12.47	144	7.20	184	4.41	144	8.18	184	6.49
25	20.28	65	16.94	105	12.33	145	7.10	185	4.36	145	8.12	185	6.46
26	20.22	66	16.84	106	12.20	146	7.01	186	4.32	146	8.05	186	6.44
27	20.15	67	16.74	107	12.07	147	6.91	187	4.27	147	7.99	187	6.42
28	20.08	68	16.64	108	11.94	148	6.82	188	4.23	148	7.93	188	6.40
29	20.01	69	16.53	109	11.81	149	6.73	189	4.18	149	7.87	189	6.38
30	19.94	70	16.43	110	11.67	150	6.64	190	4.14	150	7.81	190	6.36
31	19.87	71	16.33	111	11.54	151	6.55	191	4.09	151	7.75	191	6.35
32	19.80	72	16.22	112	11.40	152	6.46	192	4.05	152	7.69	192	6.33
33	19.73	73	16.12	113	11.26	153	6.38	193	4.01	153	7.64	193	6.31
34	19.65	74	16.01	114	11.13	154	6.30	194	3.97	154	7.59	194	6.30
35	19.58	75	15.90	115	10.99	155	6.22	195	3.93	155	7.53	195	6.28
36	19.50	76	15.79	116	10.85	156	6.14	196	3.89	156	7.48	196	6.27
37	19.42	77	15.69	117	10.71	157	6.06	197	3.85	157	7.43	197	6.26
38	19.35	78	15.58	118	10.57	158	5.98	198	3.81	158	7.39	198	6.24
39	19.27	79	15.47	119	10.43	159	5.91	199	3.77	159	7.34	199	6.23
40	19.19	80	15.36	120	10.28	160	5.83	200	3.73	160	7.29	200	6.22

* K taken as 1.0 for secondary members.   Note: $C_c = 126.1$

$S_{yp} = 36,000$ psi

TABLE C.2 ALLOWABLE STRESS 1000 psi FOR COMPRESSION MEMBERS OF 42,000 SPECIFIED YIELD STRESS STEEL

$S_{yp} = 42,000$ psi

	Main and Secondary Members KL/r not over 120						Main Members KL/r 121 to 200				Secondary Members* L/r 121 to 200			
$\frac{KL}{r}$	F/A 1000 psi	$\frac{KL}{r}$	F/A 1000 psi	$\frac{KL}{r}$	F/A 1000 psi	$\frac{KL}{r}$	F/A 1000 psi	$\frac{KL}{r}$	F/A 1000 psi	$\frac{L}{r}$	F/A 1000 psi	$\frac{L}{r}$	F/A 1000 psi
1	25.15	41	21.98	81	16.92	121	10.20	161	5.76	121	10.25	161	7.25
2	25.10	42	21.87	82	16.77	122	10.03	162	5.69	122	10.13	162	7.20
3	25.05	43	21.77	83	16.62	123	9.87	163	5.62	123	10.02	163	7.16
4	24.99	44	21.66	84	16.47	124	9.71	164	5.55	124	9.91	164	7.12
5	24.94	45	21.55	85	16.32	125	9.56	165	5.49	125	9.80	165	7.08
6	24.88	46	21.44	86	16.17	126	9.41	166	5.42	126	9.70	166	7.04
7	24.82	47	21.33	87	16.01	127	9.26	167	5.35	127	9.59	167	7.00
8	24.76	48	21.22	88	15.86	128	9.11	168	5.29	128	9.49	168	6.96
9	24.70	49	21.10	89	15.71	129	8.97	169	5.23	129	9.40	169	6.93
10	24.63	50	20.99	90	15.55	130	8.84	170	5.17	130	9.30	170	6.89
11	24.57	51	20.87	91	15.39	131	8.70	171	5.11	131	9.21	171	6.85
12	24.50	52	20.76	92	15.23	132	8.57	172	5.05	132	9.12	172	6.82
13	24.43	53	20.64	93	15.07	133	8.44	173	4.99	133	9.03	173	6.79
14	24.36	54	20.52	94	14.91	134	8.32	174	4.93	134	8.94	174	6.76
15	24.29	55	20.40	95	14.75	135	8.19	175	4.88	135	8.86	175	6.73
16	24.2	56	20.28	96	14.59	136	8.07	176	4.82	136	8.78	176	6.70
17	24.15	57	20.16	97	14.43	137	7.96	177	4.77	137	8.70	177	6.67
18	24.07	58	20.03	98	14.26	138	7.84	178	4.71	138	8.62	178	6.64
19	24.00	59	19.91	99	14.09	139	7.73	179	4.66	139	8.54	179	6.61
20	23.92	60	19.79	100	13.93	140	7.62	180	4.61	140	8.47	180	6.58
21	23.84	61	19.66	101	13.76	141	7.51	181	4.56	141	8.39	181	6.56
22	23.76	62	19.53	102	13.59	142	7.41	182	4.51	142	8.32	182	6.53
23	23.68	63	19.40	103	13.42	143	7.30	183	4.46	143	8.25	183	6.51
24	23.59	64	19.27	104	13.25	144	7.20	184	4.41	144	8.18	184	6.49
25	23.51	65	19.14	105	13.08	145	7.10	185	4.36	145	8.12	185	6.46
26	23.42	66	19.01	106	12.90	146	7.01	186	4.32	146	8.05	186	6.44
27	23.33	67	18.88	107	12.73	147	6.91	187	4.27	147	7.99	187	6.42
28	23.24	68	18.75	108	12.55	148	6.82	188	4.23	148	7.93	188	6.40
29	23.15	69	18.61	109	12.37	149	6.73	189	4.18	149	7.87	189	6.38
30	23.06	70	18.48	110	12.19	150	6.64	190	4.14	150	7.81	190	6.36
31	22.97	71	18.34	111	12.01	151	6.55	191	4.09	151	7.75	191	6.35
32	22.88	72	18.20	112	11.83	152	6.46	192	4.05	152	7.69	192	6.33
33	22.78	73	18.06	113	11.65	153	6.38	193	4.01	153	7.64	193	6.31
34	22.69	74	17.92	114	11.47	154	6.30	194	3.97	154	7.59	194	6.30
35	22.59	75	17.78	115	11.28	155	6.22	195	3.93	155	7.53	195	6.28
36	22.49	76	17.64	116	11.10	156	6.14	196	3.89	156	7.48	196	6.27
37	22.39	77	17.50	117	10.91	157	6.06	197	3.85	157	7.43	197	6.26
38	22.29	78	17.35	118	10.72	158	5.98	198	3.81	158	7.39	198	6.24
39	22.19	79	17.21	119	10.55	159	5.91	199	3.77	159	7.34	199	6.23
40	22.08	80	17.06	120	10.37	160	5.83	200	3.73	160	7.29	200	6.22

* K taken as 1.0 for secondary members.   Note: $\bar{C}_c = 116.7$

TABLE C.3 ALLOWABLE STRESS 1000 psi FOR COMPRESSION MEMBERS OF 45,000 SPECIFIED YIELD STRESS STEEL

$S_{yp} = 45,000$ psi

\multicolumn{4}{c	}{Main and Secondary Members KL/r not over 120}	\multicolumn{4}{c	}{Main Members KL/r 121 to 200}	\multicolumn{4}{c}{Secondary Members* L/r 121 to 200}
$\dfrac{KL}{r}$	F/A 1000 psi	$\dfrac{KL}{r}$	F/A 1000 psi	$\dfrac{KL}{r}$	F/A 1000 psi	$\dfrac{KL}{r}$	F/A 1000 psi	$\dfrac{KL}{r}$	F/A 1000 psi	$\dfrac{KL}{r}$	F/A 1000 psi	$\dfrac{L}{r}$	F/A 1000 psi	$\dfrac{L}{r}$	F/A 1000 psi
1	26.95	41	23.39	81	17.67	121	10.20	161	5.76	121	10.25	161	7.25
2	26.89	42	23.27	82	17.51	122	10.03	162	5.69	122	10.13	162	7.20
3	26.83	43	23.15	83	17.34	123	9.87	163	5.62	123	10.02	163	7.16
4	26.77	44	23.03	84	17.17	124	9.71	164	5.55	124	9.91	164	7.12
5	26.71	45	22.90	85	17.00	125	9.56	165	5.49	125	9.80	165	7.08
6	26.64	46	22.78	86	16.82	126	9.41	166	5.42	126	9.70	166	7.04
7	26.58	47	22.65	87	16.65	127	9.26	167	5.35	127	9.59	167	7.00
8	26.51	48	22.53	88	16.48	128	9.11	168	5.29	128	9.49	168	6.96
9	26.44	49	22.40	89	16.30	129	8.97	169	5.23	129	9.40	169	6.93
10	26.37	50	22.27	90	16.12	130	8.84	170	5.17	130	9.30	170	6.89
11	26.30	51	22.14	91	15.95	131	8.70	171	5.11	131	9.21	171	6.85
12	26.22	52	22.01	92	15.77	132	8.57	172	5.05	132	9.12	172	6.82
13	26.15	53	21.88	93	15.59	133	8.44	173	4.99	133	9.03	173	6.79
14	26.07	54	21.74	94	15.40	134	8.32	174	4.93	134	8.94	174	6.76
15	25.99	55	21.61	95	15.22	135	8.19	175	4.88	135	8.86	175	6.73
16	25.91	56	21.47	96	15.04	136	8.07	176	4.82	136	8.78	176	6.70
17	25.82	57	21.33	97	14.85	137	7.96	177	4.77	137	8.70	177	6.67
18	25.74	58	21.19	98	14.66	138	7.84	178	4.71	138	8.62	178	6.64
19	25.65	59	21.05	99	14.47	139	7.73	179	4.66	139	8.54	179	6.61
20	25.57	60	20.91	100	14.28	140	7.62	180	4.61	140	8.47	180	6.58
21	25.48	61	20.77	101	14.09	141	7.51	181	4.56	141	8.39	181	6.56
22	25.39	62	20.63	102	13.90	142	7.41	182	4.51	142	8.32	182	6.53
23	25.29	63	20.48	103	13.71	143	7.30	183	4.46	143	8.25	183	6.51
24	25.20	64	20.34	104	13.51	144	7.20	184	4.41	144	8.18	184	6.49
25	25.11	65	20.19	105	13.32	145	7.10	185	4.36	145	8.12	185	6.46
26	25.01	66	20.04	106	13.12	146	7.01	186	4.32	146	8.05	186	6.44
27	24.91	67	19.89	107	12.92	147	6.91	187	4.27	147	7.99	187	6.42
28	24.81	68	19.74	108	12.72	148	6.82	188	4.23	148	7.93	188	6.40
29	24.71	69	19.59	109	12.52	149	6.73	189	4.18	149	7.87	189	6.38
30	24.61	70	19.43	110	12.31	150	6.64	190	4.14	150	7.81	190	6.36
31	24.50	71	19.28	111	12.11	151	6.55	191	4.09	151	7.75	191	6.35
32	24.40	72	19.12	112	11.90	152	6.46	192	4.05	152	7.69	192	6.33
33	24.29	73	18.97	113	11.69	153	6.38	193	4.01	153	7.64	193	6.31
34	24.18	74	18.81	114	11.49	154	6.30	194	3.97	154	7.59	194	6.30
35	24.07	75	18.65	115	11.29	155	6.22	195	3.93	155	7.53	195	6.28
36	23.96	76	18.49	116	11.10	156	6.14	196	3.89	156	7.48	196	6.27
37	23.85	77	18.33	117	10.91	157	6.06	197	3.85	157	7.43	197	6.26
38	23.74	78	18.17	118	10.72	158	5.98	198	3.81	158	7.39	198	6.24
39	23.62	79	18.00	119	10.55	159	5.91	199	3.77	159	7.34	199	6.23
40	23.51	80	17.84	120	10.37	160	5.83	200	3.73	160	7.29	200	6.22

$S_{yp} = 45,000$ psi

* K taken as 1.0 for secondary members.   Note: $C_c = 112$ ⁸

TABLE C.4 ALLOWABLE STRESS 1000 psi FOR COMPRESSION MEMBERS OF 50,000 SPECIFIED YIELD STRESS STEEL

$S_{yp} = 50,000$ psi

	Main and Secondary Members KL/r not over 120						Main Members KL/r 121 to 200				Secondary Members* L/r 121 to 200			
	$\frac{KL}{r}$	F/A 1000 psi	$\frac{KL}{r}$	F/A 1000 psi	$\frac{KL}{r}$	F/A 1000 psi	$\frac{KL}{r}$	F/A 1000 psi	$\frac{KL}{r}$	F/A 1000 psi	$\frac{L}{r}$	F/A 1000 psi	$\frac{L}{r}$	F/A 1000 psi
$S_{yp} = 50,000$ psi	1	29.94	41	25.69	81	18.81	121	10.20	161	5.76	121	10.25	161	7.25
	2	29.87	42	25.55	82	18.61	122	10.03	162	5.69	122	10.13	162	7.20
	3	29.80	43	25.40	83	18.41	123	9.87	163	5.62	123	10.02	163	7.16
	4	29.73	44	25.26	84	18.20	124	9.71	164	5.55	124	9.91	164	7.12
	5	29.66	45	25.11	85	17.99	125	9.56	165	5.49	125	9.80	165	7.08
	6	29.58	46	24.96	86	17.79	126	9.41	166	5.42	126	9.70	166	7.04
	7	29.50	47	24.81	87	17.58	127	9.26	167	5.35	127	9.59	167	7.00
	8	29.42	48	24.66	88	17.37	128	9.11	168	5.29	128	9.49	168	6.96
	9	29.34	49	24.51	89	17.15	129	8.97	169	5.23	129	9.40	169	6.93
	10	29.26	50	24.35	90	16.94	130	8.84	170	5.17	130	9.30	170	6.89
	11	29.17	51	24.19	91	16.72	131	8.70	171	5.11	131	9.21	171	6.85
	12	29.08	52	24.04	92	16.50	132	8.57	172	5.05	132	9.12	172	6.82
	13	28.99	53	23.88	93	16.29	133	8.44	173	4.99	133	9.03	173	6.79
	14	28.90	54	23.72	94	16.06	134	8.32	174	4.93	134	8.94	174	6.76
	15	28.80	55	23.55	95	15.84	135	8.19	175	4.88	135	8.86	175	6.73
	16	28.71	56	23.39	96	15.62	136	8.07	176	4.82	136	8.78	176	6.70
	17	28.61	57	23.22	97	15.39	137	7.96	177	4.77	137	8.70	177	6.67
	18	28.51	58	23.06	98	15.17	138	7.84	178	4.71	138	8.62	178	6.64
	19	28.40	59	22.89	99	14.94	139	7.73	179	4.66	139	8.54	179	6.61
	20	28.30	60	22.72	100	14.71	140	7.62	180	4.61	140	8.47	180	6.58
	21	28.19	61	22.55	101	14.47	141	7.51	181	4.56	141	8.39	181	6.56
	22	28.08	62	22.37	102	14.24	142	7.41	182	4.51	142	8.32	182	6.53
	23	27.97	63	22.20	103	14.00	143	7.30	183	4.46	143	8.25	183	6.51
	24	27.86	64	22.02	104	13.77	144	7.20	184	4.41	144	8.18	184	6.49
	25	27.75	65	21.85	105	13.53	145	7.10	185	4.36	145	8.12	185	6.46
	26	27.63	66	21.67	106	13.29	146	7.01	186	4.32	146	8.05	186	6.44
	27	27.52	67	21.49	107	13.04	147	6.91	187	4.27	147	7.99	187	6.42
	28	27.40	68	21.31	108	12.80	148	6.82	188	4.23	148	7.93	188	6.40
	29	27.28	69	21.12	109	12.57	149	6.73	189	4.18	149	7.87	189	6.38
	30	27.15	70	20.94	110	12.34	150	6.64	190	4.14	150	7.81	190	6.36
	31	27.03	71	20.75	111	12.12	151	6.55	191	4.09	151	7.75	191	6.35
	32	26.90	72	20.56	112	11.90	152	6.46	192	4.05	152	7.69	192	6.33
	33	26.77	73	20.38	113	11.69	153	6.38	193	4.01	153	7.64	193	6.31
	34	26.64	74	20.19	114	11.49	154	6.30	194	3.97	154	7.59	194	6.30
	35	26.51	75	19.99	115	11.29	155	6.22	195	3.93	155	7.53	195	6.28
	36	26.38	76	19.80	116	11.10	156	6.14	196	3.89	156	7.48	196	6.27
	37	26.25	77	19.61	117	10.91	157	6.06	197	3.85	157	7.43	197	6.26
	38	26.11	78	19.41	118	10.72	158	5.98	198	3.81	158	7.39	198	6.24
	39	25.97	79	19.21	119	10.55	159	5.91	199	3.77	159	7.34	199	6.23
	40	25.83	80	19.01	120	10.37	160	5.83	200	3.73	160	7.29	200	6.22

* K taken as 1.0 for secondary members.   Note: $C_c = 107.0$

TABLE C.5 ALLOWABLE STRESS 1000 psi FOR COMPRESSION MEMBERS OF 55,000 SPECIFIED YIELD STRESS STEEL

$S_{yp} = 55{,}000$ psi

Main and Secondary Members KL/r not over 120						Main Members KL/r 121 to 200				Secondary Members* L/r 121 to 200			
$\dfrac{KL}{r}$	F/A 1000 psi	$\dfrac{KL}{r}$	F/A 1000 psi	$\dfrac{KL}{r}$	F/A 1000 psi	$\dfrac{KL}{r}$	F/A 1000 psi	$\dfrac{KL}{r}$	F/A 1000 psi	$\dfrac{L}{r}$	F/A 1000 psi	$\dfrac{L}{r}$	F/A 1000 psi
1	32.93	41	27.94	81	19.80	121	10.20	161	5.76	121	10.25	161	7.25
2	32.85	42	27.78	82	19.56	122	10.03	162	5.69	122	10.13	162	7.20
3	32.77	43	27.61	83	19.32	123	9.87	163	5.62	123	10.02	163	7.16
4	32.69	44	27.43	84	19.08	124	9.71	164	5.55	124	9.91	164	7.12
5	32.60	45	27.26	85	18.83	125	9.56	165	5.49	125	9.80	165	7.08
6	32.51	46	27.08	86	18.58	126	9.41	166	5.42	126	9.70	166	7.04
7	32.42	47	26.91	87	18.34	127	9.26	167	5.35	127	9.59	167	7.00
8	32.33	48	26.73	88	18.08	128	9.11	168	5.29	128	9.49	168	6.96
9	32.23	49	26.55	89	17.83	129	8.97	169	5.23	129	9.40	169	6.93
10	32.14	50	26.36	90	17.58	130	8.84	170	5.17	130	9.30	170	6.89
11	32.03	51	26.18	91	17.32	131	8.70	171	5.11	131	9.21	171	6.85
12	31.93	52	25.99	92	17.06	132	8.57	172	5.05	132	9.12	172	6.82
13	31.82	53	25.80	93	16.80	133	8.44	173	4.99	133	9.03	173	6.79
14	31.72	54	25.61	94	16.53	134	8.32	174	4.93	134	8.94	174	6.76
15	31.61	55	25.42	95	16.27	135	8.19	175	4.88	135	8.86	175	6.73
16	31.49	56	25.23	96	16.00	136	8.07	176	4.82	136	8.78	176	6.70
17	31.38	57	25.03	97	15.73	137	7.96	177	4.77	137	8.70	177	6.67
18	31.26	58	24.83	98	15.46	138	7.84	178	4.71	138	8.62	178	6.64
19	31.14	59	24.63	99	15.19	139	7.73	179	4.66	139	8.54	179	6.61
20	31.02	60	24.43	100	14.91	140	7.62	180	4.61	140	8.47	180	6.58
21	30.89	61	24.23	101	14.63	141	7.51	181	4.56	141	8.39	181	6.56
22	30.76	62	24.03	102	14.35	142	7.41	182	4.51	142	8.32	182	6.53
23	30.63	63	23.82	103	14.08	143	7.30	183	4.46	143	8.25	183	6.51
24	30.50	64	23.61	104	13.81	144	7.20	184	4.41	144	8.18	184	6.49
25	30.37	65	23.40	105	13.54	145	7.10	185	4.36	145	8.12	185	6.46
26	30.23	66	23.19	106	13.29	146	7.01	186	4.32	146	8.05	186	6.44
27	30.09	67	22.98	107	13.04	147	6.91	187	4.27	147	7.99	187	6.42
28	29.95	68	22.76	108	12.80	148	6.82	188	4.23	148	7.93	188	6.40
29	29.81	69	22.54	109	12.57	149	6.73	189	4.18	149	7.87	189	6.38
30	29.67	70	22.33	110	12.34	150	6.64	190	4.14	150	7.81	190	6.36
31	29.52	71	22.11	111	12.12	151	6.55	191	4.09	151	7.75	191	6.35
32	29.37	72	21.88	112	11.90	152	6.46	192	4.05	152	7.69	192	6.33
33	29.22	73	21.66	113	11.69	153	6.38	193	4.01	153	7.64	193	6.31
34	29.07	74	21.43	114	11.49	154	6.30	194	3.97	154	7.59	194	6.30
35	28.91	75	21.21	115	11.29	155	6.22	195	3.93	155	7.53	195	6.28
36	28.76	76	20.98	116	11.10	156	6.14	196	3.89	156	7.48	196	6.27
37	28.60	77	20.75	117	10.91	157	6.06	197	3.85	157	7.43	197	6.26
38	28.44	78	20.51	118	10.72	158	5.98	198	3.81	158	7.39	198	6.24
39	28.28	79	20.28	119	10.55	159	5.91	199	3.77	159	7.34	199	6.23
40	28.11	80	20.04	120	10.37	160	5.83	200	3.73	160	7.29	200	6.22

* K taken as 1.0 for secondary members.   Note: $C_c = 102.0$

S_{yp} 55,000 psi

TABLE C.6 ALLOWABLE STRESS 1000 psi FOR COMPRESSION MEMBERS OF 60,000 SPECIFIED YIELD STRESS STEEL

$S_{yp} = 60,000$ psi

	Main and Secondary Members KL/r not over 120						Main Members KL/r 121 to 200				Secondary Members* L/r 121 to 200			
	KL/r	F/A 1000 psi	KL/r	F/A 1000 psi	KL/r	F/A 1000 psi	KL/r	F/A 1000 psi	KL/r	F/A 1000 psi	L/r	F/A 1000 psi	L/r	F/A 1000 psi
$S_{yp}=60,000$ psi	1	35.92	41	30.15	81	20.65	121	10.20	161	5.76	121	10.25	161	7.25
	2	35.83	42	29.95	82	20.37	122	10.03	162	5.69	122	10.13	162	7.20
	3	35.74	43	29.75	83	20.09	123	9.87	163	5.62	123	10.02	163	7.16
	4	35.64	44	29.55	84	19.80	124	9.71	164	5.55	124	9.91	164	7.12
	5	35.54	45	29.35	85	19.51	125	9.56	165	5.49	125	9.80	165	7.08
	6	35.44	46	29.15	86	19.22	126	9.41	166	5.42	126	9.70	166	7.04
	7	35.34	47	28.94	87	18.93	127	9.26	167	5.35	127	9.59	167	7.00
	8	35.23	48	28.73	88	18.63	128	9.11	168	5.29	128	9.49	168	6.96
	9	35.12	49	28.52	89	18.34	129	8.97	169	5.23	129	9.40	169	6.93
	10	35.01	50	28.31	90	18.04	130	8.84	170	5.17	130	9.30	170	6.89
	11	34.89	51	28.09	91	17.73	131	8.70	171	5.11	131	9.21	171	6.85
	12	34.77	52	27.87	92	17.43	132	8.57	172	5.05	132	9.12	172	6.82
	13	34.65	53	27.66	93	17.12	133	8.44	173	4.99	133	9.03	173	6.79
	14	34.52	54	27.43	94	16.81	134	8.32	174	4.93	134	8.94	174	6.76
	15	34.40	55	27.21	95	16.50	135	8.19	175	4.88	135	8.86	175	6.73
	16	34.27	56	26.98	96	16.19	136	8.07	176	4.82	136	8.78	176	6.70
	17	34.13	57	26.76	97	15.87	137	7.96	177	4.77	137	8.70	177	6.67
	18	34.00	58	26.53	98	15.55	138	7.84	178	4.71	138	8.62	178	6.64
	19	33.86	59	26.29	99	15.24	139	7.73	179	4.66	139	8.54	179	6.61
	20	33.71	60	26.06	100	14.93	140	7.62	180	4.61	140	8.47	180	6.58
	21	33.57	61	25.82	101	14.64	141	7.51	181	4.56	141	8.39	181	6.56
	22	33.42	62	25.58	102	14.35	142	7.41	182	4.51	142	8.32	182	6.53
	23	33.27	63	25.34	103	14.08	143	7.30	183	4.46	143	8.25	183	6.51
	24	33.12	64	25.10	104	13.81	144	7.20	184	4.41	144	8.18	184	6.49
	25	32.96	65	24.86	105	13.54	145	7.10	185	4.36	145	8.12	185	6.46
	26	32.81	66	24.61	106	13.29	146	7.01	186	4.32	146	8.05	186	6.44
	27	32.65	67	24.36	107	13.04	147	6.91	187	4.27	147	7.99	187	6.42
	28	32.48	68	24.11	108	12.80	148	6.82	188	4.23	148	7.93	188	6.40
	29	32.32	69	23.86	109	12.57	149	6.73	189	4.18	149	7.87	189	6.38
	30	32.15	70	23.60	110	12.34	150	6.64	190	4.14	150	7.81	190	6.36
	31	31.98	71	23.34	111	12.12	151	6.55	191	4.09	151	7.75	191	6.35
	32	31.81	72	23.08	112	11.90	152	6.46	192	4.05	152	7.69	192	6.33
	33	31.63	73	22.82	113	11.69	153	6.38	193	4.01	153	7.64	193	6.31
	34	31.45	74	22.56	114	11.49	154	6.30	194	3.97	154	7.59	194	6.30
	35	31.28	75	22.29	115	11.29	155	6.22	195	3.93	155	7.53	195	6.28
	36	31.09	76	22.02	116	11.10	156	6.14	196	3.89	156	7.48	196	6.27
	37	30.91	77	21.75	117	10.91	157	6.06	197	3.85	157	7.43	197	6.26
	38	30.72	78	21.48	118	10.72	158	5.98	198	3.81	158	7.39	198	6.24
	39	30.53	79	21.21	119	10.55	159	5.91	199	3.77	159	7.34	199	6.23
	40	30.34	80	20.93	120	10.37	160	5.83	200	3.73	160	7.29	200	6.22

* K taken as 1.0 for secondary members.   Note: $C_c = 97.7$

TABLE C.7 ALLOWABLE STRESS 1000 psi FOR COMPRESSION MEMBERS OF 65,000 SPECIFIED YIELD STRESS STEEL

$S_{yp} = 65,000$ psi

Main and Secondary Members KL/r not over 120								Main Members KL/r 121 to 200				Secondary Members* L/r 121 to 200			
$\dfrac{KL}{r}$	F/A 1000 psi	$\dfrac{KL}{r}$	F/A 1000 psi	$\dfrac{KL}{r}$	F/A 1000 psi	$\dfrac{KL}{r}$	F/A 1000 psi	$\dfrac{KL}{r}$	F/A 1000 psi	$\dfrac{L}{r}$	F/A 1000 psi	$\dfrac{L}{r}$	F/A 1000 psi
1	38.90	41	32.30	81	21.36	121	10.20	161	5.76	121	10.25	161	7.25
2	38.81	42	32.08	82	21.03	122	10.03	162	5.69	122	10.13	162	7.20
3	38.70	43	31.85	83	20.70	123	9.87	163	5.62	123	10.02	163	7.16
4	38.59	44	31.62	84	20.37	124	9.71	164	5.55	124	9.91	164	7.12
5	38.48	45	31.39	85	20.04	125	9.56	165	5.49	125	9.80	165	7.08
6	38.37	46	31.35	86	19.70	126	9.41	166	5.42	126	9.70	166	7.04
7	38.25	47	30.92	87	19.36	127	9.26	167	5.35	127	9.59	167	7.00
8	38.13	48	30.68	88	19.02	128	9.11	168	5.29	128	9.49	168	6.96
9	38.00	49	30.43	89	18.67	129	8.97	169	5.23	129	9.40	169	6.93
10	37.87	50	30.19	90	18.32	130	8.84	170	5.17	130	9.30	170	6.89
11	37.74	51	29.94	91	17.97	131	8.70	171	5.11	131	9.21	171	6.85
12	37.61	52	29.69	92	17.62	132	8.57	172	5.05	132	9.12	172	6.82
13	37.47	53	29.44	93	17.26	133	8.44	173	4.99	133	9.03	173	6.79
14	37.32	54	29.18	94	16.90	134	8.32	174	4.93	134	8.94	174	6.76
15	37.18	55	28.92	95	16.55	135	8.19	175	4.88	135	8.86	175	6.73
16	37.03	56	28.66	96	16.20	136	8.07	176	4.82	136	8.78	176	6.70
17	36.87	57	28.40	97	15.87	137	7.96	177	4.77	137	8.70	177	6.67
18	36.72	58	28.14	98	15.55	138	7.84	178	4.71	138	8.62	178	6.64
19	36.56	59	27.87	99	15.24	139	7.73	179	4.66	139	8.54	179	6.61
20	36.40	60	27.60	100	14.93	140	7.62	180	4.61	140	8.47	180	6.58
21	36.23	61	27.33	101	14.64	141	7.51	181	4.56	141	8.39	181	6.56
22	36.06	62	27.05	102	14.35	142	7.41	182	4.51	142	8.32	182	6.53
23	35.89	63	26.78	103	14.08	143	7.30	183	4.46	143	8.25	183	6.51
24	35.71	64	26.50	104	13.81	144	7.20	184	4.41	144	8.18	184	6.49
25	35.54	65	26.21	105	13.54	145	7.10	185	4.36	145	8.12	185	6.46
26	35.36	66	25.93	106	13.29	146	7.01	186	4.32	146	8.05	186	6.44
27	35.17	67	25.64	107	13.04	147	6.91	187	4.27	147	7.99	187	6.42
28	34.99	68	25.35	108	12.80	148	6.82	188	4.23	148	7.93	188	6.40
29	34.80	69	25.06	109	12.57	149	6.73	189	4.18	149	7.87	189	6.38
30	34.60	70	24.76	110	12.34	150	6.64	190	4.14	150	7.81	190	6.36
31	34.41	71	24.47	111	12.12	151	6.55	191	4.09	151	7.75	191	6.35
32	34.21	72	24.17	112	11.90	152	6.46	192	4.05	152	7.69	192	6.33
33	34.01	73	23.87	113	11.69	153	6.38	193	4.01	153	7.64	193	6.31
34	33.81	74	23.56	114	11.49	154	6.30	194	3.97	154	7.59	194	6.30
35	33.60	75	23.25	115	11.29	155	6.22	195	3.93	155	7.53	195	6.28
36	33.39	76	22.94	116	11.10	156	6.14	196	3.89	156	7.48	196	6.27
37	33.18	77	22.63	117	10.91	157	6.06	197	3.85	157	7.43	197	6.26
38	32.96	78	22.32	118	10.72	158	5.98	198	3.81	158	7.39	198	6.24
39	32.75	79	22.00	119	10.55	159	5.91	199	3.77	159	7.34	199	6.23
40	32.53	80	21.68	120	10.37	160	5.83	200	3.73	160	7.29	200	6.22

* K taken as 1.0 for secondary members.   Note: $C_c = 93.8$

TABLE C.8 ALLOWABLE STRESS 1000 psi FOR COMPRESSION MEMBERS OF 90,000 SPECIFIED YIELD STRESS STEEL

$F_y = 90,000$ psi

	Main and Secondary Members KL/r not over 120						Main Members KL/r 121 to 200				Secondary Members* L/r 121 to 200			
$\frac{KL}{r}$	F/A 1000 psi	$\frac{KL}{r}$	F/A 1000 psi	$\frac{KL}{r}$	F/A 1000 psi	$\frac{KL}{r}$	F/A 1000 psi	$\frac{KL}{r}$	F/A 1000 psi	$\frac{L}{r}$	F/A 1000 psi	$\frac{L}{r}$	F/A 1000 psi
1	53.84	41	42.39	81	22.76	121	10.20	161	5.76	121	10.25	161	7.25
2	53.68	42	42.00	82	22.21	122	10.03	162	5.69	122	10.13	162	7.20
3	53.51	43	41.59	83	21.68	123	9.87	163	5.62	123	10.02	163	7.16
4	53.33	44	41.19	84	21.16	124	9.71	164	5.55	124	9.91	164	7.12
5	53.15	45	40.78	85	20.67	125	9.56	165	5.49	125	9.80	165	7.08
6	52.95	46	40.36	86	20.19	126	9.41	166	5.42	126	9.70	166	7.04
7	52.75	47	39.94	87	19.73	127	9.26	167	5.35	127	9.59	167	7.00
8	52.55	48	39.51	88	19.28	128	9.11	168	5.29	128	9.49	168	6.96
9	52.33	49	39.08	89	18.85	129	8.97	169	5.23	129	9.40	169	6.93
10	52.11	50	38.65	90	18.44	130	8.84	170	5.17	130	9.30	170	6.89
11	51.89	51	38.21	91	18.03	131	8.70	171	5.11	131	9.21	171	6.85
12	51.65	52	37.77	92	17.64	132	8.57	172	5.05	132	9.12	172	6.82
13	51.41	53	37.32	93	17.27	133	8.44	173	4.99	133	9.03	173	6.79
14	51.17	54	36.86	94	16.90	134	8.32	174	4.93	134	8.94	174	6.76
15	50.92	55	36.41	95	16.55	135	8.19	175	4.88	135	8.86	175	6.73
16	50.66	56	35.94	96	16.20	136	8.07	176	4.82	136	8.78	176	6.70
17	50.39	57	35.47	97	15.87	137	7.96	177	4.77	137	8.70	177	6.67
18	50.12	58	35.00	98	15.55	138	7.84	178	4.71	138	8.62	178	6.64
19	49.85	59	34.52	99	15.24	139	7.73	179	4.66	139	8.54	179	6.61
20	49.56	60	34.04	100	14.93	140	7.62	180	4.61	140	8.47	180	6.58
21	49.28	61	33.56	101	14.64	141	7.51	181	4.56	141	8.39	181	6.56
22	48.98	62	33.06	102	14.35	142	7.41	182	4.51	142	8.32	182	6.53
23	48.68	63	32.57	103	14.08	143	7.30	183	4.46	143	8.25	183	6.51
24	48.38	64	32.07	104	13.81	144	7.20	184	4.41	144	8.18	184	6.49
25	48.07	65	31.56	105	13.54	145	7.10	185	4.36	145	8.12	185	6.46
26	47.75	66	31.05	106	13.29	146	7.01	186	4.32	146	8.05	186	6.44
27	47.43	67	30.53	107	13.04	147	6.91	187	4.27	147	7.99	187	6.42
28	47.10	68	30.01	108	12.80	148	6.82	188	4.23	148	7.93	188	6.40
29	46.77	69	29.48	109	12.57	149	6.73	189	4.18	149	7.87	189	6.38
30	46.43	70	28.95	110	12.34	150	6.64	190	4.14	150	7.81	190	6.36
31	46.09	71	28.41	111	12.12	151	6.55	191	4.09	151	7.75	191	6.35
32	45.74	72	27.87	112	11.90	152	6.46	192	4.05	152	7.69	192	6.33
33	45.39	73	27.32	113	11.69	153	6.38	193	4.01	153	7.64	193	6.31
34	45.03	74	26.77	114	11.49	154	6.30	194	3.97	154	7.59	194	6.30
35	44.67	75	26.21	115	11.29	155	6.22	195	3.93	155	7.53	195	6.28
36	44.30	76	25.65	116	11.10	156	6.14	196	3.89	156	7.48	196	6.27
37	43.93	77	25.08	117	10.91	157	6.06	197	3.85	157	7.43	197	6.26
38	43.55	78	24.50	118	10.72	158	5.98	198	3.81	158	7.39	198	6.24
39	43.17	79	23.92	119	10.55	159	5.91	199	3.77	159	7.34	199	6.23
40	42.78	80	23.33	120	10.37	160	5.83	200	3.73	160	7.29	200	6.22

* K taken as 1.0 for secondary members. Note: $C_c = 79.8$

TABLE C.9 ALLOWABLE STRESS 1000 psi FOR COMPRESSION MEMBERS OF 100,000 SPECIFIED YIELD STRESS STEEL

$S_{yp} = 100,000$ psi

Main and Secondary Members KL/r not over 120						Main Members KL/r 121 to 200				Secondary Members* L/r 121 to 200			
$\frac{KL}{r}$	F/A 1000 psi	$\frac{KL}{r}$	F/A 1000 psi	$\frac{KL}{r}$	F/A 1000 psi	$\frac{KL}{r}$	F/A 1000 psi	$\frac{KL}{r}$	F/A 1000 psi	$\frac{L}{r}$	F/A 1000 psi	$\frac{L}{r}$	F/A 1000 psi
1	59.82	41	46.12	81	22.76	121	10.20	161	5.76	121	10.25	161	7.25
2	59.62	42	45.64	82	22.21	122	10.03	162	5.69	122	10.13	162	7.20
3	59.42	43	45.16	83	21.68	123	9.87	163	5.62	123	10.02	163	7.16
4	59.21	44	44.67	84	21.16	124	9.71	164	5.55	124	9.91	164	7.12
5	58.99	45	44.17	85	20.67	125	9.56	165	5.49	125	9.80	165	7.08
6	58.76	46	43.67	86	20.19	126	9.41	166	5.42	126	9.70	166	7.04
7	58.53	47	43.17	87	19.73	127	9.26	167	5.35	127	9.59	167	7.00
8	58.28	48	42.65	88	19.28	128	9.11	168	5.29	128	9.49	168	6.96
9	58.03	49	42.14	89	18.85	129	8.97	169	5.23	129	9.40	169	6.93
10	57.77	50	41.61	90	18.44	130	8.84	170	5.17	130	9.30	170	6.89
11	57.50	51	41.08	91	18.03	131	8.70	171	5.11	131	9.21	171	6.85
12	57.22	52	40.55	92	17.64	132	8.57	172	5.05	132	9.12	172	6.82
13	56.93	53	40.00	93	17.27	133	8.44	173	4.99	133	9.03	173	6.79
14	56.64	54	39.46	94	16.90	134	8.32	174	4.93	134	8.94	174	6.76
15	56.34	55	38.90	95	16.55	135	8.19	175	4.88	135	8.86	175	6.73
16	56.03	56	38.35	96	16.20	136	8.07	176	4.82	136	8.78	176	6.70
17	55.72	57	37.78	97	15.87	137	7.96	177	4.77	137	8.70	177	6.67
18	55.39	58	37.21	98	15.55	138	7.84	178	4.71	138	8.62	178	6.64
19	55.06	59	36.63	99	15.24	139	7.73	179	4.66	139	8.54	179	6.61
20	54.72	60	36.05	100	14.93	140	7.62	180	4.61	140	8.47	180	6.58
21	54.38	61	35.46	101	14.64	141	7.51	181	4.56	141	8.39	181	6.56
22	54.03	62	34.87	102	14.35	142	7.41	182	4.51	142	8.32	182	6.53
23	53.67	63	34.26	103	14.08	143	7.30	183	4.46	143	8.25	183	6.51
24	53.30	64	33.66	104	13.81	144	7.20	184	4.41	144	8.18	184	6.49
25	52.93	65	33.04	105	13.54	145	7.10	185	4.36	145	8.12	185	6.46
26	52.55	66	32.42	106	13.29	146	7.01	186	4.32	146	8.05	186	6.44
27	52.17	67	31.80	107	13.04	147	6.91	187	4.27	147	7.99	187	6.42
28	51.78	68	31.16	108	12.80	148	6.82	188	4.23	148	7.93	188	6.40
29	51.38	69	30.52	109	12.57	149	6.73	189	4.18	149	7.87	189	6.38
30	50.97	70	29.88	110	12.34	150	6.64	190	4.14	150	7.81	190	6.36
31	50.56	71	29.22	111	12.12	151	6.55	191	4.09	151	7.75	191	6.35
32	50.15	72	28.56	112	11.90	152	6.46	192	4.05	152	7.69	192	6.33
33	49.72	73	27.90	113	11.69	153	6.38	193	4.01	153	7.64	193	6.31
34	49.29	74	27.22	114	11.49	154	6.30	194	3.97	154	7.59	194	6.30
35	48.86	75	26.54	115	11.29	155	6.22	195	3.93	155	7.53	195	6.28
36	48.42	76	25.85	116	11.10	156	6.14	196	3.89	156	7.48	196	6.27
37	47.97	77	25.19	117	10.91	157	6.06	197	3.85	157	7.43	197	6.26
38	47.51	78	24.54	118	10.72	158	5.98	198	3.81	158	7.39	198	6.24
39	47.05	79	23.93	119	10.55	159	5.91	199	3.77	159	7.34	199	6.23
40	46.59	80	23.33	120	10.37	160	5.83	200	3.73	160	7.29	200	6.22

$S_{yp} = 100,000$ psi

* K taken as 1.0 for secondary members.   Note: $C_c = 75.7$

Appendix D

Properties of pipe

*Reprinted with the permission of *Fluid Mechanics for Engineering Technology* by Irving Granet, Prentice-Hall, Inc., 1971.

Table D.1 Properties of pipe[†]

Schedules, Wall Thicknesses, and Weights
Conforming to ASA Standard B36.10, 1950, for Wrought Steel and Wrought Iron Pipe

Nominal Pipe Size (in.)	Outside Diameter, D(in.)	Wall Thickness, t(in.)	Inside Diameter d(in.)	Inside Diameter d^2(squared)	Inside Diameter d^5(fifth power)	Area of Metal (in.^{2})	Internal Cross-Sectional Area in.^{2}	Internal Cross-Sectional Area ft^2	External Surface (ft^2)	Moment of Inertia (in.^{4})	Weight (Pounds) of Pipe (per ft)	Weight (Pounds) of Water (per ft of pipe)
\multicolumn{13}{c}{Schedule 10}												
14 o.d.	14.0	0.250	13.50	182.25	448,403	10.80	143.14	0.994	3.665	255.3	36.71	62.03
16 o.d.	16.0	0.250	15.50	240.25	894,660	12.37	188.69	1.310	4.189	383.7	42.05	81.74
18 o.d.	18.0	0.250	17.50	306.25	1,641,309	13.94	240.53	1.670	4.712	549.1	47.39	104.21
20 o.d.	20.0	0.250	19.50	380.25	2,819,505	15.51	298.65	2.074	5.236	756.4	52.73	129.42
24 o.d.	24.0	0.250	23.50	552.25	7,167,030	18.65	433.74	3.012	6.283	1,315.0	63.41	187.95
30 o.d.	30.0	0.312	29.376	862.95	21,875,768	29.10	677.76	4.707	7.854	3,206.0	98.93	293.72
\multicolumn{13}{c}{Schedule 20}												
8	8.625	0.250	8.125	66.02	35,409	6.57	51.85	0.3601	2.258	57.72	22.36	22.47
10	10.75	0.250	10.25	105.06	113,141	8.24	82.52	0.5731	2.814	113.7	28.04	35.76
12	12.75	0.250	12.25	150.06	275,855	9.82	117.86	0.8185	3.338	191.8	33.38	51.07
14 o.d.	14.0	0.312	13.376	178.92	428,185	13.42	140.52	0.975	3.665	314.4	45.68	60.89
16 o.d.	16.0	0.312	15.376	236.42	859,442	15.38	185.69	1.290	4.189	473.2	52.36	80.50
18 o.d.	18.0	0.312	17.376	301.92	1,583,978	17.34	237.13	1.647	4.712	678.2	59.03	102.77
20 o.d.	20.0	0.375	19.25	370.56	2,643,344	23.12	291.04	2.021	5.236	1113	78.60	125.67
24 o.d.	24.0	0.375	23.25	540.56	6,793,832	27.83	424.56	2.948	6.283	1942	94.62	183.95
30 o.d.	30.0	0.500	29.00	841.0	20,511,149	46.34	660.52	4.587	7.854	5,042	157.53	286.23
\multicolumn{13}{c}{Schedule 30}												
8	8.625	0.277	8.071	65.14	34,248	7.26	51.16	0.3553	2.258	63.35	24.70	22.17
10	10.75	0.307	10.136	102.74	106,987	10.07	80.69	0.5603	2.814	137.4	34.24	34.96

Appendix D 469

12	12.75	0.330	12.09	146.17	258,304	12.87	114.80	0.7972	3.338	248.4	43.77	49.74
14 o.d.	14.0	0.375	13.25	175.56	408,394	16.05	137.88	0.9575	3.665	372.8	54.57	59.75
16 o.d.	16.0	0.375	15.25	232.56	824,801	18.41	182.65	1.268	4.189	562.1	62.58	79.12
18 o.d.	18.0	0.437	17.126	293.30	1,473,261	24.11	230.36	1.600	4.712	930.3	82.06	99.84
20 o.d.	20.0	0.500	19.0	361.00	2,476,099	30.63	283.53	1.969	5.236	1,457	104.13	122.87
24 o.d.	24.0	0.562	22.876	523.31	6,264,703	41.39	411.00	2.854	6.283	2,843	140.80	178.09
30 o.d.	30.0	0.625	28.75	826.56	19,642,160	57.68	649.18	4.508	7.854	6,224	196.08	281.30

Schedule 40

1/8	0.405	0.068§	0.269	0.0724	0.00141	0.072	0.0569	0.00040	0.106	0.001064	0.24	0.0250
1/4	0.540	0.088§	0.364	0.1325	0.00639	0.125	0.1041	0.00072	0.141	0.003312	0.42	0.0449
3/8	0.675	0.091§	0.493	0.2430	0.02912	0.167	0.1909	0.00133	0.177	0.007291	0.57	0.0830
1/2	0.840	0.109§	0.622	0.3869	0.09310	0.250	0.3039	0.00211	0.220	0.01709	0.85	0.1317
3/4	1.050	0.113§	0.824	0.679	0.3799	0.333	0.5333	0.00371	0.275	0.03704	1.13	0.2315
1	1.315	0.133§	1.049	1.100	1.270	0.494	0.8639	0.00600	0.344	0.08734	1.68	0.3744
1 1/4	1.660	0.140§	1.380	1.904	5.005	0.669	1.495	0.01040	0.435	0.1947	2.27	0.6490
1 1/2	1.900	0.145§	1.610	2.592	10.82	0.799	2.036	0.01414	0.497	0.3099	2.72	0.8823
2	2.375	0.154§	2.067	4.272	37.72	1.075	3.356	0.02330	0.622	0.666	3.65	1.454
2 1/2	2.875	0.203§	2.469	6.096	91.75	1.704	4.788	0.03322	0.753	1.530	5.79	2.073
3	3.5	0.216§	3.068	9.413	271.8	2.228	7.393	0.05130	0.916	3.017	7.58	3.201
3 1/2	4.0	0.226§	3.548	12.59	562.2	2.680	9.888	0.06870	1.047	4.788	9.11	4.287
4	4.5	0.237§	4.026	16.21	1,058	3.173	12.73	0.08840	1.178	7.233	10.79	5.516
5	5.563	0.258§	5.047	25.47	3,275	4.304	20.01	0.1390	1.456	15.16	14.62	8.674
6	6.625	0.280§	6.065	36.78	8,206	5.584	28.89	0.2006	1.734	28.14	18.97	12.52
8	8.625	0.322§	7.981	63.70	32,380	8.396	50.03	0.3474	2.258	72.49	25.55	21.68
10	10.75	0.365§	10.02	100.4	101,000	11.90	78.85	0.5475	2.814	160.7	40.48	34.16
12	12.75	0.406	11.938	142.5	242,470	15.77	111.93	0.7773	3.338	300.3	53.53	48.50
14 o.d.	14.0	0.437	13.126	172.3	389,638	18.61	135.32	0.9397	3.665	429.1	63.37	58.64
16 o.d.	16.0	0.500	15.000	225.0	759,375	24.35	176.72	1.2272	4.189	731.9	82.77	76.58
18 o.d.	18.0	0.562	16.876	284.8	1,368,820	30.79	223.68	1.5533	4.712	1172	104.75	96.93

Nominal Pipe Size (in.)	Outside Diameter D(in.)	Wall Thickness t(in.)	Inside Diameter d(in.)	Inside Diameter d^2(squared)	Inside Diameter d^5(fifth power)	Area of Metal (in.^{2})	Internal Cross-Sectional Area in.^{2}		External Surface (ft^2)	Moment of Inertia (in.^{4})	Weight (Pounds) of Pipe (per ft)	Weight (Pounds) of Water (per ft of pipe)
							in.^{2}	ft^2				
20 o.d.	20.0	0.593	18.814	354.0	2,357,244	36.15	278.00	1.9305	5.236	1703	122.91	120.46
24 o.d.	24.0	0.687	22.626	511.9	5,929,784	50.31	402.07	2.7921	6.283	3424	171.17	174.23
Schedule 60												
8	8.625	0.406	7.813	61.04	29,113	10.48	47.94	0.3329	2.258	88.73	35.64	20.77
10	10.75	0.500	9.75	95.06	88,110	16.10	74.66	0.5185	2.814	212.0	54.74	32.35
12	12.75	0.562	11.626	135.16	212,399	21.52	106.16	0.7372	3.338	400.4	73.16	46.00
14 o.d.	14.0	0.593	12.814	164.20	345,480	24.98	128.96	0.8956	3.665	562.3	84.91	55.86
16 o.d.	16.0	0.656	14.688	215.74	683,618	31.62	169.44	1.1766	4.189	932.4	107.50	73.42
18 o.d.	18.0	0.750	16.500	272.25	1,222,981	40.64	213.83	1.4849	4.712	1,515	138.17	92.80
20 o.d.	20.0	0.812	18.376	337.68	2,095,342	48.95	265.21	1.8417	5.236	2,257	166.40	114.92
24 o.d.	24.0	0.968	22.064	486.82	5,229,029	70.04	382.35	2.6552	6.283	4,654	238.11	165.94
Schedule 80												
$\frac{1}{8}$	0.405	0.095	0.215	0.0462	0.000459	0.093	0.0363	0.00025	0.106	0.001216	0.31	0.0157
$\frac{1}{4}$	0.540	0.119	0.302	0.0912	0.002513	0.157	0.0716	0.00050	0.141	0.003766	0.54	0.031
$\frac{3}{8}$	0.675	0.126	0.423	0.1789	0.01354	0.217	0.1405	0.00098	0.177	0.008619	0.74	0.0609
$\frac{1}{2}$	0.840	0.147	0.546	0.2981	0.04852	0.320	0.2341	0.00163	0.220	0.02008	1.09	0.1013
$\frac{3}{4}$	1.050	0.154	0.742	0.5506	0.2249	0.433	0.4324	0.00300	0.275	0.04479	1.47	0.1875
1	1.315	0.179	0.957	0.9158	0.8027	0.639	0.7193	0.00499	0.344	0.1056	2.17	0.3112
$1\frac{1}{4}$	1.660	0.191	1.278	1.633	3.409	0.881	1.283	0.00891	0.435	0.2418	3.00	0.5553
$1\frac{1}{2}$	1.900	0.200	1.500	2.250	7.594	1.068	1.767	0.01225	0.498	0.3912	3.63	0.7648
2	2.375	0.218	1.939	3.760	27.41	1.477	2.953	0.02050	0.622	0.8679	5.02	1.279
$2\frac{1}{2}$	2.875	0.276	2.323	5.396	67.64	2.254	4.238	0.02942	0.753	1.924	7.66	1.834
3	3.5	0.300	2.900	8.410	205.1	3.016	6.605	0.04587	0.917	3.894	10.25	2.859
$3\frac{1}{2}$	4.0	0.318	3.364	11.32	430.8	3.678	8.891	0.06170	1.047	6.280	12.51	3.847

Appendix D   471

4	4.5	0.337	3.826	14.64		819.8	4.407	11.50	0.07986	1.178	9.610	14.98	4.976
5	5.563	0.375	4.813	23.16		2,583	6.112	18.19	0.1263	1.456	20.67	20.78	7.875
6	6.625	0.432	5.761	33.19		6,346	8.405	26.07	0.1810	1.734	40.49	28.57	11.29
8	8.625	0.500	7.625	58.14		25,775	12.76	45.66	0.3171	2.257	105.7	43.39	19.79
10	10.75	0.593	9.564	91.47		80,020	18.92	71.84	0.4989	2.817	244.8	64.33	31.13
12	12.75	0.687	11.376	129.41		190,523	26.03	101.64	0.7958	3.338	475.1	88.51	44.04
14 o.d.	14.0	0.750	12.500	156.25		305,176	31.22	122.72	0.8522	3.665	687.3	106.13	53.18
16 o.d.	16.0	0.843	14.314	204.89		600,904	40.14	160.92	1.1175	4.189	1,156	136.46	69.73
18 o.d.	18.0	0.937	16.125	260.05		1,090,518	50.23	204.24	1.4183	4.712	1,833	170.75	88.50
20 o.d.	20.0	1.031	17.938	321.77		1,857,248	61.44	252.72	1.7550	5.236	2,772	208.87	109.51
24 o.d.	24.0	1.218	21.564	465.01		4,662,798	87.17	365.22	2.5362	6.283	5,672	296.36	158.26

Schedule 100

8	8.625	0.593	7.439	55.34		22,781	14.96	43.46	0.3018	2.258	121.3	50.87	18.83
10	10.75	0.718	9.314	86.75		69,357	22.63	68.13	0.4732	2.814	286.1	76.93	29.53
12	12.75	0.843	11.064	122.41		165,791	31.53	96.14	0.6677	3.338	561.6	107.20	41.66
14 o.d.	14.0	0.937	12.126	147.04		262,173	38.45	115.49	0.8020	3.665	824.4	130.73	50.04
16 o.d.	16.0	1.031	13.938	194.27		526,020	48.48	152.58	1.0596	4.189	1,364	164.83	66.12
18 o.d.	18.0	1.156	15.688	246.11		950,250	61.17	193.30	1.3423	4.712	2,180	207.96	83.76
20 o.d.	20.0	1.281	17.438	304.08		1,612,398	75.34	238.82	1.6585	5.236	3,316	256.10	103.65
24 o.d.	24.0	1.531	20.938	438.40		4,024,179	108.07	344.32	2.3911	6.283	6,853	367.40	149.43

Schedule 120

4	4.5	0.438	3.625	13.15		626.8	5.578	10.33	0.0717	1.178	11.65	19.01	4.47
5	5.563	0.500	4.563	20.82		1,978	7.953	16.35	0.1136	1.456	25.73	27.04	7.09
6	6.625	0.562	5.501	30.26		5,037	10.705	23.77	0.1650	1.734	49.61	36.39	10.30
8	8.625	0.718	7.189	51.68		19,202	17.84	40.59	0.2819	2.257	140.5	60.63	17.59
10	10.75	0.843	9.064	82.16		61,179	26.24	64.53	0.4481	2.817	324.2	89.20	27.96
12	12.75	1.000	10.750	115.56		143,563	36.91	90.76	0.6303	3.338	641.6	125.49	39.33
14 o.d.	14.0	1.093	11.814	139.57		230,134	44.32	109.62	0.7612	3.665	929.8	150.67	47.57
16 o.d.	16.0	1.218	13.564	183.98		459,133	56.56	144.50	1.0035	4.189	1,555	192.29	62.62

Nominal Pipe Size (in.)	Outside Diameter D(in.)	Wall Thickness t(in.)	Inside Diameter d(in.)	Inside Diameter d²(squared)	Inside Diameter d⁵(fifth power)	Area of Metal (in.²)	Internal Cross-Sectional Area in.²	Internal Cross-Sectional Area ft²	External Surface (ft²)	Moment of Inertia (in.⁴)	Weight (Pounds) of Pipe (per ft)	Weight (Pounds) of Water (per ft of pipe)
18 o.d.	18.0	1.375	15.250	232.56	824,783	71.82	182.65	1.2684	4.712	2,499	244.14	79.27
20 o.d.	20.0	1.500	17.000	289.00	1,419,857	87.18	226.98	1.5762	5.236	3,754	296.37	98.35
24 o.d.	24.0	1.812	20.376	415.18	3,512,301	126.31	326.08	2.2644	6.283	7,827	429.39	141.52
Schedule 140											
8	8.625	0.812	7.001	49.01	16,819	19.93	38.50	0.2673	2.257	153.7	67.76	16.68
10	10.75	1.000	8.750	76.56	51,291	30.63	60.13	0.4176	2.817	367.8	104.13	26.06
12	12.75	1.125	10.500	110.25	127,628	41.08	86.59	0.6013	3.338	700.5	139.68	37.52
14 o.d.	14.0	1.250	11.500	132.25	201,136	50.07	103.87	0.7213	3.665	1,027	170.22	45.01
16 o.d.	16.0	1.438	13.125	172.29	389,670	65.74	135.32	0.9397	4.189	1,760	223.50	58.64
18 o.d.	18.0	1.562	14.876	221.30	728,502	80.66	173.80	1.2070	4.712	2,749	274.23	75.32
20 o.d.	20.0	1.750	16.500	272.25	1,222,981	100.33	213.82	1.4849	5.236	4,216	341.10	92.66
24 o.d.	24.0	2.062	19.876	395.09	3,102,022	142.11	310.28	2.1547	6.283	8,625	483.13	134.45
Schedule 160											
½	0.840	0.187	0.466	0.2172	0.002197	0.3836	0.1706	0.00118	0.220	0.02212	1.30	0.074
¾	1.050	0.218	0.614	0.3770	0.08726	0.5698	0.2961	0.00206	0.275	0.05269	1.94	0.130
¾	1.315	0.250	0.815	0.6642	0.3596	0.8365	0.5217	0.00362	0.344	0.1251	2.84	0.230
1	1.660	0.250	1.160	1.346	2.100	1.107	1.057	0.00734	0.435	0.2839	3.76	0.46
1¼	1.900	0.281	1.338	1.790	4.288	1.429	1.406	0.00976	0.498	0.4824	4.86	0.61
1½	2.375	0.343	1.689	2.853	13.74	2.190	2.241	0.01556	0.622	1.162	7.44	0.97
2	2.875	0.375	2.125	4.516	43.33	2.945	3.546	0.02463	0.753	2.353	10.01	1.54
2½	3.5	0.438	2.625	6.896	124.9	4.205	5.416	0.03761	0.917	5.032	14.32	2.35
3½	4.0											
4	4.5	0.531	3.438	11.82	480.3	6.621	9.283	0.06447	1.178	13.27	22.51	4.02
5	5.563	0.625	4.313	18.60	1,492	9.696	14.61	0.1015	1.456	30.03	32.96	6.33

6	6.625	0.718	5.189	26.93	3,762	13.32	21.15	0.1469	1.734	58.97	45.30	9.16
8	8.625	0.906	6.813	46.42	14,679	21.97	36.46	0.2532	2.257	165.9	74.69	15.80
10	10.75	1.125	8.500	72.25	44,371	34.02	56.75	0.3941	2.817	399.3	115.65	24.59
12	12.75	1.312	10.126	102.54	106,461	47.14	80.53	0.5592	3.338	781.1	160.27	34.89
14 o.d.	14.0	1.406	11.188	125.17	175,292	55.63	98.31	0.6827	3.665	1,117	189.12	42.60
16 o.d.	16.0	1.593	12.814	164.20	345,486	72.10	128.96	0.8955	4.189	1,894	245.11	55.97
18 o.d.	18.0	1.781	14.438	208.46	627,412	90.75	163.72	1.1369	4.712	3,021	308.51	71.05
20 o.d.	20.0	1.968	16.064	258.05	1,069,699	111.49	202.67	1.4074	5.236	4,586	379.0l̄	87.96
24 o.d.	24.0	2.343	19.314	373.03	2,687,570	159.41	292.98	2.0345	6.283	9,458	541.94	127.15

† Data taken with permission from *Catalog #57*, the Walworth Co., New York, 1957.
‡ This column also represents the contents in cubic feet per foot of length.
§ These thicknesses are identical with those listed in ASA B36.10—1950 for standard wall pipe.
¶ These thicknesses are identical with those listed in ASA B36.10—1950 for extra strong wall pipe.

Appendix E

Miscellaneous tables

COEFFICIENTS OF EXPANSION

The coefficient of linear expansion (ε) is the change in length, per unit of length, for a change of one degree of temperature. The coefficient of surface expansion is approximately two times the linear coefficient, and the coefficient of volume expansion, for solids, is approximately three times the linear coefficient.

A bar, free to move, will increase in length with an increase in temperature and will decrease in length with a decrease in temperature. The change in length will be $\varepsilon t l$, where ε is the coefficient of linear expansion, t the change in temperature, and l the length. If the ends of a bar are fixed, a change in temperature (t) will cause a change in the unit stress of $E\varepsilon t$, and in the total stress of $AE\varepsilon t$, where A is the cross sectional area of the bar and E the modulus of elasticity.

The following table gives the coefficient of linear expansion for 100°, or 100 times the value indicated above.

Example: A piece of medium steel is exactly 40 feet long at 60°F. Find the length at 90°F, assuming the ends free to move.

$$\text{Change of length} = \varepsilon t l = \frac{0.00065 \times 30 \times 40}{100} = 0.0078 \text{ ft}$$

The length at 90°F is 40.0078 feet.

Example: A piece of medium steel is exactly 40 feet long and the ends are fixed. If the temperature increases 30°F, what is the resulting change in the unit stress?

---

*Reprinted with the Permission of the *American Institute of Steel Construction Manual of Steel Construction,* Seventh Edition, 1970.

Change in unit stress $= E\varepsilon t = \dfrac{29{,}000{,}000 \times 0.00065 \times 30}{100}$

$= 5655$ lbs per sq. inch

COEFFICIENTS OF EXPANSION FOR 100 DEGREES — 100ε

Materials	Linear Expansion		Materials	Linear Expansion	
	Celsius	Fahrenheit		Celsius	Fahrenheit
METALS AND ALLOYS			**STONE AND MASONRY**		
Aluminum, wrought	0.00231	0.00128	Ashlar masonry	0.00063	0.00035
Brass	0.00188	0.00104	Brick masonry	0.00061	0.00034
Bronze	0.00181	0.00101	Cement, portland	0.00126	0.00070
Copper	0.00168	0.00093	Concrete	0.00099	0.00055
Iron, cast, gray	0.00106	0.00059	Granite	0.00080	0.00044
Iron, wrought	0.00120	0.00067	Limestone	0.00076	0.00042
Iron, wire	0.00124	0.00069	Marble	0.00081	0.00045
Lead	0.00286	0.00159	Plaster	0.00166	0.00092
Magnesium, various alloys	0.0029	0.0016	Rubble masonry	0.00063	0.00035
			Sandstone	0.00097	0.00054
Nickel	0.00126	0.00070	Slate	0.00080	0.00044
Steel, mild	0.00117	0.00065			
Steel, stainless, 18-8	0.00178	0.00099			
Zinc, rolled	0.00311	0.00173			
TIMBER			**TIMBER**		
Fir (parallel to fiber)	0.00037	0.00021	Fir (perpendicular to fiber)	0.0058	0.0032
Maple (parallel to fiber)	0.00064	0.00036	Maple (perpendicular to fiber)	0.0048	0.0027
Oak (parallel to fiber)	0.00049	0.00027	Oak (perpendicular to fiber)	0.0054	0.0030
Pine (parallel to fiber)	0.00054	0.00030	Pine (perpendicular to fiber)	0.0034	0.0019

EXPANSION OF WATER
Maximum Density $= 1$

°C	Volume	°C	Volume	°C	Volume	°C	Volume	°C	Volume
0	1.000126	10	1.000257	30	1.004234	50	1.011877	70	1.022384
4	1.000000	20	1.001732	40	1.007627	60	1.016954	80	1.029003

°C	Volume
90	1.035829
100	1.043116

WIRE AND SHEET METAL GAGES
In decimals of an inch

Name of Gage	*United States Standard Gage		The United States Steel Wire Gage	American or Brown & Sharpe Wire Gage	New Birmingham Standard Sheet & Hoop Gage	British Imperial or English Legal Standard Wire Gage	Birmingham or Stubs Iron Wire Gage	Name of Gage
Principal Use	Uncoated Steel Sheets and Light Plates		Steel Wire except Music Wire	Non-Ferrous Sheets and Wire	Iron and Steel Sheets and Hoops	Wire	Strips, Bands, Hoops and Wire	Principal Use
Gage No.	Weight Oz. per Sq. Ft.	Approx. Thickness Inches	Thickness, Inches					Gage No.
7/0's			.4900		.6666	.500		7/0's
6/0's			.4615	.5800	.625	.464		6/0's
5/0's			.4305	.5165	.5883	.432	.500	5/0's
4/0's			.3938	.4600	.5416	.400	.454	4/0's
3/0's			.3625	.4096	.500	.372	.425	3/0's
2/0's			.3310	.3648	.4452	.348	.380	2/0's
1/0			.3065	.3249	.3964	.324	.340	1/0
1			.2830	.2893	.3532	.300	.300	1
2			.2625	.2576	.3147	.276	.284	2
3	160	.2391	.2437	.2294	.2804	.252	.259	3
4	150	.2242	.2253	.2043	.250	.232	.238	4
5	140	.2092	.2070	.1819	.2225	.212	.220	5
6	130	.1943	.1920	.1620	.1981	.192	.203	6
7	120	.1793	.1770	.1443	.1764	.176	.180	7
8	110	.1644	.1620	.1285	.1570	.160	.165	8
9	100	.1495	.1483	.1144	.1398	.144	.148	9
10	90	.1345	.1350	.1019	.1250	.128	.134	10
11	80	.1196	.1205	.0907	.1113	.116	.120	11
12	70	.1046	.1055	.0808	.0991	.104	.109	12
13	60	.0897	.0915	.0720	.0882	.092	.095	13
14	50	.0747	.0800	.0641	.0785	.080	.083	14
15	45	.0673	.0720	.0571	.0699	.072	.072	15
16	40	.0598	.0625	.0508	.0625	.064	.065	16
17	36	.0538	.0540	.0453	.0556	.056	.058	17
18	32	.0478	.0475	.0403	.0495	.048	.049	18
19	28	.0418	.0410	.0359	.0440	.040	.042	19
20	24	.0359	.0348	.0320	.0392	.036	.035	20
21	22	.0329	.0317	.0285	.0349	.032	.032	21
22	20	.0299	.0286	.0253	.0313	.028	.028	22
23	18	.0269	.0258	.0226	.0278	.024	.025	23
24	16	.0239	.0230	.0201	.0248	.022	.022	24
25	14	.0209	.0204	.0179	.0220	.020	.020	25
26	12	.0179	.0181	.0159	.0196	.018	.018	26
27	11	.0164	.0173	.0142	.0175	.0164	.016	27
28	10	.0149	.0162	.0126	.0156	.0148	.014	28
29	9	.0135	.0150	.0113	.0139	.0136	.013	29
30	8	.0120	.0140	.0100	.0123	.0124	.012	30
31	7	.0105	.0132	.0089	.0110	.0116	.010	31
32	6.5	.0097	.0128	.0080	.0098	.0108	.009	32
33	6	.0090	.0118	.0071	.0087	.0100	.008	33
34	5.5	.0082	.0104	.0063	.0077	.0092	.007	34
35	5	.0075	.0095	.0056	.0069	.0084	.005	35
36	4.5	.0067	.0090	.0050	.0061	.0076	.004	36
37	4.25	.0064	.0085	.0045	.0054	.0068		37
38	4	.0060	.0080	.0040	.0048	.0060		38
39			.0075	.0035	.0043	.0052		39
40			.0070	.0031	.0039	.0048		40

*U. S. Standard Gage is officially a weight gage, in oz. per sq. ft. as tabulated. The Approx. Thickness shown is the "Manufacturers' Standard" of the American Iron and Steel Institute, based on steel as weighing 501.81 lb. per cu. ft. (489.6 true weight plus 2.5 per cent for average over-run in area and thickness). The AISI standard nomenclature for flat rolled carbon steel is as follows:

Thickness (Inches)	Width (Inches)					
	To 3½ incl.	Over 3½ To 6	Over 6 To 8	Over 8 To 12	Over 12 To 48	Over 48
0.2300 & thicker	Bar	Bar	Bar	Plate	Plate	Plate
0.2299 to 0.2031	Bar	Bar	Strip	Strip	Sheet	Plate
0.2030 to 0.1800	Strip	Strip	Strip	Strip	Sheet	Plate
0.1799 to 0.0449	Strip	Strip	Strip	Strip	Sheet	Sheet
0.0448 to 0.0344	Strip	Strip				
0.0343 to 0.0255	Strip		Hot rolled sheet and strip not generally produced in these widths and thicknesses			
0.0254 & thinner						

Index

Acceleration of gravity, 4
Allowable stress:
  bolts, 83
  columns, 360
  rivets, 81, 83, 85
  shear, 81, 169, 185
  tension, 83, 89
  welds, 105, 106
Angle of curvature, 307
Angle of twist:
  circular shafts, 169, 174
  non-circular shafts, 189
Apparent elastic limit, 64
Apparent stress, 60
Arc welding, 83
Areas:
  centroids, 121, 134
  moment of inertia, 121, 134, 140
  percent reduction of, 67
  polar moment of inertia, 171
  radius of gyration, 149
  section modulus, 261
  table of properties, 134, 221, 318

Axial loads:
  combined with bending, 381, 392
  combined with torsion, 416
Axis, neutral (*See* neutral axis)

Balanced reinforcement, 275
Bar, torsion, 181
Beam stresses, 255
Beams:
  bending, 212
  bending moments, 211, 224
  bolt spacing, 286
  built in, 208
  built up, 286
  cantilever, 207, 215, 221, 313
  composite, 286
  continuous, 209, 215, 341
  deflections, 304, 312, 326
  design, 275
  elastic curve, 305
  flexural stresses, 256
  loads, 207, 209
  moments, 206
  moving loads, 235

478   Index

Beams *(Continued)*
  neutral axis, 258
  neutral plane, 256
  non-uniform loads, 210
  of three or more supports, 209, 341
  of several materials, 266
  overhanging, 208
  propped, 209
  radius of curvature, 306
  reinforced concrete, 273
  rotating test, 70
  section modulus, 261
  shear forces in, 211, 224
  shear stresses in, 278
  simple, 207
  slopes, 310, 326
  statically determinate, 207
  statically indeterminate, 209, 334
  stresses in, 255, 257, 278
  supports, 207
  three moment equation, 341
  transformed section, 266
  uniform loading, 210
Bearing, 14, 84, 86, 89
Bearing failure in riveted joints, 80, 89
Bearing of rivets, 84, 89
Bending:
  combined, 381, 391, 392, 413
  moment diagrams, 206, 215, 224
  of beams, 212, 256
  sign of, 213
  stresses, 257
Biaxial stress, 26, 400
Bolted couplings, 177
Bolts, 83
Breaking strength, 66
Brinell hardness, 71
Brittle failure, 397
Brittle material, 67, 397
Buckling of columns, 360, 363
Built-in beams, 208

Butt joints:
  riveted, 80
  welded, 101, 102, 103

Cantilever beams, 207, 214, 220
Center of gravity, 12, 97, 121, 258
Centric loads, 12, 97
Centroid, (*See* center of gravity)
Charpy test, 72
Chicago column formula, 369
Circle, Mohr's (*See* Mohr's circle)
Circumferential stress, 40
Coefficient of expansion, 32, Appendix E
Columns:
  buckling, 360, 363
  design, 376
  eccentrically loaded, 360, 381, 392
  Euler's formula, 364
  historical, 368
  intermediate, 360, 368
  long, 360, 363
  new AISC Method, 376
  parabolic formula, 369
  Rankine-Gordon formula, 369, 374
  slender, 361
  slenderness ratio, 361
  straight line formula, 369
Combined stress, 28, 381, 391, 392, 413
Composite beams, 286
Compression, 11, 68
Compressive stress, 12
Concentrated load, 207
Concrete, reinforced, (*See* reinforced concrete)
Continuous beams, 209, 341
Continuous loading, 11
Couplings, 173
Conversion factors, 6
Creep, 68
Critical column load, 363, 364

Crushing of rivets and plate, 84, 89
Curvature of beams, 306
Cyclic stresses, 70
Cylinder, 40
Curvature, angle of, 307

Deflection:
   angular, 169, 174, 182, 189
   beam, (See beam deflections)
   springs, 180
Deformation:
   angular, 18, 169, 174
   axial, 17
Design of riveted joints, 89
Design stress, (See allowable stress)
Deviation, 312
Diagonal failure, 86
Distributed loads, (See uniformly distributed loads)
Double shear, 16, 84, 89, 90
Ductility, 55, 67
Dynamic loads, 11

Eccentric loading:
   columns, 381, 392
   riveted joints, 97
   welds, 108
Edge failure of riveted joints, 86
Effective length, 365
Efficiency of riveted joints, 93
Elastic curve, 305
Elastic limit,
   Johnson's apparent, 64
Elasticity, modulus of:
   tension, 20, 45
   shear, 21, 29
Elevated temperature testing, 68
Elongation, 17, 66
   gage length effect, 55
Elongation, percentage, 55, 66
End conditions for columns, 365
Endurance limit, 70

Endurance strength, (See fatigue failure)
Engineering stress-strain diagram, 56
Equivalent bending moment, 414
Equivalent shearing stress, 397
Equivalent torque, 414
Euler's formula, 364
Expansion coefficient, 32 Appendix E
Extensometer, 56

Factor of safety, 366
Failure, high temperature, 68
Failure of columns, 361
Failure of welded joints, 101
Fatigue failure, 70
Fatigue testing, 70
Fiber stress, 255, 257, 258
Fillet welds, 101, 104
First moment of area, (See center of gravity and centroid)
Flexure formula, 256, 259, 306
Flexural rigidity, 310
Floor framing, 264
Flow, shear, 192
Force, 3
Forming of rivets, 78
Formulas for deflection of beams, 326
Fracture, brittle, 72
Fracture, strength, 66
Friction type riveted joint, 82

Gas welding, 100
Gage length, 55
Gage, sheet metal, Appendix E
Generalized stress, 28, 408
Gordon-Rankine formula, (See Rankine-Gordon formula)
Gravity, center of, (See center of gravity and centroid)
Gyration, radius of, 149, 361

Hardness, 71
Helical springs, 183
High temperature effect on strength, 68
Hinged ends of colums, 365
Hollow shaft, torsion of, 171
Hooke's law, 20, 28, 63
Horizontal shearing stress, 212, 278
Horsepower, 165, 167

Inertia, moment of, (*See* moment of inertia)
Indeterminate stress, 29, 208,

Johnson's apparent elastic limit, 64
Joints:
  bolted, 83
  butt, 80, 101, 102, 103
  eccentrically loaded, 97, 108
  fillet, 101, 104
  lap, 80, 101, 102, 103
  riveted, 77, 80
  welded, 77, 100, 101, 102

Lap joint:
  riveted, 77, 80
  welded, 77, 101, 102, 103
Lateral strain, (*See* Poisson's ratio)
Length, gage, 55
Limit:
  creep, 68
  elastic, 20, 64
  endurance, 70
  proportional, 20, 64
Linear expansion, coefficient of, 32, Appendix E
Load:
  centric, 12, 97
  concentrated, 209
  dynamic, 11
  eccentric, 97, 108, 381, 392
  moving, 235
  non-uniform, 210
  uniformly distributed, 210
Load, critical, 363, 364
Load-shear relationship in beams, 211, 224
Load terms, three moment equation, 344
Long columns, 363
Longitudonal stress:
  in cylinders, 41
  in spheres, 42
  shear in beams, 278
Lower yield point, 65

Mass, 3
Materials testing, 54
Mechanical properties of materials, 54
Modulus:
  of elasticity, 20, 64
  of rigidity (shear), 21, 29, 174
  of rupture, 66
  section, 261
  spring, 181
  youngs, 20
  toughness, 67
Mohr's circle, 403, 417
Moment, bending in beams, 206, 210, 211, 224
Moment area method, 308
Moment of area:
  first, (*See* center of gravity)
  second, (*See* moment of inertia)
Moment of inertia:
  rectangular, 121, 140
  polar, 108, 141, 171
Moving loads, 235

Necking down, 67
Neutral axis, 258
Neutral plane, 256
Non-circular sections in torsion, 189
Normal stress, 12, 397, 399
Notched specimen, 72

Offset yield, 65
Overhanging beams, 208

Parabolas, properties of, 256
Parabolic column formula, 369
Parallel axis transfer theorem, 142
Percent elongation, 55, 66
Percent reduction in area, 67
Permanent stress, 65
Permissible stress, (See allowable stress)
Pipe, properties of, Appendix E
Plane, neutral, 256
Plane, stress, 408, (See also biaxial stress)
Plate tension, 85
Poisson's ratio, 23, 24
Polar moment of inertia, 108, 141, 171
Power transmission, 165
Pressure vessels, 40
Primary creep, 69
Primary shear in welds, 105
Principal stress, 395
Properties, mechanical, 54
Properties of areas, 134, 221, 318
Properties of structural shapes, Appendix B, 436
Proportional limit, 20, 63
Propped beam, 209, 335
Pure shear, 394

Radius of curvature, 306
Radius of gyration, 149, 361
Rankine-Gordon formula, 369, 374
Ratio:
  Poisson's, 23, 24
  slenderness, 361
Recorder, stress strain, 56, 60
Rectangular, moment of inertia, 121, 140, 141
Reduction of area, 67
Reinforced concrete, 273
Relaxation, 68

Repeated loading, 11, 70
Resistance welding, 100
Reversal of stress, 70
Rigidity, modulus of, (See modulus of rigidity)
Rivet:
  shear, 83, 89
  bending, 83
  crushing, 84, 89
Riveted joints:
  allowable stresses, 83
  butt, 80
  design, 89
  eccentric load on, 97
  efficiency, 93
  lap, 77, 80
  types of failure, 83
Rockwell hardness, 71
Rotating beam test, 70
Rupture, modulus of, (See modulus of rupture)
Rupture strength, 66
Rupture test, 68

SI system, 1
S-N curve, 70
Safety factor, (See factor of safety)
Scleroscope, 71
Secant formula, 382
Second moment of area, (See moment of inertia)
Secondary creep, 70
Secondary stress:
  riveted joints, 98
  welded joints, 105
Section modulus, 261
Section, transformed, 266
Set, permanent, 65
Set-up stress, 85
Shafts:
  angle of twist, 169, 174
  combined stresses in, 413
  couplings, 177
  torsion in circular, 169
  torsion in non-circular, 189

Shear:
  flow, 192
  horizontal, 212, 279
  in beams, 206, 211, 278
  in rivets, 83
  in shafts, 165, 169
  in welds, 104, 106
  longitudonal, 190
  modulus, 21, 29
  primary, 104
  pure, 394
  secondary, 98, 105
  strain, 17
  transverse, 190
  vertical, 212, 279
Shear diagrams, 212, 215, 224
Shear modulus, 21, 29, 174
Shore scleroscope, 71
Simply supported beams, 207
Single shear, 18, 83, 89, 90
Slenderness ratio, 361
Slope in beams, 310, 326
Sphere, 40
Spring modulus, 189
Springs:
  helical, 180, 183
  torsion, 181
Static loading, 11
Statically determinate beams, 206
Statically indeterminate beams, 208, 334
Statically indeterminate stresses, 29
Stiffness, (See Modulus of elasticity)
Straight line formula, 369
Strain, 10, 17, 56
  generalized, 28
  shear, 18
  volumetric, 24
Strength:
  buckling, 360, 363
  compressive, 65
  fatigue, 70
  fracture, 66
  of rivets, 81
  rupture, 66
  tensile, 65
  ultimate, 66
  yield, 65
Stress:
  allowable, (See allowable stress)
  bearing, 14
  biaxial, 399
  breaking, 66
  combined, 28, 381, 391, 392, 413
  compression, 11, 68, 381
  engineering, 56
  fatigue, 70
  fiber, 257, 278
  indeterminate, 29, 208, 334
  in cylinders, 40
  in rivets, 81
  in spheres, 40
  in welds, 104, 106
  normal, 12, 397, 399, 408
  shear, 18, 81, 169, 185, 278
  simple, 12
  temperature, 32, 81
  tension, 11, 56, 83, 89, 381
  triaxial, 28, 408
Stress strain curve, 20
  in tension, 56
  true, 60
Stress strain recorder, 56, 60
Structural loads, 10
Superposition, 233, 325

Tearing, 85, 86
Temperature stress, 32, 81
Tensile strength, effect of temperature on, 66
Tensile test, 54, 61
Tension, 11, 83, 89
Tertiary creep, 69

Testing:
    charpy, 72
    compression, 68
    creep, 68
    hardness, 71
    impact, 72
    stress-rupture, 68
    temperature, 68
    tensile, 54
Testing machine:
    impact, 72
    universal, 56
Thermal expansion, coefficient of, 32, Appendix E
Thick-walled cylinders, 41
Thin:
    cylinders, 40
    spheres, 40
    torsion members, 183
Three moment equation, 341
Throat, 104
Torque, 165
Torsion, 165
    circular sections, 169
    non-circular sections, 189
Torsion bars, 165, 181
Toughness, 67
Transformed section, 266
Transition point, creep, 69
Travelling loads, (See moving loads)
Triaxial stress, 28, 408

True stress strain curve, 60
Twist, angle of, 169, 174, 182, 189

Ultimate strength, 66
Uniform stress distribution, 12
Uniformly distributed load, 210
Units, 1
Universal testing machine, 56
Upper yield point, 65
Upsetting, 78

Vertical shear, 212, 279
Vickers hardness, 71
Volumetric strain, 24

Wahl spring factor, 185
Weight, 3
Welded joints, 78
    butt, 101, 102, 103
    design of, 99, 104
    fillet, 101, 102, 103
    lap, 101, 102, 103
    stresses in, 106, 107
    throat of, 104
Work, 166
Working stress, (See allowable stress)

Yield point, 65
Yield strength, 65
Young's modulus, 20, 64